Topics in Geobiology

Volume 48

The Topics in Geobiology series covers the broad discipline of geobiology that is devoted to documenting life history of the Earth. A critical theme inherent in addressing this issue and one that is at the heart of the series is the interplay between the history of life and the changing environment. The series aims for high quality, scholarly volumes of original research as well as broad reviews.

Geobiology remains a vibrant as well as a rapidly advancing and dynamic field. Given this field's multidiscipline nature, it treats a broad spectrum of geologic, biologic, and geochemical themes all focused on documenting and understanding the fossil record and what it reveals about the evolutionary history of life. The Topics in Geobiology series was initiated to delve into how these numerous facets have influenced and controlled life on Earth.

Recent volumes have showcased specific taxonomic groups, major themes in the discipline, as well as approaches to improving our understanding of how life has evolved.

Taxonomic volumes focus on the biology and paleobiology of organisms—their ecology and mode of life—and, in addition, the fossil record—their phylogeny and evolutionary patterns—as well as their distribution in time and space.

Theme-based volumes, such as predator-prey relationships, biomineralization, paleobiogeography, and approaches to high-resolution stratigraphy, cover specific topics and how important elements are manifested in a wide range of organisms and how those dynamics have changed through the evolutionary history of life.

Comments or suggestions for future volumes are welcomed.

Neil H. Landman Department of Paleontology American Museum of Natural History New York, USA E-mail: landman@amnh.org Peter J. Harries Department of Marine, Earth and Atmospheric Sciences North Carolina State University Raleigh, USA E-mail: pjharrie@ncsu.edu

More information about this series at http://www.springer.com/series/6623

Sergio Martínez · Alejandra Rojas ·
Fernanda Cabrera
Editors

Actualistic Taphonomy in South America

Springer

Editors
Sergio Martínez
Departamento de Paleontología
Facultad de Ciencias, Universidad de la República
Montevideo, Uruguay

Alejandra Rojas
Departamento de Paleontología
Facultad de Ciencias, Universidad de la República
Montevideo, Uruguay

Fernanda Cabrera
Departamento de Paleontología
Facultad de Ciencias, Universidad de la República
Montevideo, Uruguay

ISSN 0275-0120
Topics in Geobiology
ISBN 978-3-030-20624-6 ISBN 978-3-030-20625-3 (eBook)
https://doi.org/10.1007/978-3-030-20625-3

This Springer imprint is published by the registered company Springer Nature Switzerland AG
The registered company address is: Gewerbestrasse 11, 6330 Cham, Switzerland

Preface

Taphonomy has certainly come a long way. Although the publication of the neologism coined by Efremov in 1940 is considered—not without reason—the foundational event of the discipline, it is also true that its basic concepts had been used since the dawn of Paleontology. Classic broad-spectrum books such as Darwin's "On the Origin of Species" and early paleontological treatises as Zittel's incorporated important sections on preservation.

Among the antecedents immediately before 1940, the scientific production of the Aktuo-Paläontologie school in Senckenberg am Meer under the direction of Rudolf Richter is of utmost importance. Posterior to 1940, taphonomical studies developed irregularly, being mostly considered subsidiary to paleoecological research, in times when Paleoecology became a high profile discipline, especially from the 1960 decade until the mid-1980s.

A symbolic inflection point was the launching of PALAIOS in 1986, a journal representative of a new and influential generation, which emphasized the importance of obtaining new data through the study of patterns and processes of fossilization, instead of worrying about the loss of (biological) information. This revitalization of the discipline included the Actualistic Taphonomy branch, as was later named. This kind of approach has dealt with a wide range of organisms, and incorporated quantitative techniques that have broadened and deepened the investigations.

During the past years, Actualistic Taphonomy had a big boost in South America, especially in the so-called "southern cone" (Argentina, Brazil, Uruguay), as shown by numerous papers, books, talks, and theses. The state of the art in the region deserved a focused meeting in order to exchange experiences and to promote the development of the discipline into the future. This book is a consequence of the Workshop "Actualistic Taphonomy in South America", that took place in Montevideo (Uruguay) during October 9–11, 2017, organized by the Invertebrate Paleontology & Ichnology Laboratory (Paleontology Department, Facultad de Ciencias, Universidad de la República). The book chapters show the amplitude of the actualistic taphonomic studies in South America. Subjects of study comprise plants, invertebrates, vertebrates, ichnites, and human artifacts (zooarcheological and lithic), showing the diverse specializations of the authors. According to the nature and spirit of the event and book, the chapters exhibit different amplitude, in which the reader will find revision and original papers, new ideas, qualitative and quantitative approaches. Consequently, this book pretends to be as ample and diverse as our continent is.

Montevideo, Uruguay

Sergio Martínez
Alejandra Rojas
Fernanda Cabrera

Acknowledgments

The chapters of this book were revised by Julio Aguirre (Universidad de Granada), Luis Borrero (CONICET/Universidad de Buenos Aires), Cristian Favier Dubois (CONICET/Universidad del Centro de la Provincia de Buenos Aires), Ari Iglesias (Universidad Nacional del Comahue), Julieta Martinelli (University of Washington), Mariana Mondini (CONICET/Universidad Nacional de Córdoba), James Nebelsick (Universität Tübingen), Sandro Petró (Universidade Federal do Rio Grande do Sul), Rodrigo Scalise Horodyski (Universidade do Vale do Rio dos Sinos), Rafael Suárez (Universidad de la República, Montevideo), Adam Tomašových (Slovak Academy of Sciences), and Martín Ubilla (Universidad de la República, Montevideo). Facultad de Ciencias (UdelaR), PEDECIBA, and ANII supported the work of the editors. Thank you very much to all of them.

Contents

Contributors

María C. Álvarez Instituto de Investigaciones Arqueológicas y Paleontológicas del Cuaternario Pampeano (INCUAPA-CONICET), Facultad de Ciencias Sociales, Universidad Nacional del Centro de la Provincia de Buenos Aires, Olavarría, Buenos Aires, Argentina

Eloisa García Añino CONICET/División Arqueología, Facultad de Ciencias Naturales y Museo, Universidad Nacional de La Plata, La Plata, Buenos Aires, Argentina

Diego Antuña Departamento de Paleontología, Facultad de Ciencias, Universidad de la República, Montevideo, Uruguay

Fernando M. Archuby Instituto de Investigación en Paleobiología y Geología (IIPG), UNRN-CONICET, General Roca, Río Negro, Argentina

Laura Beovide Centro de Investigación Regional Arqueológica y Territorial (CIRAT), Dirección Para el Desarrollo de la Ciencia y el Conocimiento, MEC, San José, Uruguay

Karen Borrazzo CONICET, Instituto Multidisciplinario de Historia y Ciencias Humanas (CONICET-IMHICIHU), Buenos Aires, Argentina;
Facultad de Filosofía y Letras, Universidad de Buenos Aires (FFyL-UBA), Buenos Aires, Argentina

Fernanda Cabrera Departamento de Paleontología, Facultad de Ciencias, Universidad de la República, Montevideo, Uruguay

Paula A. Cristini Instituto de Investigaciones en Ciencia y Tecnología de Materiales (INTEMA), CONICET, UNMdP, Mar del Plata, Argentina

Claudio G. De Francesco Instituto de Investigaciones Marinas y Costeras (IIMyC), CONICET, UNMdP, Mar del Plata, Argentina

Paula Dentzien-Dias Laboratório de Geologia e Paleontologia, Instituto de Oceanografia, Universidade Federal do Rio Grande, Rio Grande, RS, Brazil

Marisel C. Díaz Instituto de Investigaciones Marinas y Costeras (IIMyC), CONICET, UNMdP, Mar del Plata, Argentina

Matias do Nascimento Ritter Centro de Estudos Costeiros, Limnológicos e Marinhos, Campus Litoral Norte, Universidade Federal do Rio Grande do Sul, Imbé, Rio Grande do Sul, Brazil

Fernando Erthal Departamento de Paleontologia e Estratigrafia, Instituto de Geociências, Universidade Federal do Rio Grande do Sul, Porto Alegre, Rio Grande do Sul, Brazil

Heitor Francischini Laboratório de Paleontologia de Vertebrados, Programa de Pós-Graduação em Geociências, Instituto de Geociências, Universidade Federal do Rio Grande do Sul, Porto Alegre, RS, Brazil

Renato Pirani Ghilardi Faculdade de Ciências, Universidade Estadual Paulista, UNESP, Bauru, SP, Brazil

Mariela E. González Instituto de Investigaciones Arqueológicas y Paleontológicas del Cuaternario Pampeano (INCUAPA-CONICET), Facultad de Ciencias Sociales, Universidad Nacional del Centro de la Provincia de Buenos Aires, Olavarría, Buenos Aires, Argentina

María A. Gutiérrez Instituto de Investigaciones Arqueológicas y Paleontológicas del Cuaternario Pampeano (INCUAPA-CONICET), Facultad de Ciencias Sociales, Universidad Nacional del Centro de la Provincia de Buenos Aires, Olavarría, Buenos Aires, Argentina

Gabriela S. Hassan Instituto de Investigaciones Marinas y Costeras (IIMyC), CONICET, UNMdP, Mar del Plata, Argentina

Darío O. Hermo CONICET/División Arqueología, Facultad de Ciencias Naturales y Museo, Universidad Nacional de La Plata, La Plata, Buenos Aires, Argentina

Cristian A. Kaufmann Instituto de Investigaciones Arqueológicas y Paleontológicas del Cuaternario Pampeano (INCUAPA-CONICET), Facultad de Ciencias Sociales, Universidad Nacional del Centro de la Provincia de Buenos Aires, Olavarría, Buenos Aires, Argentina

Spencer G. Lucas New Mexico Museum of Natural History and Science, Albuquerque, NM, USA

Lucía A. Magnín CONICET/División Arqueología, Facultad de Ciencias Naturales y Museo, Universidad Nacional de La Plata, La Plata, Buenos Aires, Argentina

Laura Marchionni CONICET/División Arqueología, Facultad de Ciencias Naturales y Museo, Universidad Nacional de La Plata, La Plata, Buenos Aires, Argentina

Thays Paganella Marcondes Faculdade de Ciências, Universidade Estadual Paulista, UNESP, Bauru, SP, Brazil

Sergio Martínez Departamento de Paleontología, Facultad de Ciencias, Universidad de la República, Montevideo, Uruguay

Agustina Massigoge Instituto de Investigaciones Arqueológicas y Paleontológicas del Cuaternario Pampeano (INCUAPA-CONICET), Facultad de Ciencias Sociales, Universidad Nacional del Centro de la Provincia de Buenos Aires, Olavarría, Buenos Aires, Argentina

Laura L. Miotti CONICET/División Arqueología, Facultad de Ciencias Naturales y Museo, Universidad Nacional de La Plata, La Plata, Buenos Aires, Argentina

Bruno H. Mosquera CONICET/División Arqueología, Facultad de Ciencias Naturales y Museo, Universidad Nacional de La Plata, La Plata, Buenos Aires, Argentina

Sueli Yoshinaga Pereira Instituto de Geociências Universidade Estadual de Campinas-UNICAMP, Campinas, SP, Brazil

Maria Clara Silva Pinto Faculdade de Ciências, Universidade Estadual Paulista, UNESP, Bauru, SP, Brazil

Daniel J. Rafuse Instituto de Investigaciones Arqueológicas y Paleontológicas del Cuaternario Pampeano (INCUAPA-CONICET), Facultad de Ciencias Sociales, Universidad Nacional del Centro de la Provincia de Buenos Aires, Olavarría, Buenos Aires, Argentina

Fresia Ricardi-Branco Instituto de Geociências Universidade Estadual de Campinas-UNICAMP, Campinas, SP, Brazil

Francisco Santiago Rios Instituto de Geociências Universidade Estadual de Campinas-UNICAMP, Campinas, SP, Brazil

Andrea Roche Escuela Superior de Ciencias Marinas (ESCiMar) and Centro de Investigación Aplicada y Transferencia Tecnológica en Recursos Marinos Almirante Storni (CIMAS), Río Negro, Argentina

Sabrina Coelho Rodrigues Instituto de Ciências Exatas e Naturais do Pontal, Universidade Federal de Uberlândia, Ituiutaba, MG, Brazil

Alejandra Rojas Departamento de Paleontología, Facultad de Ciencias, Universidad de la República, Montevideo, Uruguay

Nahuel A. Scheifler Instituto de Investigaciones Arqueológicas y Paleontológicas del Cuaternario Pampeano (INCUAPA-CONICET), Facultad de Ciencias Sociales, Universidad Nacional del Centro de la Provincia de Buenos Aires, Olavarría, Buenos Aires, Argentina

Cesar L. Schultz Laboratório de Paleontologia de Vertebrados, Programa de Pós-Graduação em Geociências, Instituto de Geociências, Universidade Federal do Rio Grande do Sul, Porto Alegre, RS, Brazil

Eleonor Tietze Instituto de Investigaciones Marinas y Costeras (IIMyC), CONICET, UNMdP, Mar del Plata, Argentina

Chapter 1
Taphonomy of Recent Bioclastic Deposits from the Southern Brazil Shelf: Stratigraphic Potential

Fernando Erthal and Matias do Nascimento Ritter

Abstract In the Southern Brazil Shelf (SBS), surface bioclastic concentrations are associated with putative paleo-shorelines formed where wave ravinement surfaces are probably present. From the late Last Glacial Maximum, the SBS can be considered a sediment-starved passive margin continental shelf, with its morphostructural development fairly known. There, fourteen molluscan shell samples from near shelf-break deposits ("distal shell-rich"), eleven from proximal, low depth bioclastic deposits ("proximal shell-rich") and ten samples from sandy substrate ("shell-poor") were evaluated for taphonomic damage accordingly to updated protocols. Multivariate statistical analysis showed significant differences between the three groups of shelly samples. Low-intensity damage states (such as natural bright and ornamentation) dominate samples from the distal shell-rich deposit, whereas the inverse occurs in the proximal deposit (samples from the shell-poor locations present an intermediate damage pattern). This pattern is consistent either with onlap/toplap and backlap shell bed formation, according to characteristics determined in the literature. The condition of these three areas may reflect degrees of exposure at the taphonomically-active zone, the magnitude of time averaging and duration of shell accumulation, and even the lack of shelf accommodation space, which in turn is related to glacioeustatic sea-level oscillations.

Keywords Time-averaging · Stratigraphic paleobiology · Coquina · Shell preservation · Onlap deposits

F. Erthal (✉)
Departamento de Paleontologia e Estratigrafia, Instituto de Geociências, Universidade Federal do Rio Grande do Sul, Porto Alegre, Rio Grande do Sul 91501-970, Brazil
e-mail: fernando.erthal@ufrgs.br

M. do Nascimento Ritter
Centro de Estudos Costeiros, Limnológicos e Marinhos, Campus Litoral Norte, Universidade Federal do Rio Grande do Sul, Imbé, Rio Grande do Sul 95625-000, Brazil
e-mail: matias.ritter@ufrgs.br

S. Martínez et al. (eds.), *Actualistic Taphonomy in South America*,
Topics in Geobiology 48, https://doi.org/10.1007/978-3-030-20625-3_1

1.1 Conceptual Background

In the marine realm, besides from storm deposits or other allogenic processes, shell beds frequently develop in stratigraphically predictable areas through a combination of a series of sedimentary factors (Brett 1995; Kidwell 1989, 1991). Whereas the sedimentological and stratigraphic significance of fossil concentrations is well recognized (e.g., Kidwell 1989; Patzkowsky and Holland 2012), these have not commonly been used for ecological studies because of obvious issues of temporal and spatial averaging. However, recent works have suggested that shell beds can accurately record broad-scale ecological changes through geologic time (Li and Droser 1999; Kidwell 2013). Transgressive shell beds, with complex internal microstratigraphy (sensu Kidwell 1991), are common in Plio-Pleistocene shallow marine sedimentary cycles around the world (some examples include Abbot 1997; Kondo et al. 1998; Kidwell 1989; Fürsich and Oschmann 1993; Fürsich and Pandey 2003; Simões and Kowalewski 1998). Although cycles of coastal retrogradation and regression may punctuate transgressions, bioclastic deposits within the TST context probably encompass a high degree of time-averaging (Kidwell 1989; Fürsich and Pandey 2003; Scarponi et al. 2013). Good estimates and direct measures of the time enclosed within shell deposits are available for a variety of modern marine depositional settings (Krause et al. 2010; Scarponi et al. 2013; Kidwell 2013; Dexter et al. 2014; Dominguez et al. 2016; Ritter et al. 2017).

Naturally occurring bioclastic death assemblages are useful in determining the environmental control on the style of post-mortem damage to skeletal hard parts (Tomašových and Zuschin 2009; Parsons-Hubbard et al. 2011). Marine molluscan assemblages are likely to be ecologically mixed in composition and have taphonomically complex histories, because of shallowing and deepening cycles and the continuous shallow-water erosional reworking (Kidwell 1991). Intrinsic properties such as the mineralogy of carbonatic bioclasts (proportion of aragonite, high-Mg calcite, and low-Mg), may modulate the rate of taphonomic loss (Glover and Kidwell 1993). The taphonomic patterns of present-day developing deposits are well documented, especially for tropical marine areas. Hendy et al. (2006) described the taphonomic properties of several shell-beds from Miocene-Pliocene deposits of New Zealand under a sequence stratigraphy framework. They emphasized that closer attention should be paid to the taphonomic condition of shells within bioclastic concentrations, to refine environmental and stratigraphic interpretation. In this contribution, the naturally-occurring mollusk death assemblages from the Southern Brazil Shelf are considered a potential analogous for transgressive deposits preserved in the fossil record. The aim is to test if the damage profile of mollusk shells from several areas of shell accumulation allows predicting the properties of shell beds in similar transgressive system tracts from the deep-time fossil record.

1.2 Study Area

Despite considerable clastic contribution by the Plata River, the Southern Brazil Shelf (SBS, roughly within latitudes 27°S and 34°S) can be presently characterized as a sediment-starved passive margin continental shelf (Corrêa 1996). Deposition of a large amount of post-rift, primarily clastic sediment produced a wide (100–200 km), shallow (100–140 m), and gently sloping (0.03–0.08°) continental shelf (Dillenburg and Barboza 2014; Weschenfelder et al. 2014). Sandy sediments dominate the adjacent continental shelf out to a water depth of circa 60 m while muddy sediments predominate in areas of deeper water (Corrêa 1996; Dillenburg and Barboza 2014). In the southern area, sand bodies, sand waves and elongated bioclastic deposits (between isobaths of 10 and 50 m) interrupt the usually smooth morphology (Figueiredo 1975; Villwock et al. 1986; Corrêa 1996).

The present physiography of the coastal plain and shelf is mainly the result of Quaternary high-frequency (4th order to 5th order), glacioeustatic sea level changes, that affected the sedimentary systems along the coast. The post-glacial sea-level history of this area extends back circa 15.5 kyr ago when the sea-level was 120–130 m below the present position (Villwock et al. 1986; Corrêa 1996). After that time, the sea-level rose at an average rate of 1.2 cm/yr, beginning after the onset of the Postglacial Marine Transgression (PMT). Sea level stabilized at 9 kyr (between depths of 32 and 45 m) and 8 kyr (between depths of 20 and 25 m; Corrêa 1996; Nagai et al. 2014). Sparse data exists on the sea-level behavior during middle to late Holocene, and sea-level curves for areas to the north indicate that at the culmination of the PMT, the sea-level was 1–3 m (perhaps up to 4 m) above its present level, which occurred between 4.1 and 2.7 kyrs ago (Corrêa 1996; Angulo et al. 2006; Martínez and Rojas 2013).

1.3 Materials and Methods

A sampling of top marine sediment, rich in molluscan dead remains was conducted during several oceanographic expeditions promoted by Brazil Navy in cooperation with universities (Kowsmann et al. 1977; Figueiredo and Tessler 2004). Given their assumed reworked nature (Figueiredo 1975; Corrêa 1996), shell samples from 35 bottom-level locations (Fig. 1.1) can be evaluated in terms of developing condensed deposits within a (fourth to fifth order) transgressive sequence, such as the apparent stage of SBS. At least 150 shells or fragments of multiple species of bivalves larger than 4 mm by location were inspected for taphonomic damage, following standard taphonomic protocols such as the presented in Ritter et al. (2019). Briefly, we quantified the intensity of taphonomic variables considering damage states varying from 0 (zero, no damage present) and 1 (damage present). Based on location and sediment type, samples were grouped into three regions or so-called provinces: one proximal, with high carbonate content (>50%) and low depths (<60 m); one distal, with high

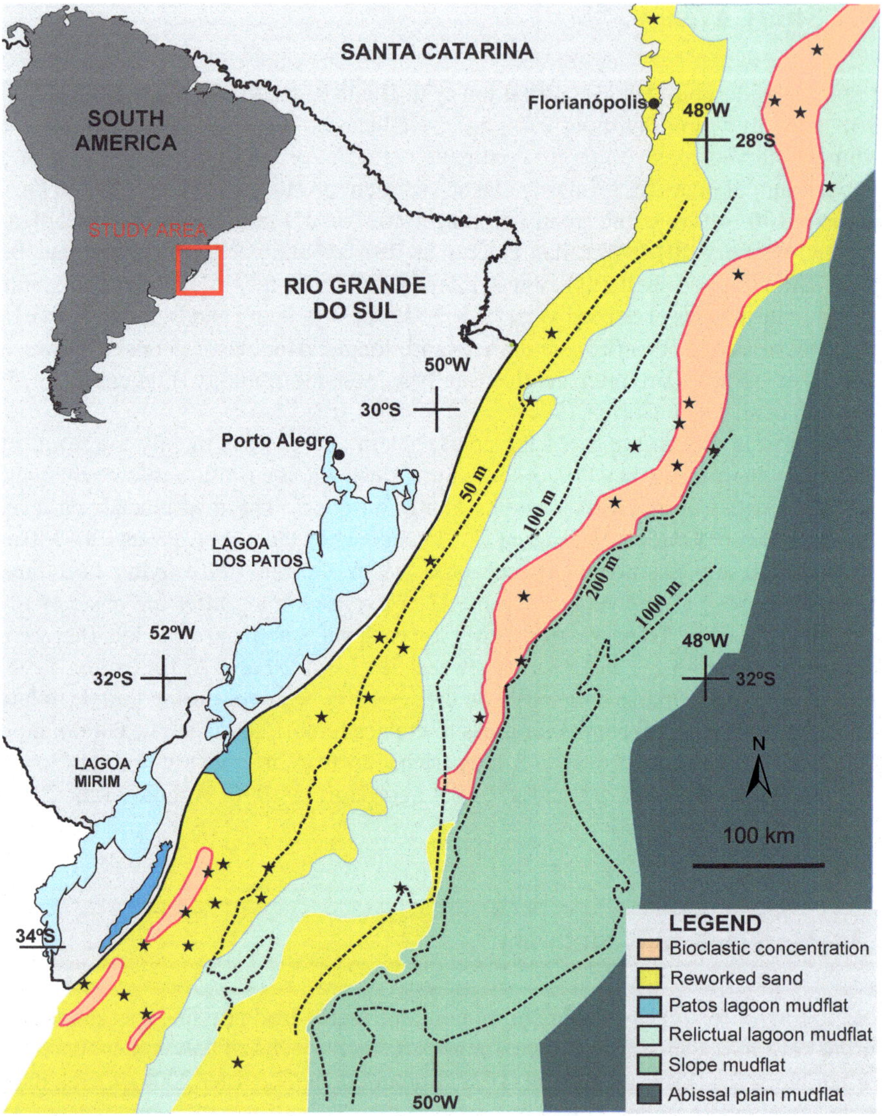

Fig. 1.1 Location of the study area and sampling sites, with indication of main geomorphological and sedimentary features (map modified from Figueiredo 1975; Vicalvi 1977; Kowsmann and Costa 1979; Kowsmann et al. 1977; Villwock and Tomazelli 1995; Corrêa 1996; Tomazelli and Vilwock 2000; Corrêa et al. 2014; Dillenburg and Barboza 2014; Weschenfelder et al. 2014)

carbonate content (>50%) and great depth (>100 m); and a third area including very low carbonate ($\ll$50%) and high sand contents.

Non-metrical multidimensional scaling (NMDS) was used to visualize the distribution of sampling units accordingly to their taphonomic profile. To search for differences in the taphonomic profiled between sedimentary provinces (proximal and distal shell-rich, and sandy shell-poor) and depth ranges (shallow: less than 100 m; deep: more than 100 m), we used multivariate analysis of variance with permutation (PERMANOVA; Anderson 2001). The relationship between taphonomic profile and sediment grain-size and composition (environmental variables) was explored using Constrained Analysis of Proximities (CAP, also known as distance-based Redundancy Analysis; Legendre and Legendre 2012). NMDS, PERMANOVA, and CAP were conducted based on Manhattan distance, which is better suited for taphonomic data (Tomašových and Zuschin 2009). All statistical analysis and graphics generation were performed using statistical software R (R Development Core 2018).

1.4 Results

The multivariate analysis evaluates the total variation present in the data at one time, allowing discriminating bioclastic provinces in the SBS, with potentially better interpretations. Damage profiles within shell-rich and shell-poor provinces are strongly variable. The dispersion of sampling units from SBS according to the variation in their damage condition is shown in Fig. 1.2. The three sampling areas (proximal and distal shell-rich and sandy shell-poor) showed significantly different taphonomic profile, according to PERMANOVA (Table 1.1). This difference persists even when separating the damage profile into the biological and physical origin and color alteration, and the taphonomic variation follows the depth gradient, as samples shallower than 100 m are significantly different from samples deeper than 100 m (Table 1.1). A neat variation of the taphonomic profile according the sedimentary carbonate and gravel content is observed in the NMDS (Fig. 1.2).

Table 1.1 Results of PERMANOVA examining the multivariate difference between three groups of sampling units (provinces) considering presence of shell-rich and shell-poor areas and groups of sampling sites according to depth (shallow × deep samples)

	F value	P value
Province		
Total damage profile	5.0211	<0.01
Biological damage	4.2356	<0.01
Physical damage	7.5111	<0.01
Color alteration	3.1103	<0.01
Depth		
Total damage profile	7.7297	<0.01
Biological damage	7.2194	<0.01
Physical damage	13.308	<0.01
Color alteration	2.7113	>0.05

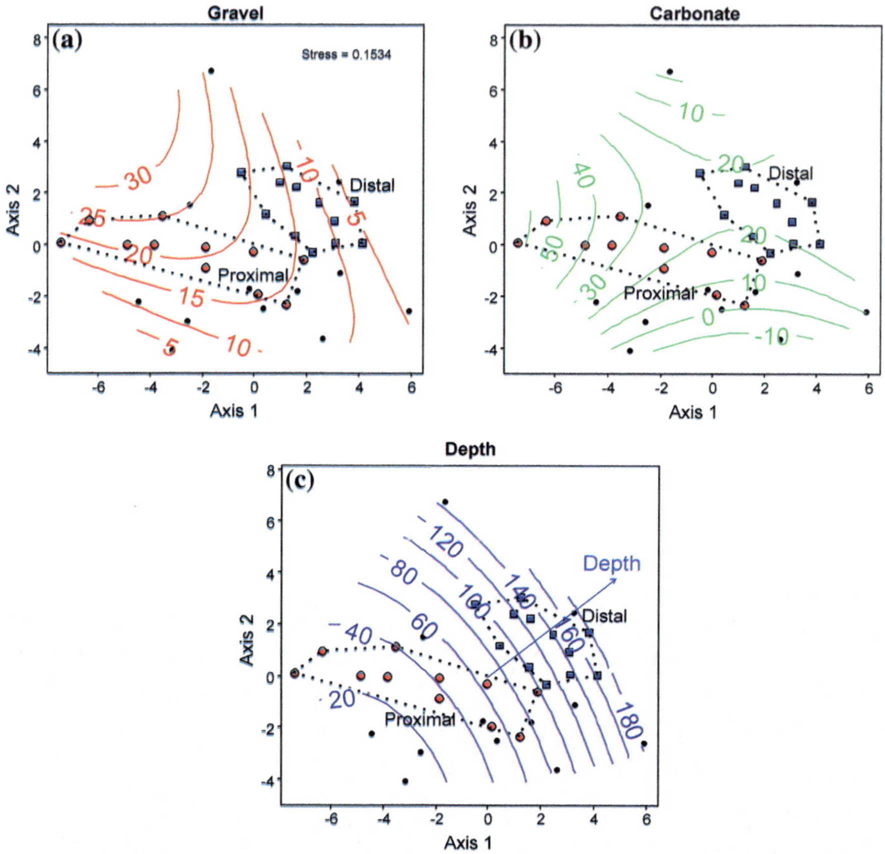

Fig. 1.2 NMDS (based in Manhattan distance) showing dispersion of the studied mollusk samples from the SBS based on the taphonomic profile. Function ordisurf (in the "vegan" package; Oksanen et al. 2019) allowed fitting thin plate splines (gradients) showing proportion of sedimentary carbonate (**a**), gravel content (**b**) and depth gradient (**c**). Dotted hulls show sites contained within the proximal and distal shell-rich zones, which form significantly different groups accordingly to PERMANOVA (Squares = distal shell-rich province; Circles = proximal shell-rich province; Smaller dots = dispersed shell-poor samples). The proportion of damage for each taphonomic variable at each sampling site was standardized (scaling to zero mean and unit variance) and normalized (making the sum of squares equal to one) prior to analysis

Constrained analysis of proximities based on Manhattan distance of taphonomic data shows that the environmental factors explain almost 35% of the entire taphonomic profile, with 73% of variation indicated by Canonical Axis 1 (53.2%) and Canonical Axis 2 (19.7%) (Fig. 1.3a). In the proximal shell-rich province, the damage is mainly controlled by the increase in carbonate and gravel content of the sediment, and secondarily by a decrease in depth and mud (clay + silt) (Fig. 1.3a). Figure 1.3b–q show the variation of all taphonomic conditions measures in the samples across the SBS. Most biologically-originated damages are found in the distal shell-rich province, where depth and gravel and sand proportions affected most (Fig. 1.3b–g). Depth and sand content, which are inversely proportional, and the variation in gravel and mud content appears to influence the damage pattern in the proximal shell-rich samples, an area with the strongest signal for physical origin damage (Fig. 1.3h–q). It is noteworthy that a small group of taphonomic signatures (drilling, large pits, natural ornamentation and bright) appears to influence more the distal shell-rich area. Based on the constrained analysis of color alteration and sedimentary factors, the shell-rich provinces overlap (Fig. 1.3d), despite being significantly different (Table 1.1).

1.5 Discussion

1.5.1 Taphonomic Profile of Bioclastic Deposits from SBS

Although several damage states lump at the origin of canonical axes, key variables relate to increases and/or decreases in the measured sedimentary variables. The clustering of low-damage (near pristine) signatures such as natural ornamentation and natural bright, which occurs preferentially in the distal shell-rich province, may attest the lower influence of high-energy environmental conditions in that region. This pattern was also found in comparable Cenozoic depositional settings, and may reflect the intense reworking caused by the action of waves and currents (Kidwell 1986; Fürsich and Oschmann 1993; Kondo et al. 1998; Hendy et al. 2006). Nevertheless, damage reflecting high-energy environments, such as ornamentation loss, secondary luster, and accentuated ornamentation are common in SBS. Bioclastic remains at the proximal province might have faced harsh environmental conditions, during a longer time span, consistent with the reworked nature of the SBS deposits.

Physically originated signatures discriminate the shell-rich provinces better, implying that transport by waves and currents act coordinately over sedimentary sequences and damage condition of bioclastic deposits. Sediment bypassing and starvation (Kidwell 1991) can often lead to intense mechanical wearing of shells, as well as colonization by bioeroding and encrusting organisms (Young and Nelson 1985; Cutler and Flessa 1995). The intense bioerosion found in all sites and provinces of SBS may be due to the constant exposure of shells to the sediment-water interface, caused either by the low clastic sedimentation or little accommodation space in the studied area during post-LGM times.

A mild bathymetric pattern of taphonomic alteration is present, based on the results of NMDS. Attritional-related signatures are more frequent in the proximal (shallow) shell-rich province while dissolution-related signatures prevail in the distal shell-rich area (Figs. 1.2 and 1.3). It means that some bathymetric conditioning might have induced the observed taphonomic profile of shell samples from the SBS. On the other hand, many thousands of years must separate the genesis of the two provinces, regardless of whether they relate to early episodes of shoreline stabilization

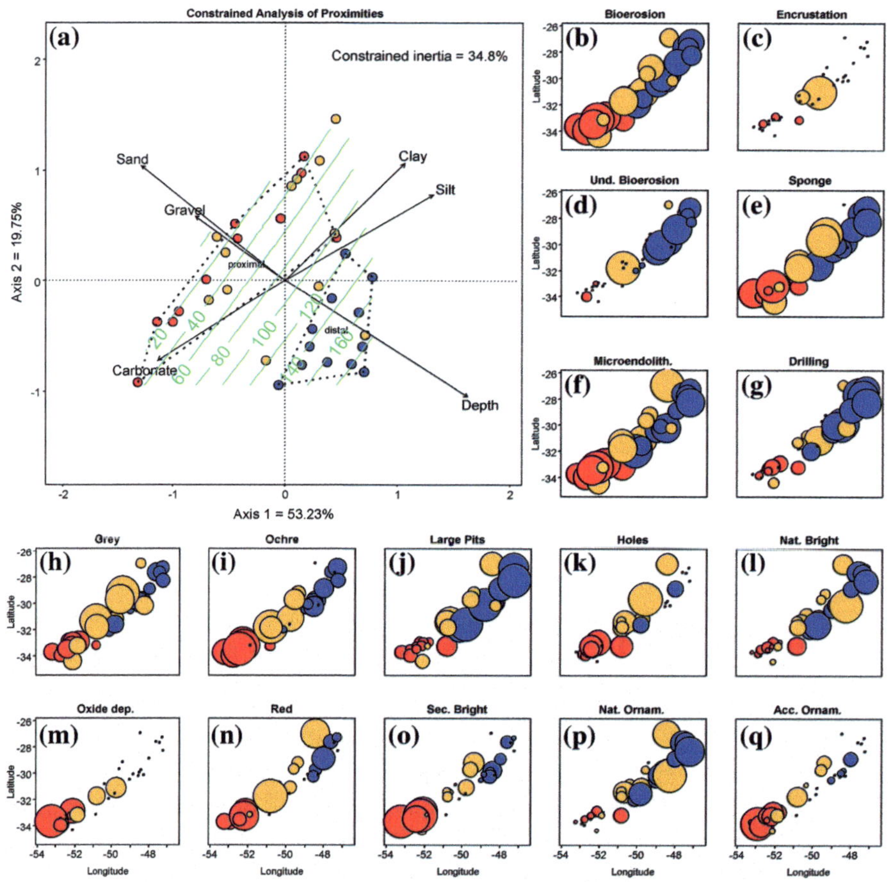

Fig. 1.3 Constrained Analysis of Proximities of the shelly samples from SBS. **a** Entire taphonomic profile. **b** Damage of biological origin. **c** Damage of physical origin. **d** Color alteration. Environmental factors are in capital letters when significant after forward selection, and taphonomic signatures are in italics. Sampling units were omitted from the dispersion plot for clarity. The intensity of distinctive taphonomic variables through sampling units are displayed in graphs (**b**–**p**). Abbreviations are: Und. Bioerosion = Undetermined Bioerosion, Microendolith = Microendolithic Traces, Nat. Bright = Natural Bright, Oxide dep. = Oxide deposits (crusts), Sec. Bright = Secondary Bright, Nat. Ornam. = Natural Ornamentation, Acc. Ornam = Accentuated Ornamentation. Gray, Ochre and Red refers to color alteration (secondary colors acquired by the shells)

or not. Ritter et al. (2017, 2018) have shown that the age-variation of these death-assemblages can reach three-order of magnitude.

The bioclasts in the shell-rich areas might have been affected by abiotic and biotic factors taking place during significant shifts in environmental conditions, from below storm wave base level to nearshore and foreshore (Fig. 1.4), and probably under lagoon influence (Ritter et al. 2013). In addition, the longer the shells remain exposed in the upper limit of the benthic TAZ, the stronger will be the environmental imprint on the taphonomic pattern preserved (Davies et al. 1989; Olszewski 2004; Smith and Nelson 2003; Parsons-Hubbard et al. 2011). Although the prevalence of biological damage in the distal province, and physical damage characterizing better the proximal province, the biologically originated signatures are also known to relate to specific environmental settings or conditions (Fürsich and Oschmann 1993; Lescinsky et al. 2002). The color alteration in bioclastic remains associated with the type of sedimentary geochemical properties, including but not limited to inhibiting factors such as sedimentary iron (e.g., Cai et al. 2006; Best et al. 2007; Powell et al. 2011). The color of shells altered to red hues (Fig. 1.3) was implied with subaerial exposure of bioclastic remains, with precipitation of reduced iron and manganese (Anjos-Zerfass et al. 2008; Corrêa et al. 2008). The effect of intrinsic factors such as species composition and shell mineralogy and microstructure on the relation between damage profile and depositional properties in the SBS is an interesting issue to detail in a future work.

1.5.2 *Influence of Sedimentary Regime and Discontinuity Surfaces*

Condensed deposits of all types are important and complex repositories of paleoecological and taphonomic data. Densely fossiliferous coquinas mark the base of marine depositional sequences: each shell deposit onlaps erosional (subaerial) disconformities (Kidwell 1991; Hendy et al. 2006), and are traceable over large areas (tens to hundreds of km^2). That is the case of assemblages examined by Kidwell (1989; see also Kondo et al. 1998), including those presently found in SBS (Figs. 1.1 and 1.4). The ratio between sedimentation and erosion rates controls most of the taphonomic signatures observed in molluscan assemblages (Kidwell 1986, 1991; Tomašových et al. 2006). Whereas low sedimentation and reworking favor abrasion (as seen in the proximal area; Fig. 1.3), the winnowing of sediment enhances bioerosion.

Shell beds are dense (bioclast supported) concentrations of shelly fossils. It is a synonym of coquinas, lumachellas, shell gravels and bioclastic calcirudites, which may form transgressive lags, where shell supply is abundant, detrital sediment supply is low, in areas with significant sediment bypass or even lack of accommodation space in nearshore environments (Kidwell 1989, 1991). All these sedimentary features may be inferable from the bioclastic deposits present in the SBS. Despite low-to-null clastic supply in SBS, shell deposits may not be as dense or voluminous in this area when compared to other Quaternary deposits around the world. The estimated thickness for the shell deposits shown in Figs. 1.1 and 1.4 may vary from little more than 5 cm to 2.5 m (Figueiredo 1975). Emerged shell concentrations in the coastal plain of Southern Brazil reach thickness up to 6 m (Dillenburg and Barboza 2014). That bioclastic facies is highly reworked, reflecting high-energy environments (Kowsmann et al. 1977).

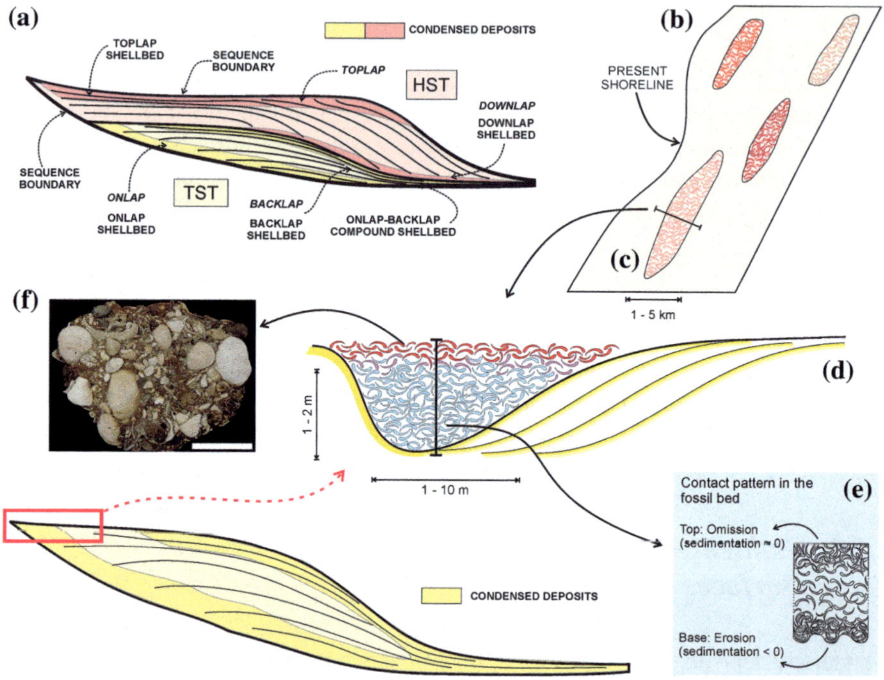

Fig. 1.4 Summary of the possible nature of developing onlap and backlap condensed deposits present in the SBS. **a** Generalized stratigraphic architecture showing transgressive and high-stand system tracts. **b** Generalized map and morphology of bioclastic bodies **c** in proximal continental shelf. **d** Hypothetical cross-cut of a single bioclastic deposit. **e** Contact pattern of the bioclastic deposit, based mostly upon classification of Kidwell (1985). **f** Example of a coquina found in the nearshore area of SBS, that is probably reworked from the top of an underwater bioclastic body (scale bar = 6 cm). Based mostly upon this and previously published work by Kidwell (1991), Fürsich and Oschmann (1993), Corrêa (1996), Fürsich and Pandey (2003), Hendy et al. (2006), Dillenburg and Barboza (2014), Weschenfelder et al. (2014)

In shallow marine accumulations, shell beds usually develop associated with changes in net sedimentation involving rates of deposition, omission (bypass) and erosion (Kidwell 1989, 1991). The degree of alteration of bioclasts may be consistently variable according to the sedimentary regime, although the sedimentation rate and shell input determine the relative exposure time of bioclasts at the sediment-water interface (Speyer and Brett 1986; Fürsich and Oschmann 1993; Smith and Nelson 2003; Tomašových and Zuschin 2009; Parsons-Hubbard et al. 2011).

The importance of contact-based discontinuities increases as they can reflect boundary surfaces used in sequence stratigraphy, especially where no other sedimentological or paleontological evidence can be determined (Courville and Collin 2002). Bioclastic lithosomes may develop above wave ravinement surface (or transgressive surface) in low gradient shelf, where sedimentation rate is low. Figure 1.4 explores a possible scenario for the formation of bioclastic-rich deposits under the framework of sequence stratigraphy. Wave ravinement, at the base of a transgressive system tract, together with low sediment supply creates small valleys. This is the place of formation of onlap shell beds, where reworked shells of marine mollusks are deployed. While the entire bioclastic body probably encompasses intense time-averaging and a complex taphonomic pattern, the topmost shell layers probably reflect much of present-day environmental conditions. Marine benthos can colonize transgressive shell lags, creating taphonomically complex assemblages of potentially non-coeval fossils. Also, changes in sedimentation rate may produce discontinuity surfaces necessary for the recognition of stratigraphic sequences.

The overall result is that the type and pattern of damage present in marine (and coastal/transitional) assemblages offers a valuable information for the interpretative power of sequence stratigraphy (Brett 1995, 1998; Fürsich and Pandey 2003; Hendy et al. 2006). Transgressive deposits are relatively easy to trace in outcrop and seismic logs, for they are frequently bounded by erosive or omission surfaces (Cattaneo and Steel 2003; Catuneanu et al. 2011; Hendy et al. 2006). Moreover, transgressive system tracts typically preserve mixed and within-habitat transported assemblages (Zecchin and Catuneanu 2013), reflecting high-energy environments with variable degrees of inferred time-averaging and species richness, however (Fürsich and Pandey 2003). Therefore, transgressive fossil deposits (i.e., transgressive lags, but especially condensed onlap deposits) hardly represent paleocommunities.

1.5.3 Exposure Time of Bioclasts in Sediment-Water Interface

Combined sedimentological and taphonomic features from shell-rich deposits in the context of transgressive system tracts reveal that bioclasts persisted on the seafloor for a considerably long time, with intense post-mortem alteration prior to final burial (Kidwell 1989; Fürsich and Oschmann 1993; Abbott 1997; Abbott et al. 2005; Cantalamessa et al. 2005; Hendy et al. 2006). Skeletons included in those shell beds may be derived either from inhumation of indigenous living mollusks, previously buried and erosionally exhumed fossils or from shells transported from immediately adjacent, coeval environments, resulting in spatial mixing and remaniés (sensu Craig and Hallam 1963). Therefore, the fauna composition in those particular settings (such as the shell-rich provinces in SBS; Figs. 1.1 and 1.4) does not reflect the previous productivity history in the area. However, the topmost shell constituents may reflect the later environmental conditions, depending on the ultimate rate of sedimentation. Direct age dating of bioclastic material coupled to damage state discrimination indicated that at least two distinctive assemblages of taphonomically-altered remains are present in the SBS, with a clear shallow vs. deep patterning (Ritter et al. 2018).

Most actualistic studies conducted at facies-level scale found no general environmental trend that could be applied for fossil interpretation (Parsons-Hubbard et al. 2011; Powell et al. 2012). Those events occurred during Late Pleistocene-Early

Holocene (Ritter et al. 2017), although most of the shells are recent (97% younger than 6 ka). The result is a taphonomically altered sum of all those events and is thought to separate each of those processes. Environmental factors depending on depth may affect both the duration of bioclastic accumulation and the taphonomic profile of bioclasts (Ritter et al. 2018). Postmortem alteration is, by turn, a function of both exposure time (to the sediment-water interface) and the rate of bioclast production. Under low sedimentation rates, shells accumulate at or just below the SWI, with long exposure periods to post-mortem processes. Therefore, a positive correlation between shelliness and taphonomic alteration is predictable (Kidwell 1986; Fürsich and Aberhan 1990; Olszewski 2004), a pattern also observed in this study.

Caution is needed to understand the variety of circumstances involved in the rate of shell production and destruction, and some shelly deposits are better suited for paleoecological studies than others (Kidwell 1986, 1991). Condensed deposits may be taphonomically irregular, both within a single deposit and between deposits in a single basin, for shells may suffer countless cycles of burial and exhumation during periods of significant environmental change (e.g., distal vs. proximal shell-rich deposits in SBS; Fig. 1.4; Davies et al. 1989; Kidwell 1991).

1.6 Final Remarks

Several fossil concentrations preserve strongly biased ecological and environmental information, sometimes impairing further paleobiological reconstruction. On the other hand, the formation and preservation of a shell bed may reveal the intenser effect of environmental data than the actual fossils it contains. It is not well known whether rates of shell production in nearshore environments of Southern Brazil Shelf are high; therefore, shells may be considerably old, despite high rates of destruction in SWI. There is no common-sense in using damage state as an (even indirect) indicator for time-averaging (e.g., Ritter et al. 2017, 2018) since most taphonomic features become highly modified during burial and early diagenesis. On the other hand, different hard parts degrade at different rates. The cumulative—time averaged assemblages—may not represent the average environmental condition that prevailed during the duration of its formation.

The low present-day sedimentation in the Southern Brazil Shelf, allied to any rate of hard part productivity enabled the formation of highly time-averaged bioclastic deposits (low productivity + low sedimentation = high time averaging), with the formation of *remaniés* (bioclastic material with complex depositional and taphonomic histories, accumulated during wide time intervals and possibly subject to several burial and re-exposure cycles). Shells with highly variable taphonomic conditions compose the distal and proximal shelly provinces, implying intense and prolonged exposure to harmful agents in the taphonomically active zone. Despite this variability in postmortem alteration, the damage profile of these bioclastic bodies reflects the history of sea level variations occurred in the area, and this pattern can be applied to fossil concentrations present in the older deposits, allowing better interpretation of marine sedimentary sequences.

The proposition here is that the bioclastic deposits from shell-rich areas in SBS formed during early, proximal transgressive system tracts (regions of onlap), underlain by wave ravinement surfaces. It would imply that the lower boundary of such deposits is erosive, and the upper limit of those deposits consists of an omission to erosive surfaces. The better preservation state of shells from the distal province are consistent with the taphonomic pattern exhibited by fossils deposited at regions of apparent truncation (backlap, or condensed section).

Acknowledgements This paper greatly benefited from comments and discussion with several people: I. C. S. Corrêa and E. R. Beltram (Universidade Federal do Rio Grande do Sul), R. S. Horodyski (Universidade do Vale do Rio dos Sinos), M. L. Assine (Universidade Estadual Paulista at Rio Claro), and M. G. Simões (Universidade Estadual Paulista at Botucatu). Brazil funding agencies (Coordenação de Aperfeiçoamento de Pessoal de Nível Superior—CAPES, Conselho Nacional de Desenvolvimento Científico e Tecnológico—CNPq), and Petroleum National Agency (ANP) partly funded this research.

References

Abbott ST (1997) Mid-cycle condensed shellbeds from mid-Pleistocene cyclothems, New Zealand: implications for sequence architecture. Sedimentology 44:805–824

Abbott ST, Naish TR, Carter RM, Pillans BJ (2005) Sequence stratigraphy of the Nukumaruan Stratotype (Pliocene–Pleistocene, c. 2.08–1.63 Ma), Wanganui Basin, New Zealand. J R Soc NZ 35:123–150

Anderson MJ (2001) A new method for non-parametric multivariate analysis of variance. Austral Ecol 26:32–46

Angulo RJ, Lessa GC, Souza MC (2006) A critical review of Mid- to Late Holocene sea-level fluctuations on the eastern Brazilian coastline. Quat Sci Rev 25:486–506

Anjos-Zerfass GS, Souza PA, Chemale Jr F (2008) Biocronoestratigrafia da Bacia de Pelotas: estado atual e aplicação na geologia do petróleo. Rev Bras Geocienc 38:47–62

Best MMR, Ku TCW, Kidwell SM, Walter LM (2007) Carbonate preservation in shallow marine environments: unexpected role of tropical siliciclastics. J Geol 115:437–456

Brett CE (1995) Sequence stratigraphy, biostratigraphy and taphonomy in shallow marine environments. Palaios 10:597–616

Brett CE (1998) Sequence stratigraphy, paleoecology and evolution: biotic clues and responses to sea-level fluctuations. Palaios 13:241–262

Cai WJ, Chen F, Powell EN, Walker SE, Parsons-Hubbard KM, Staff GM, Wang Y, Aston-Alcox KA, Callender WR, Brett CE (2006) Preferential dissolution of carbonate shells driven by petroleum seep activity in the Gulf of Mexico. Earth Planet Sci Lett 248:227–243

Cantalamessa G, Di Celma C, Ragaini L (2005) Sequence stratigraphy of the Punta Ballena Member of the Jama Formation (Early Pleistocene, Ecuador): insights from integrated sedimentologic, taphonomic and paleoecologic analysis of molluscan shell concentrations. Paleogeogr Paleoclimatol Paleoecol 216:1–25

Cattaneo A, Steel RJ (2003) Transgressive deposits: a review of their variability. Earth-Sci Rev 62:187–228

Catuneanu O, Galloway WE, Kendall CGSC, Miall AD, Posamentier HW, Strasser A, Tucker ME (2011) Sequence stratigraphy: methodology and nomenclature. Newsl Stratigr 44:173–245

Courville P, Collin PY (2002) Taphonomic sequences—a new tool for sequence stratigraphy. Geology 30:511–514
Corrêa ICS (1996) Les variations du niveau de la mer durant les derniers 17.500 ans BP: exemple de la plate-forme continentale du Rio Grande do Sul-Brésil. Mar Geol 130:163–178
Corrêa ICS, Ayup-Zouain RN, Weschenfelder J, Tomazelli LJ (2008) Áreas fonte dos minerais pesados e sua distribuição sobre a plataforma continental sul-brasileira, Uruguaia e norte-argentina. Pesquisas 35:137–150
Corrêa ICS, Medeanic S, Weschenfelder J, Nunes JC, Baitelli R (2014) The palaeo-drainage of the La Plata River in Southern Brazil continental shelf. Rev Bras Geofis 32:259–271
Craig GY, Hallam A (1963) Size-frequency and growth-ring analyses of *Mytius edulis* and *Cardium edule* and their paleoecological significance. Paleontology 6:731–750
Cutler AH, Flessa KW (1995) Bioerosion, dissolution and precipitation as taphonomic agents at high and low latitudes. Senckenb Marit 25:115–121
Davies DJ, Powell EN, Stanton RJ Jr (1989) Relative rates of shell dissolution and net sediment accumulation—a commentary: can shell beds form by the gradual accumulation of biogenic debris on the sea floor? Lethaia 22:207–212
Dexter TA, Kaufman DS, Krause RA Jr, Barbour Wood SL, Simões MG, Huntley JW, Yanes Y, Romanek CS, Kowalewski M (2014) A continuous multi-millennial record of surficial bivalve mollusk shells from the São Paulo Bight, Brazilian shelf. Quatern Res 81:274–283
Dillenburg SR, Barboza EG (2014) The strike-fed sandy coast of Southern Brazil. In: Martini IP, Wanless HR (eds) Sedimentary coastal zones from high to low latitudes: similarities and differences. Geological Society, London, Special Publications, 388, pp 333–352
Dominguez JG, Kosnik MA, Allen AP, Hua Q, Jacob DE, Kaufman DS, Whitacre K (2016) Time-averaging and stratigraphic resolution in death assemblages and Holocene deposits: Sydney Harbour's molluscan record. Palaios 31:563–574
Figueiredo AG (1975) Geologia dos depósitos calcários biodetríticos da Plataforma Continental do Rio Grande do Sul. Programa de Pós-Graduação em Geociências, Universidade Federal do Rio Grande do Sul, M.Sc. dissertation
Figueiredo AG, Tessler MG (2004) Topografia e composição do substrato marinho da Região Sudeste-Sul do Brasil. Série Documentos REVIZEE: Score Sul. Instituto Oceanografico, São Paulo
Fürsich FT, Aberhan M (1990) Significance of time-averaging for palaeocommunity analysis. Lethaia 23(2):143–152
Fürsich FT, Oschmann W (1993) Shell beds as tools in basin analysis: the Jurassic of Kachchh, western India. J Geol Soc 150:169–185
Fürsich FT, Pandey DK (2003) Sequence stratigraphic significance of sedimentary cycles and shell concentrations in the Upper Jurassic-Lower Cretaceous of Kachchh, western India. Paleogeogr Paleoclimatol Paleoecol 193:285–309
Glover CP, Kidwell SM (1993) Influence of organic matrix on the post-mortem destruction of molluscan shells. J Geol 101:729–747
Hendy AJW, Kamp PJJ, Vonk AJ (2006) Cool-water shell bed taphofacies from Miocene-Pliocene shelf sequences in New Zealand: utility of taphofacies in sequence stratigraphic analysis. In: Pedley HM, Carannante G (eds) Cool-water carbonates: depositional systems and palaeoenvironmental controls. Geological Society, London, Special Publications 255, pp 283–305
Kidwell SM (1985) Palaeobiological and sedimentological implications of fossil concentrations. Nature 318:457–460
Kidwell SM (1986) Models for fossil concentrations: paleobiologic implications. Paleobiology 12:6–24
Kidwell SM (1989) Stratigraphic condensation of marine transgressive records: origin of major shell deposits from the Miocene of Maryland. J Geol 97:1–24
Kidwell SM (1991) Condensed deposits in siliciclastic sequences: expected and observed feature. In: Einsele G, Ricken W, Seilacher A (eds) Cycles and events in stratigraphy. Springer, Berlin, pp 682–695

Kidwell SM (2013) Time-averaging and fidelity of modern death assemblages: building a taphonomic foundation for conservation palaeobiology. Palaeontology 56:487–522
Kondo Y, Abbott ST, Kitamura A, Kamp PJJ, Naish TR, Kamataki T, Saul GS (1998) The relationship between shellbed type and sequence architecture: examples from Japan and New Zealand. Sediment Geol 122:109–127
Kowsmann RO, Costa MPA (1979) Sedimentação Quaternária da margem continental Brasileira e das áreas oceânicas adjacentes. PETROBRAS, Rio de Janeiro, Projeto REMAC Series 8
Kowsmann RO, Costa MPA, Vicalvi MA, Coutinho MGM, Gambôa LAP (1977) Modelo da sedimentação holocênica na plataforma continental sul brasileira. Série Projeto REMAC 2:7–26
Krause RA, Barbour SL, Kowalewski M, Kaufman DS, Romanek CS, Simões MG, Wehmiller JF (2010) Quantitative comparisons and models of time-averaging in bivalve and brachiopod shell accumulations. Paleobiology 36:428–452
Legendre P, Legendre L (2012) Numerical ecology, 3rd edn. Elsevier, Amsterdam
Lescinsky HL, Edinger E, Risk MJ (2002) Mollusc shell encrustation and bioerosion rates in a modern epeiric sea: taphonomy experiments in the Java Sea, Indonesia. Palaios 17:171–191
Li X, Droser ML (1999) Lower and Middle Ordovician shell beds from the Basin and Range province of the western United States (California, Nevada, and Utah). Palaios 14:215–233
Martínez S, Rojas A (2013) Relative sea level during the Holocene in Uruguay. Paleogeogr Paleoclimatol Paleoecol 374:123–131
Nagai RH, Sousa SHM, Mahiques MM (2014) The southern Brazil shelf. In: Chiocci FL, Chivas AR (eds) Continental shelves of the world: their evolution during the last glacio-eustatic cycle. Geological Society, London, Memoirs, 41, pp 305–313
Oksanen J, Blanchet FG, Friendly M, Kindt R, Legendre P, McGlinn D, Minchin PR, O'Hara RB, Simpson GL, Solymos P, Henry M, Stevens H, Szoecs E, Wagner H (2019) Vegan: community ecology package. R package version 2.5-5. https://CRAN.R-project.org/package=vegan
Olszewski TD (2004) Modeling the influence of taphonomic destruction, reworking, and burial on time-averaging in fossil accumulations. Palaios 19:39–50
Parsons-Hubbard KM, Brett CE, Walker SE (2011) Taphonomic field experiments and the role of the shelf and slope experimental taphonomy initiative. Paleogeogr Paleoclimatol Paleoecol 312:195–208
Patzkowsky ME, Holland SM (2012) Stratigraphic paleobiology: understanding the distribution of fossil taxa in time and space. The University of Chicago Press, Chicago
Powell EN, Staff GM, Callender WR, Ashton-Alcox KA, Brett CE, Parsons-Hubbard KM, Walker SE, Raymond A (2011) Taphonomic degradation of molluscan remains during thirteen years on the continental shelf and slope of the northwestern Gulf of Mexico. Paleogeogr Paleoclimatol Paleoecol 312:209–232
Powell EN, Hu X, Cai WJ, Ashton-Alcox KA, Parsons-Hubbard KM, Walker SE (2012) Geochemical controls on carbonate shell taphonomy in northern gulf of Mexico Continental Shelf and Slope sediments. Palaios 27:571–584
R Core Team (2018) R: A language and environment for statistical computing. R Foundation for Statistical Computing, Vienna, Austria. http://www.R-project.org/
Ritter MN, Erthal F, Coimbra JC (2013) Taphonomic signatures in molluscan fossil assemblages from the Holocene lagoon system in the northern part of the coastal plain, Rio Grande do Sul State, Brazil. Quat Int 305:5–14
Ritter MN, Erthal F, Coimbra JC (2019) Depth as an overarching environmental variable that modulates the preservation potential and temporal resolution of shelly taphofacies. Lethaia 52:44–56
Ritter MN, Erthal F, Kosnik MA, Coimbra JC, Kaufman DS (2017) Spatial variation in the temporal resolution of subtropical shallow-water molluscan death assemblages. Palaios 32:559–571
Scarponi D, Kaufman DS, Amorosi A, Kowalewski M (2013) Sequence stratigraphy and the resolution of the fossil record. Geology 41:239–242
Simões MG, Kowalewski M (1998) Shell beds as paleoecological puzzles: a case study from the Upper Permian of the Paraná Basin, Brazil. Facies 38:175–196

Smith AM, Nelson C (2003) Effects of early sea-floor processes on the taphonomy of temperate shelf skeletal carbonate deposits. Earth-Sci Rev 63:1–31
Speyer SE, Brett CE (1986) Trilobite taphonomy and Middle Devonian Tafofacies. Palaios 1:312–327
Tomašových A, Zuschin M (2009) Variation in brachiopod preservation along a carbonate shelf-basin transect (Red Sea and Gulf of Aden): environmental sensitivity and tafofacies. Palaios 24:697–716
Tomašových A, Fürsich FT, Olszewski TD (2006) Modeling shelliness and alteration in shell beds: variation in hardpart input and burial rates leads to opposing predictions. Paleobiology 32:278–298
Tomazelli LJ, Villwock JA (2000) O Cenozoico no Rio Grande do Sul: Geologia da Planície Costeira. In: Holz M, De Ros LF (eds) Geologia do Rio Grande do Sul, pp 375–406, CIGO/UFRGS
Vicalvi MA (1977) Sedimentos quaternários da plataforma continental e talude do sul do Brasil: estratigrafia e curvas paleoclimáticas. PETROBRAS, Rio de Janeiro, Série Projeto REMAC 2:27–76
Villwock JA, Tomazelli LJ (1995) Geologia costeira do Rio Grande do Sul. Notas Técnicas 8:1–45
Villwock JA, Tomazelli LJ, Loss EL, Dehnhardt EA, Horn Filho NO, Bachi FA, Dehnhardt BA (1986) Geology of the Rio Grande do Sul Coastal Province. Quat South Am Antarct Pen 4:79–97
Weschenfelder J, Baitelli R, Corrêa ICS, Bortolin EC, Santos CB (2014) Quaternary incised valleys in Southern Brazil coastal zone. J South Am Earth Sci 55:83–93
Young HR, Nelson CR (1985) Biodegradation of temperate-water skeletal carbonates by boring sponges on the Scott shelf, British Columbia, Canada. Mar Geol 65:33–45
Zecchin M, Catuneanu O (2013) High-resolution sequence stratigraphy of clastic shelves I: units and bounding surfaces. Mar Pet Geol 39:1–25

Chapter 2
The Fossil, the Dead, the Living: Beach Death Assemblages and Molluscan Biogeography of the Uruguayan Coast

Alejandra Rojas and Sergio Martínez

Abstract Shelly death assemblages are major sources of information for Actualistic Taphonomy studies. They preserve a sample, often biased, of the living communities, provide biostratinomic evidence useful for taphonomic analyses, allow one to establish shell time averaging and residence time, and are valuable resources in conservation paleobiology studies. Besides these paleontologically related approaches, death assemblages have traditionally been a source for species records in a given location or area. Although a species' geographic range should be established through the collection of living specimens, there are numerous records in the literature based on dead shells from beaches. For the Uruguayan coast, there are some examples of this situation, and they are especially important when considering distributional endpoints. One cause that accounts for dead shells on a beach is the reworking of fossiliferous deposits. Increased knowledge about Uruguayan Quaternary marine deposits can provide explanations for why some species are found in modern beach death assemblages. For example, the bivalve *Anomalocardia flexuosa* has been recorded in several Quaternary outcrops, and its shells have been found on modern beaches, but no living specimens have been recorded on the Uruguayan coast. A shell collected from the Parque del Plata beach (Canelones county) yielded an accelerator mass spectrometry (AMS) radiocarbon age of circa 6,600 years before present (BP), a finding that suggests that the dated shell came from a nearby Holocene deposit. This contribution provides an insight for future research that integrates information from Quaternary and modern marine death assemblages to discuss the molluscan biogeography of the Uruguayan coast.

Keywords Mollusks · Range endpoints · Quaternary · Fossil assemblage · Death assemblage

---

A. Rojas (✉) · S. Martínez
Departamento de Paleontología, Facultad de Ciencias, Universidad de la República, Iguá 4225, 11400 Montevideo, Uruguay
e-mail: alejandra@fcien.edu.uy

S. Martínez
e-mail: smart@fcien.edu.uy

S. Martínez et al. (eds.), *Actualistic Taphonomy in South America*,
Topics in Geobiology 48, https://doi.org/10.1007/978-3-030-20625-3_2

2.1 Introduction

Shelly death assemblages constitute a valuable resource for Actualistic Taphonomy research. They provide evidence of biostratinomic processes, have been useful in time averaging and residence time studies, allow temporal, spatial, and ecological fidelity studies between living and dead counterparts, and represent a tool for applied taphonomy approaches such as conservation paleobiology (e.g., Fürsich and Flessa 1987; Flessa et al. 1993; Kowalewski et al. 1994; Kidwell 1998; 2013; Zuschin and Oliver 2003; Kowalewski and LaBarbera 2004; Kotzian and Simões 2006; Lokwood and Work 2006; Weber and Zuschin 2013; Cristini and De Francesco 2017; Dietl and Flessa 2017). The most diverse and abundant group in modern marine and transitional setting death assemblages are mollusks, due to their high representation in invertebrate communities and their resistant hard parts. Although the biogeographic range of a species should be established from living specimen records, molluscan remains found on beaches and emerged settings have been used for the detection of species in a particular region or area, because they are an accessible place for sampling. On the Uruguayan coast, empty shells (with a variable grade of preservation) had been considered as evidence for the presence of several molluscan species (see examples in Barattini and Ureta 1961; Figueiras and Sicardi 1968, 1971). These records are frequently cited in subsequent contributions, such as taxonomic lists, catalogs, or online databases, in which the original nature of the record may be lost. This situation is especially concerning when considering whether the Uruguayan coast is or may be the endpoint of a taxon's distribution.

Scarabino and Zaffaroni (2004) discussed the status of 20 molluscan species cited in the literature as Uruguayan coastal residents and excluded them from the list of its living fauna. They considered several causes, such as fossil specimen inclusion, misidentified material, ballast specimens, incorrect locality labeling, and species recorded during anomalous oceanographic conditions. Of these possible explanations, our present contribution focuses on those records that may be based on fossil shells found in death assemblages. Paleontological knowledge of Quaternary littoral deposits is essential for this evaluation. Thus, our aim is to discuss how fossil shells can be recognized in Uruguayan coastal shelly death assemblages and how this information may be used to avoid misleading biogeographic assumptions in modern death assemblage studies.

2.2 The Uruguayan Coastal Quaternary Marine Molluscan Assemblages

The Uruguayan Quaternary marine deposits crop out patchily along a thin strip parallel to the present coastline, approximately from the mouth of the Río Negro River to the margin of the Merín Lagoon (Fig. 2.1). They are a consequence of global climatic oscillations and sea level changes that occurred during this period (e.g., Lisiecki and Raymo 2005; Jouzel et al. 2007). The formal lithostratigraphic units used to recognize these deposits are the Chuy and Villa Soriano formations (see Ubilla and Martínez 2016). Molluscan shells are the most diverse and abundant fossils, although there are occasional remains of other taxa such as crustaceans, bryozoans, corals, echinoderms, annelids, and fish (e.g., de Mata 1947; Figueiras 1961, 1962; Sprechmann 1978; Rojas 2007; Ubilla and Martínez 2016).

Quaternary marine mollusks were mentioned for Uruguay since the works of Larrañaga (1819, but published in 1894), d'Orbigny (1842), and Darwin (1846). In the twentieth century, research centered on the taxonomic composition of the deposits mostly located on the coasts of Montevideo, Colonia, and Soriano (e.g., von Ihering 1907, 1923; Teisseire 1928; Frenguelli 1930; Roselli 1939, 1976; de Mata 1947; Calcaterra 1971; Figueiras 1961, 1962, 1967). Some authors took into account the molluscan ecological preferences and made interesting paleoenvironmental inferences (von Ihering 1907; Teisseire 1928; Parodiz 1962), and new information was obtained from deposits located on the eastern coast (Figueiras 1975; Sprechmann

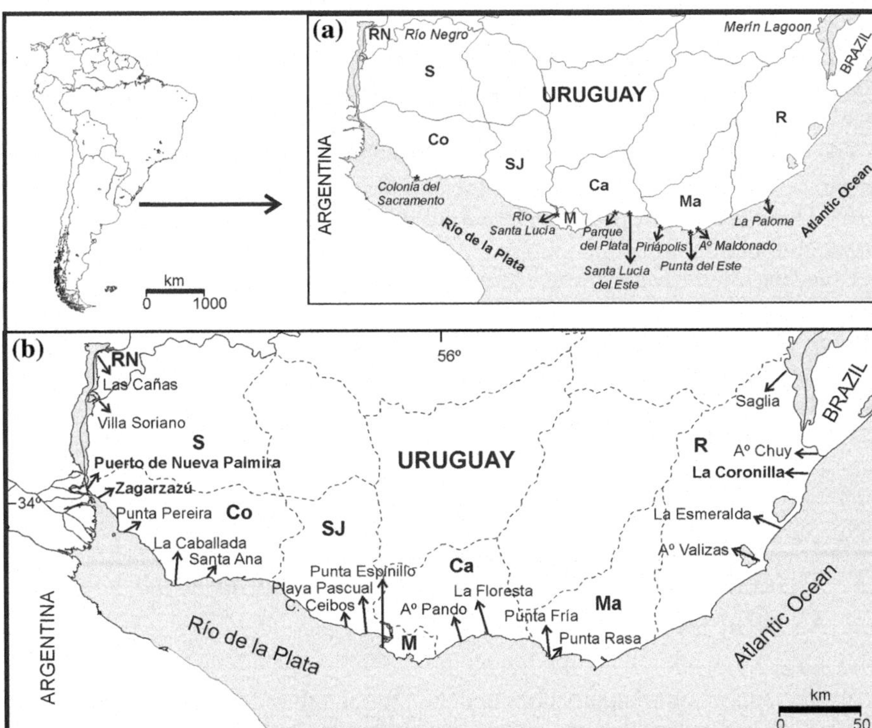

Fig. 2.1 Coastal area of Uruguay. **a** Geographic features mentioned in the text, **b** location of Quaternary fossiliferous marine deposits (bold letters correspond to Pleistocene assemblages and regular letters to Holocene assemblages). Counties are: RN Río Negro, S Soriano, Co Colonia, SJ San José, M Montevideo, Ca Canelones, Ma Maldonado, R Rocha

1978; Martínez 1988, 1990; Piñeiro et al. 1992). More recent approaches include radiocarbon dating of shells to provide a precise chronology of shell concentrations (Martínez et al. 2001, 2006; Rojas 2002, 2007; Rojas and Urteaga 2011; Martínez and Rojas 2013), an effort that led to the recognition of three Late Pleistocene assemblages and numerous Holocene shell beds (see Martínez et al. 2001; Martínez and Rojas 2013; Rojas and Martínez 2016; Ubilla and Martínez 2016). The malacological content and paleoecology of the Pleistocene assemblages were studied by Martínez et al. (2001), Rojas (2007, 2016), Demicheli (2015), Rojas and Martínez (2016), and Rojas et al. (2018a, b). In turn, paleontological knowledge of Holocene marine localities has increased from the report of specific deposits (e.g., Teisseire 1928; Calcaterra 1971; Alonso 1978; Martínez 1983, 1988, 1990; Piñeiro et al. 1992) to a more geographically comprehensive and integrative approach of the malacofauna and paleoecology of precise temporal frameworks (Martínez et al. 2006; Martínez and Rojas 2013; Ubilla and Martínez 2016). Taxonomic lists of Quaternary mollusks were compiled by Figueiras (1961, 1962, 1967) and Clavijo et al. (2005).

The Quaternary beds are characterized by a lithological composition that includes clay, silt, fine to coarse sands, and even conglomerates (Preciozzi et al. 1988). Most deposits are unconsolidated and exposed to the coastal dynamics, since they usually occur on the abrasion platform of sandy beaches, low cliffs, or in river margins and mouths. Due to this fact, fossil shells can potentially reenter the taphonomic active zone and mix into modern littoral death assemblages.

The Zagarzazú and La Coronilla marine assemblages in Colonia and Rocha counties are among the Late Pleistocene deposits that could be a potential source of fossil shells. These fossil beds have been found covered by sand or occasionally underwater during storms (Fig. 2.2).

Holocene molluscan-rich deposits that could provide shells to modern death assemblages are known for all littoral areas from the Río de la Plata to the Atlantic coast. Fossiliferous localities occur in Colonia county (Arroyo La Caballada and Santa Ana-Artilleros), San José county (Playa Pascual), Montevideo (Punta Espinillo), Canelones county (Arroyo Pando, La Floresta, Arroyo Sarandí, Punta Fría, and Punta Rasa), Rocha county (La Esmeralda, Arroyo Valizas, and Arroyo Chuy), and in the margin of the Merín Lagoon (Saglia) (Fig. 2.3).

2.3 Detection of Fossil Shells in Modern Uruguayan Coastal Death Assemblages

Researchers use several approaches to detect fossil valves in modern death assemblages, such as numerical dating, comparison between the taxonomic composition of fossil and modern assemblages in a given area, or by the taphonomic grade exhibited by the collected shells. All approaches have advantages and disadvantages, and the suitability of each for the detection of old shells in modern death assemblages depends on the similarity or dissimilarity in their taxonomic composition, the depositional

Fig. 2.2 Late Pleistocene fossiliferous deposits located at the abrasion platform of the beach. **a** La Coronilla exposed, **b** La Coronilla covered by sand, **c** Zagarzazú exposed, **d** Zagarzazú predominantly covered

and environmental settings involved, the taphonomic history of fossil assemblages, among others.

2.3.1 Taphonomic Grade

The taphonomic signatures of fossil shells in modern death assemblages will depend on their original taphonomic conditions before their reworking into modern death assemblages (e.g., from autochthonous versus parautochthonous to allochthonous deposits), the redeposition environment (e.g., high versus low energy conditions, high or low encrustation and bioturbation incidence, and high versus low sedimentation rate), the rate of freshly dead shell incorporation, and the shell time residence in different environmental settings. For example, in a study based on *Chione* spp. from modern death assemblages of the tidal flats of Bahía la Choya, Mexico, Flessa et al. (1993) concluded that the taphonomic condition did not reliably indicate the time since death. On the other hand, De Francesco and Hassan (2008) found that in an Argentine estuary modern death assemblages were mostly composed of fossil shells

Fig. 2.3 Holocene fossiliferous deposits exposed to erosive processes. **a** Santa Ana-Artilleros, **b** Cañada de los Ceibos, Kiyú, **c** Playa Pascual, **d** Arroyo Pando, **e** La Floresta, **f** La Esmeralda

and were worse preserved than fossil assemblages, due to reworking and exposition of these shells by modern biostratinomic processes.

2.3.2 Numerical Dating

Fossil shells may be detected in modern death assemblages if old ages are obtained by direct dating techniques, such as accelerator mass spectrometry (AMS) radiocarbon dating or radiocarbon calibrated amino acid racemization. One limitation for this approach is that not every collected shell will be dated. Among the dated shells, those derived from Pleistocene terraces may be successfully recognized due to the expected age differences (e.g., Wehmiller et al. 1995). However, the ages of shells that come from young (Holocene) fossil deposits may overlap with the residence time of modern time-averaged shelly assemblages. Several studies have dealt with shell survival and estimates of time-averaging in different marine subenvironments. Kidwell and Bosence (1991) estimated that maximum scales of time-averaging in coastal environments would be on the order of a few thousand years, while in open shelf habitats it would tens-of-thousands of years. Flessa et al. (1993) found that surficial shells may be 3,500 years old in Bahia La Choya accumulations, and their long survival is explained by frequent shallow burial. Flessa and Kowalewski (1994) compiled data on radiocarbon ages from the literature and found that the median age of 128 nearshore shells was 2,465 years, and that of 158 shells from the shelf was 8,870 years. The authors interpreted that the lower shell survival in nearshore deposits (<10 m) is linked to a more active sedimentary environment where shells may be rapidly destroyed or buried, in contrast to deeper environments with a lower sedimentation rate and lower taphonomic destruction rate. Recently, Kidwell (2013) summarized the literature on shell dating from death assemblages. Dating from modern-day outer shelfs yielded maximum ages of 10,000–20,000 years, whereas shells from inner shelves and beaches are typically younger, with ages of a few thousand years.

2.3.3 Comparison Between the Taxonomic Composition of Fossil Assemblages and Modern Communities

This approach may be useful if the taxonomic composition between the fossil assemblages and the adjacent living fauna from a given area differ. For example, de Francesco and Hassan (2008) found that death assemblages in the Mar Chiquita coastal lagoon were dominated by reworked fossil shells derived from Holocene beds at the lagoon margins, a finding that consequently demonstrates low ecological fidelity to the living fauna.

On the Uruguayan coast, several species are shared between the malacological fossil assemblages and the living communities. However, environmental parameters of some coastal areas have changed in the last thousand years, as well as the taxonomic composition and geographic distribution of mollusks.

2.3.3.1 Small-Scale Shifts in Geographical Distribution Due to Salinity Changes

The most noticeable difference between past and present environmental conditions is salinity. While there is currently an increasing salinity gradient from west to east (e.g., Urien 1972; Nagy et al. 1998), fossil assemblages indicate the prevalence of marine conditions into what is now the Rio de la Plata estuary (Martínez et al. 2001, 2006; Rojas 2007). In the present Atlantic region, Quaternary assemblages indicate,

Fig. 2.4 Mollusks from Late Pleistocene and Holocene assemblages from Uruguay. **a** *Erodona mactroides*, **b** *Tagelus plebeius*, **c** *Mactra isabelleana*, **d** *Pitar rostratus*, **e** *Ostrea stentina*, **f** *Heleobia australis*, **g** *Stramonita haemastoma*, **h** *Tellina gibber*, **i** *Bulla occidentalis*, **j** *Anadara brasiliana*, **k** *Anomalocardia flexuosa*. Scale bars equal 2 mm except in b that equal 5 mm

as expected, that marine littoral environments developed (Martínez et al. 2001, 2006; Rojas and Urteaga, 2011; Rojas et al. 2018b).

If past and present species geographical distributions with specific salinity preferences do not overlap, this criterion may allow for the detection of fossil specimens in modern death assemblages. For example, *Erodona mactroides* and *Tagelus plebeius* are estuarine bivalves which currently reside in the Río de la Plata, coastal lagoons, and river mouths of the Canelones, Maldonado, and Rocha counties (Scarabino et al. 2006b, 2015). Both species are commonly found in Holocene deposits (Fig. 2.4a, b), in which, for example, *E. mactroides* is sometimes dominant (Martínez et al. 2006). Contrarily, both species are rare in Pleistocene assemblages (Rojas et al. 2018a).

Marine euryhaline species such as *Mactra isabelleana* and *Pitar rostratus* are also found in Río de La Plata estuarine environments and in unconsolidated coastal and shelf area substrates. Fossil specimens for these species are present in both Pleistocene and Holocene coastal deposits, where *M. isabelleana* is more abundant than *P. rostratus* (Fig. 2.4c, d). While *Ostrea stentina* is another abundant species in the Pleistocene assemblages, and frequently found in Holocene deposits (Martínez et al. 2006; Rojas et al. 2018a), it is currently recorded in consolidated Atlantic coast substrates (Scarabino et al. 2006b) (Fig. 2.4e). A common microgastropod of the Río de La Plata and other estuarine environments is *Heleobia*. Specimens are often referred to the species *Heleobia australis*, as found in the Quaternary records. This gastropod is common and usually abundant in Holocene assemblages from different coastal sectors (Martínez et al. 2006) (Fig. 2.4f). For Pleistocene deposits, it is abundant in the Zagarzazú assemblage but rarely occurs in Puerto de Nueva Palmira and La Coronilla (Rojas et al. 2018a). All of these frequently or abundantly represented species in Quaternary assemblages demonstrate partially overlapping ranges with their current distribution in the Uruguayan coastal zone. Thus, in the latest areas, it is difficult to know whether the shells correspond to fossil specimens or members of modern death assemblages. However, where the current and fossil geographic distribution does not overlap, it is probable to identify shells reworked from Quaternary deposits. The reason is that mollusks responded to environmental changes during the last thousand years by modifying their biogeographic distribution on the Uruguayan coast.

The salinity inferences along the coast during the Quaternary, and especially in the Late Pleistocene, show a high marine influence in the western sector, where marine species are recorded (Martínez et al. 2001, 2006; Rojas and Martínez 2016). For instance, species found in the Pleistocene deposits of Puerto de Nueva Palmira, such as *Diodora patagonica*, *Lottia subrugosa*, *Lunarca ovalis*, and *Plicatula gibbosa*, are currently restricted to the Atlantic coast, about 400 km to the southeast of these deposits. *Stramonita haemastoma*, currently found to the east of Punta del Este, was recorded both in Puerto de Nueva Palmira and Zagarzazú (Fig. 2.4g); a similar situation occurs with *Tellina gibber* (Rojas and Martínez 2016) (Fig. 2.4h). Regarding marine species found in western Holocene deposits, *Buccinanops globulosus*, *Olivella tehuelcha*, and *T. gibber* were reported in La Caballada (Colonia del Sacramento). Even westwards, *L. ovalis* was recorded from the Villa Soriano locality, close to the Río Negro (Martínez et al. 2006).

Similar range displacements occur for estuarine species, such as *E. mactroides* and *T. plebeius*, since they were found westward of their current distribution boundary in the Río de La Plata. Colonia del Sacramento is nowadays the western endpoint for *E. mactroides* (Olazarri 1966), and Montevideo for *T. plebeius* (Scarabino et al. 1975). Both species are rare in the Puerto de Nueva Palmira and Zagarzazú Pleistocene deposits. However, *E. mactroides* is very abundant in most Holocene assemblages (Martínez et al. 2006). In deposits to the west of Colonia del Sacramento, *E. mactroides* is a dominant species. For instance, in Las Cañas (Río Negro county) the assemblage is monospecific. In Punta Pereira, this species represents more than 80% of the recorded specimens, and in Villa Soriano it comprises about the 60% of the recorded mollusks (see Martínez et al. 2006). The presence of *E. mactroides* shells in modern death assemblages to the west of Colonia del Sacramento is probably due to the erosion of fossil deposits, as commented by Olazarri (1966), based on the finding of highly abraded shells.

For other species, such as euryhaline taxa, a similar pattern was verified since they are recorded as displaced from their present distribution. Meanwhile, *M. isabelleana* decreases its abundance to the west of the mouth of the Río Santa Lucía (Scarabino et al. 2006b), but it is recorded in several fossil assemblages towards the northwest, reaching Villa Soriano (Martínez et al. 2006). Similarly, *P. rostratus* is found in the eastern coast up to Montevideo, while in the Holocene assemblages it is also recorded in Villa Soriano (Martínez et al. 2006).

2.3.3.2 Out-of-Range Warm Water Taxa Currently Displaced Northwards from the Uruguayan Coast

Besides the small-scale biogeographic modifications within the Uruguayan coast due to salinity changes, Quaternary molluscan assemblages record species not currently living at this latitude. These warm-water taxa are extralimitals (see Roy et al. 1995), since their southern biogeographic boundary is located at the Brazilian coast. Extralimitals were reported in both Pleistocene and Holocene assemblages (Martínez et al. 2001, 2006; Rojas and Martínez 2016; Rojas et al. 2018b) and corresponded to micro- and macromollusks. Several of these extralimitals were traditionally included among the living Uruguayan molluscan fauna and their status has been a matter of uncertainty since the records come from dead shells collected on beaches. Scarabino and Zaffaroni (2004) clarified the status of *Bulla occidentalis* (as *Bulla striata*), *Anadara chemnitzii* (as *Scapharca chemnitzii*), *Anadara brasiliana* (as *Scapharca brasiliana*), *Anomalocardia flexuosa* (as *Anomalocardia brasiliana*), *Littoraria flava*, and *Melampus coffeus*. The last species was not reported from fossil assemblages, unlike the others.

The *B. occidentalis* record in Rocha county by Barattini and Ureta (1961) and Figueiras and Sicardi (1974) was considered to be due to their presence in Holocene assemblages from the same region, as reported by Figueiras (1967) and later by Martínez et al. (2006) from Saglia (Fig. 2.4i). Further, this species was recently found in the La Coronilla Pleistocene assemblage (Rojas et al. 2018b).

Incomplete *A. brasiliana* specimens were reported by Barattini and Ureta (1961) on the La Coronilla beach, while *A. chemnitzii* loose valves were found by Figueiras and Sicardi (1968) between La Coronilla and Chuy. Despite these findings, and the fact that the latter authors mention the presence of these two arcids in Holocene deposits (see Figueiras 1961), they cited the eastern Uruguayan coast as the geographic boundary for these species. Later, Scarabino and Zaffaroni (2004) excluded both records from the living Uruguayan malacofauna and indeed further commented that the living distribution of *Anadara* (as *Scapharca*) in southern Brazil was not yet been established by the record of living specimens. The shells from *A. brasiliana* and *A. chemnitzii* found in death assemblages of La Coronilla and eastward localities can be more confidently regarded as fossils due to the record of both species in the La Coronilla Pleistocene deposit (Rojas and Martínez 2016; Rojas et al. 2018b) (Fig. 2.4j).

L. flava was included among the Uruguayan coast living mollusks by Parodiz (1962) without providing any distribution detail. Later, Figueiras and Sicardi (1971) mentioned this species (as *Littorina nebulosa flava*) on the basis of a deteriorated specimen found by the authors in the Puerto de La Paloma in Rocha and by records in private collections. These authors attributed the occasional presence of this species in the Uruguayan coast to special oceanographic conditions. Based on a preliminary communication by Rojas (2003) on the presence of *L. flava* in the Zagarzazú Pleistocene deposit, Scarabino and Zaffaroni (2004) considered that the record of empty shells could be fossil specimens since it was never found alive on the Uruguayan coast.

Another example is *Finella dubia*, a microgastropod reported by Figueiras and Sicardi (1980) and Layerle and Scarabino (1984) in Rocha and Maldonado counties, and therefore considered among the living Uruguayan coast malacofauna (e.g., Scarabino 2003). Only empty and abraded shells from this species were found by Scarabino et al. (2006a), and for this reason the authors cast doubt on the living status of *F. dubia*. Moreover, Forcelli and Narosky (2015) excluded the Uruguayan coast from the southern boundary of this species' distribution. Similar to the examples cited above, *F. dubia* was found in all Pleistocene assemblages (Martínez et al. 2001; Rojas and Martínez 2016) and in the Arroyo Chuy Holocene deposit (Martínez et al. 2006). Therefore, these records can explain the presence of its shells on the coast.

Eurytellina angulosa was regarded as a living Uruguayan species by Boss (1968) on the basis of material from the Arroyo Pando in Canelones. Scarabino and Zaffaroni (2004) considered the record as erroneous because no specimens were reported from the Uruguayan Atlantic area. Moreover, the Arroyo Pando Holocene deposit, mentioned by Broggi (1967) and more recently studied by Rojas (2007), does not contain any similar species. However, *E. angulosa* was recently found in the La Coronilla Pleistocene assemblage (Rojas et al. 2018b), and consequently it cannot be completely ruled out that the record reported by Boss (1968) corresponds to a fossil specimen.

Barattini and Ureta (1961) regarded the Uruguayan coast as the southern boundary of the *A. flexuosa* distribution, because complete specimens were found at La Coronilla beach and Arroyo Maldonado. Figueiras and Sicardi (1969) referenced the previous work and mentioned that *A. flexuosa* is very common in Holocene assemblages, and thus the reported shells could come from nearby fossil deposits. Despite this assertion, the authors still considered that the mouth of the Río de la Plata was the boundary of this species' distribution. *A. flexuosa* is indeed a common species in Quaternary assemblages along the Uruguayan coast (Fig. 2.4k). It was found in the Pleistocene marine assemblages (Martínez et al. 2001; Rojas and Martínez 2016) and Holocene beds from Soriano, Colonia, San José, Montevideo, Canelones, and Rocha counties (Figueiras 1961; Martínez et al. 2006). Besides considering that *A. flexuosa* shells could come from fossil assemblages, Scarabino and Zaffaroni (2004) also allowed for the possibility that the record by Barattini and Ureta (1961) could

Table 2.1 Radiocarbon ages obtained for *A. flexuosa* from a modern death assemblage and ages from nearby Holocene deposits. Calib 7.10 was used for calibration (Stuiver et al. 2018)

Species dated	Locality	^{14}C age	95.4% (2δ) cal. age BP	Lab. Nr.	Age source
Anomalocardia flexuosa	Modern death assemblage on Parque del Plata beach	6,650 ± 50	7,418–7,618	AA93851	This work
Mactra isabelleana	La Floresta (Holocene deposit)	4,790 ± 80	5,316–5,700	LP-904	Martínez et al. (2006)
Anomalocardia flexuosa (in life position)	Arroyo Pando (Holocene deposit)	6,630 ± 60	7,398–7,625	LP-1464	Rojas (2007)

be due to incorrect labeling or the existence of anomalous oceanographic conditions that favored the transport of *A. flexuosa* to the Uruguayan coast. Recently, Rojas et al. (2018a, b) reported articulated *A. flexuosa* specimens from the La Coronilla fossil assemblage, a finding which may explain the shell record of Barattini and Ureta (1961) at the same beach. The absence of living *A. flexuosa* records at the Uruguayan coast, and even from the adjacent Rio Grande do Sul State in Brazil, has been well established (see Scarabino et al. 2015). Despite this fact, recent research based on modern death assemblages still included *A. flexuosa* in analyses regarding molluscan composition and diversity of Maldonado beaches (as *A. brasiliana*; Boretto et al. 2018).

We found *A. flexuosa* shells on the Parque del Plata and Santa Lucía del Este beaches, and in Piriápolis in Maldonado. One valve of this species from Parque del Plata was AMS dated 6,650 ± 50 radiocarbon years before present (BP) (Table 2.1). Considering the calibrated age interval (median probability of 7,520 years BP), and taking into account that the range of ages represented by shells in recent beach death assemblages is a few thousand years (e.g., Kidwell 2013), it is probable that the dated *A. flexuosa* shell was eroded from a nearby Holocene deposit. This hypothesis is plausible since Holocene assemblages are known to occur at a short distance from Parque del Plata. Approximately 5 km to the east, the La Floresta assemblage was studied by Martínez et al. (2006) in which a radiocarbon age of 4,790 ± 80 years BP (cal. age of 5,316–5,700 years BP) was obtained. Although the mollusks originally reported from La Floresta deposit were *H. australis* and *M. isabelleana*, we recently recorded the presence of *A. flexuosa* (unpublished). Additionally, about 14 km to the west of Parque del Plata, the beds exposed at the margin of the Arroyo Pando were studied by Rojas (2007). Among the molluscan species recorded, life position specimens of *A. flexuosa* were used for standard ^{14}C dating. The obtained radiocarbon age, 6,630 ± 60 years BP (calibrated age range between 7,398–7,625 years BP), is almost identical to *A. flexuosa* shell age collected from the death assemblage of Parque del Plata.

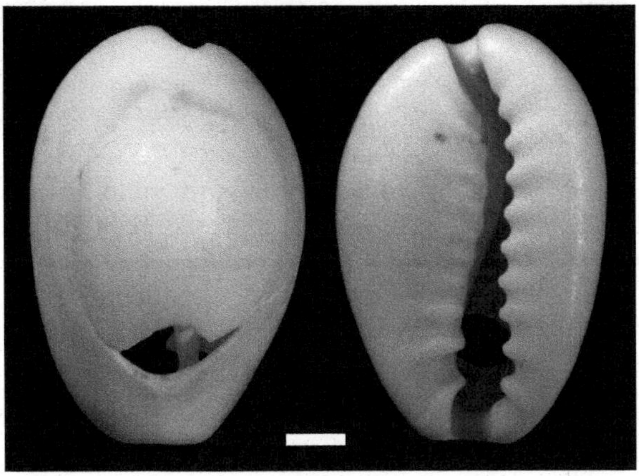

Fig. 2.5 Specimen of the alien species *Monetaria annulus* found in the modern death assemblage of Piriápolis beach, Maldonado county

2.4 Alien Molluscan Shells in Beach Death Assemblages

In January 2018, during an inspection of the death assemblage from the Piriápolis beach (Maldonado county), a shell from the Cypraeidae family was found. Despite its small size and abraded condition, it was identified as *Monetaria annulus*, a species from the central and western Pacific and Indian Oceans (Simone 2004) (Fig. 2.5). This taxon, and other alien cypreids such as *Monetaria moneta*, and western Atlantic representatives such as *Erosaria acicularis* and *Luria cinerea*, were found on several Uruguayan beaches (Juan Carlos Zaffaroni pers. comm. 2018). According to compilations by Ríos (2009) and Rosenberg (2009), the southernmost reaching species of this family in the southwestern Atlantic is *Macrocypraea zebra*, which occurs off the Santa Catarina coast in Brazil. Considering the living specimen records, *E. acicularis* was reported at the northern Saõ Paulo coast (Simone and Gonçalves 2006). Among the Uruguayan Quaternary assemblages, no Cypraeidae family members have been reported. However, as mentioned above, extralimital warm water species are common in fossil deposits. The presence of the alien *M. annulus* in Uruguayan coastal death assemblages may be explained by diverse anthropic activities, such as religious rituals that take place on beaches, as a part of pieces of craftsmanship or ballast remains. Beyond the finding of alien species shells, the extralimital shell record of western Atlantic taxa must then be carefully evaluated to avoid misleading biogeographic interpretations when studying death assemblages.

2.5 Final Remarks

The geographic distribution of the Uruguayan Quaternary marine assemblages allow for the possibility that fossil mollusk shells may be reworked into modern beach death assemblages. This has complicated the literature concerning the living Uruguayan molluscan fauna, since species recorded by empty shells were assumed to be part of living communities. Inaccurate information was afterwards amplified in subsequent publications, taxonomic lists, inventories, and webpages. The status of several species was clarified and referred to the possibility that living records may be based on fossil shells (Scarabino and Zaffaroni 2004). The current increased and improved knowledge on the taxonomic composition, location, and age of littoral Uruguayan Quaternary molluscan assemblages becomes a relevant source of information when studying death assemblages. Detection of fossil shells as part of modern death assemblages by means of numerical dating or their geographical displacement may help recognize them and assess the importance of fossil deposit erosion. Being aware of this phenomenon will contribute to the correct interpretation of the taxonomic composition of modern beach death assemblages, and their use in different actualistic approaches.

Acknowledgements We deeply thank Juan Carlos Zaffaroni for his comments on alien molluscan species found on Uruguayan beaches, and for his help in the identification of the *M. annulus* specimen.

References

Alonso C (1978) La fauna de moluscos del yacimiento de Playa Pascual con referencia a otros yacimientos estuáricos y marinos del Cuaternario de Uruguay. Com Soc Malac Uruguay 4(34):365–380

Barattini LP, Ureta EH (1961) La fauna de las costas uruguayas del este (invertebrados). Publicaciones del Museo Dámaso Antonio Larrañaga, Montevideo

Boretto GM, Rouzaut S, Cioccale M, Gordillo S, Benitez Y (2018) Dinámica costera y antropización en playas uruguayas. Un análisis integrado para su conservación. Rev Mex Ciencias Geol 35(3):291–306

Boss KJ (1968) The subfamily Tellininae in the Western Atlantic. The Genera *Tellina* (Part II) and *Tellidora*. Johnsonia 4(46):273–344

Broggi J (1967) Primera cita de *Pholas* (*Thovana*) *campechiensis* Gmelin para el Querandino uruguayo. Com Soc Malac Uruguay 2(12):47–48

Calcaterra A (1971) Algunas observaciones sobre la fauna de moluscos de la Formación Querandina (Deptos de Colonia y Soriano, Uruguay). Com Soc Malac Uruguay 3(20):79–82

Clavijo C, Scarabino F, Rojas A, Martínez S (2005) Lista sistemática de los moluscos marinos y estuarinos del Cuaternario del Uruguay. Com Soc Malac Uruguay 9(88):381–411

Cristini PA, de Francesco CG (2017) Molluscan taphonomic patterns below the sediment-water interface in freshwater shallow lakes from the southeastern Pampa plain, Argentina. Palaios 32(8):528–542

Darwin C (1846) Geological observations on South America, being the third part of the geology of the voyage of the Beagle, under the command of capt. Fitzroy RN during the years 1832–1836. Smith, Elder & Co., London

de Francesco CG, Hassan GS (2008) Dominance of reworked fossil shells in modern estuarine environments: implications for paleoenvironmental reconstructions based on biological remains. Palaios 23(1):14–23

de Mata O (1947) La formación holocena en el Departamento de Montevideo (República Oriental del Uruguay). Imprenta Nacional, Montevideo

Demicheli M (2015) Moluscos fósiles del Puerto de Nueva Palmira, Pleistoceno Tardío de Uruguay. Bachelor dissertation, Facultad de Ciencias, Universidad de la República
Dietl GP, Flessa KW (2017) Conservation paleobiology. Science and practice. The University of Chicago Press, Chicago
d'Orbigny A (1836–1843) Paléontologie. In: Voyage dans l'Amérique méridionale (le Brésil, la République orientale de l'Uruguay, la République Argentine, la Patagonie, la République du Chili, la République de Bolivia, la République du Pérou), exécuté pendant les années 1826–1833, t.3(4). París/Estrasburgo
Figueiras A (1961) Contribución al conocimiento de la malacofauna holocena del Uruguay. Com Soc Malac Uruguay 1(1):15–21
Figueiras A (1962) Sobre nuevos hallazgos de moluscos subfósiles de la Transgresión Querandina. Com Soc Malac Uruguay 1(3):53–68
Figueiras A (1967) Contribución al conocimiento de los moluscos marinos del Holoceno uruguayo. Com Soc Malac Uruguay 2(12):61–76
Figueiras A (1975) Sobre la existencia de un banco de ostréidos de probable edad pleistocénica media en el departamento de Rocha, Uruguay. Com Soc Malac Uruguay 4(28):59–92
Figueiras A, Sicardi OE (1968) Catálogo de los moluscos marinos del Uruguay. Parte II. Com Soc Malac Uruguay 2(15):255–275
Figueiras A, Sicardi OE (1969) Catálogo de los moluscos marinos del Uruguay. Parte III. Com Soc Malac Uruguay 2(16):355–378
Figueiras A, Sicardi OE (1971) Catálogo de los moluscos marinos del Uruguay. Parte VI. Com Soc Malac Uruguay 3(21):101–130
Figueiras A, Sicardi OE (1974) Catálogo de los moluscos marinos del Uruguay. Parte IX. Com Soc Malac Uruguay 3(26):323–360
Figueiras A, Sicardi OE (1980) Catálogo de los moluscos marinos del Uruguay. Parte X. Revisión actualizada de los moluscos marinos del Uruguay con descripción de las especies agregadas. Sección II—Gastropoda y Cephalopoda. Com Soc Malac Uruguay 5(38):179–272
Flessa KW, Kowalewski M (1994) Shell survival and time-averaging in nearshore and shelf environments: estimates from the radiocarbon literature. Lethaia 27:153–165
Flessa KW, Cutler AH, Meldahl KH (1993) Time and taphonomy: quantitative estimates of time-averaging and stratigraphic disorder in a shallow marine habitat. Paleobiology 19(2):266–286
Forcelli D, Narosky T (2015) Uruguayan seashells. Moluscos marinos Argentina, Uruguay, Brasil. Vázquez Manzini Editores, Buenos Aires
Frenguelli J (1930) Apuntes de Geología Uruguaya. Bol Inst Geol Perf 11:1–47
Fürsich FT, Flessa KW (1987) Taphonomy of tidal flat molluscs in the Northern Gulf of California: paleoenvironmental analysis despite the perils of preservation. Palaios 2(6):543–559
Jouzel J, Masson-Delmotte V, Cattani O, Dreyfus G, Falourd S, Hoffmann G, Minster B, Nouet J, Barnola JM, Chappellaz J, Fischer H, Gallet JC, Johnsen S, Leuenberger M, Loulergue L, Luethi D, Oerter H, Parrenin F, Raisbeck G, Raynaud D, Schilt A, Schwander J, Selmo E, Souchez R, Spahni R, Stauffer B, Steffensen JP, Stenni B, Stocker TF, Tison JL, Werner M, Wolff EW (2007) Orbital and millennial Antarctic climate variability over the past 800,000 years. Science 317:793–796
Kidwell SM (1998) Time-averaging in the marine fossil record: overview of strategies and uncertainties. Geobios 30(7):977–995
Kidwell SM (2002) Time-averaged molluscan death assemblages: palimpsests of richness, snapshots of abundance. Geol Soc Am 30(9):803–806
Kidwell SM (2013) Time-averaging and fidelity of modern death assemblages: building a taphonomic foundation for conservation palaeobiology. Palaeontology 56(3):487–522
Kidwell SM, Bosence DW (1991) Taphonomy and time averaging of marine shelly faunas. In: Allison PA, Briggs DEG (eds) Taphonomy: releasing the data locked in the fossil record. Plenum Press, New York
Kotzian CB, Simões MG (2006) Taphonomy of recent freshwater molluscan death assemblages, Touro Passo stream, southern Brazil. Rev Bras Paleontol 9(2):243–260

Kowalewski M, Labarbera M (2004) Actualistic taphonomy: death, decay, and disintegration in contemporary settings. Palaios 19(5):423–427
Kowalewski M, Flessa KW, Aggen JA (1994) Taphofacies analysis of recent shelly cheniers (beach ridges), Northeastern Baja California, Mexico. Facies 31:209–242
Larrañaga DA (1894) Memoria geológica sobre la formación del Río de la Plata, deducida de sus conchas fósiles, escrita por los años 1819. Anales del Museo Nacional de Montevideo 1(2):3–12
Layerle C, Scarabino V (1984) Moluscos del frente marítimo uruguayo entre 9 y 78 m de profundidad: análisis biocenológico. Contribuciones (Departamento de Oceanografía, Facultad de Humanidades y Ciencias) 1(9):1–17
Lisiecki LE, Raymo ME (2005) A Pliocene-Pleistocene stack of 57 globally distributed benthic $\delta^{18}O$ records. Paleoceanography 20:PA1003
Lockwood R, Work LA (2006) Quantifying taphonomic bias in molluscan death assemblages from the Upper Chesapeake Bay: patterns of shell damage. Palaios 21:442–450
Martínez S (1983) Invertebrados fósiles holocenos del Departamento de Rocha, Uruguay (Mollusca y Crustacea). Resúmenes y Comunicaciones de las Jornadas de Ciencias Naturales 3:46–47
Martínez S (1988) Moluscos fósiles holocenos de la margen occidental de la Laguna Merín (Formación Villa Soriano, Uruguay). Com Soc Malacol Uruguay 6:409–418
Martínez S (1990) Taphonomy and Palaeoecology of Holocene mollusks from the western margin of the Merin Lagoon (Villa Soriano Fm., Uruguay). Quat South Am Ant Pen 7:121–135
Martínez S, Rojas A (2013) Relative sea level during the Holocene in Uruguay. Palaeogeogr Palaeoclimatol Palaeoecol 374:123–131
Martínez S, Ubilla M, Verde M, Perea D, Rojas A, Guérèquiz R, Piñeiro G (2001) Paleoecology and geochronology of Uruguayan coastal marine Pleistocene deposits. Quat Res 55:246–254
Martínez S, Rojas A, Ubilla M, Verde M, Perea D, Piñeiro G (2006) Molluscan assemblages from the marine Holocene of Uruguay: composition, geochronology, and paleoenvironmental signals. Amegh 43(2):385–397
Nagy GJ, Martínez CM, Caffera RM, Pedrosa G, Forbes EA, Perdomo AC, López Laborde J (1998) Capítulo 2. Marco hidrológico y climático del Río de la Plata. In: Wells PG, Daborn GR (eds) El Río de la Plata –Una revisión ambiental. Un informe del Proyecto ECOPLATA. Dalhousie University, Halifax, Nova Scotia
Olazarri J (1966) Los moluscos de agua dulce del Depto. de Colonia, Uruguay. Parte I: Pelecypoda. Com Soc Malac Uruguay 2(11):15–37
Parodiz JJ (1962) Los moluscos marinos del pleistoceno rioplatense. Com Soc Malac Uruguay 1(2):29–46
Piñeiro G, Scarabino F, Verde M (1992) Una nueva localidad fosilífera del Holoceno marino de Uruguay (Punta Rasa, Departamento de Maldonado). Bol Soc Zool Uruguay 7:61–62
Preciozzi F, Spoturno J, Heinzen W, Rossi P (1988) Memoria explicativa de la Carta Geológica del Uruguay a la escala 1: 500.000. DINAMIGE, Montevideo
Ríos E (2009) Compendium of Brazilian sea shells. Evangraf, Rio Grande
Rojas A (2002) Tafonomía y cronología de nuevos depósitos fosilíferos del Holoceno marino de Uruguay. Actas de las II Jornadas Uruguayas del Cenozoico 2:80–82
Rojas A (2003) Moluscos marinos cuaternarios del Balneario Zagarzazú, Departamento de Colonia, Uruguay. Pub Esp Soc Zool Urug 1:83
Rojas A (2007) Moluscos de aguas cálidas del Cuaternario Marino del Uruguay. Master thesis dissertation, Universidad de la República
Rojas A (2016) Asociaciones fósiles del Pleistoceno marino de Uruguay. PhD thesis dissertation, Universidad de la República
Rojas A, Urteaga D (2011) Late Pleistocene and Holocene chitons (Mollusca, Polyplacophora) from Uruguay: palaeobiogeography and palaeoenvironmental reconstruction in mid latitudes of the Southwestern Atlantic. Geobios 44:377–386
Rojas A, Martínez S (2016) Marine Isotope Stage 3 (MIS 3) versus Marine Isotope Stage 5 (MIS 5) fossiliferous marine deposits from Uruguay. In: Gasparini GM, Rabassa J, Deschamps MC, Tonni

EP (eds) Marine Isotope Stage 3 in Southern South America 60 KA B.P.-30 KA B.P. Springer Earth System Sciences, Cham

Rojas A, Demicheli M, Martínez S (2018a) Taphonomy of the Late Pleistocene marine molluscan assemblages from Uruguay. N Jahrb Geol Paläontol - Abhandlungen 289(2):217–235

Rojas A, Zaffaroni JC, Martínez S (2018b) New molluscan records and palaeoecology of the Late Pleistocene marine assemblage from La Coronilla (Rocha, Uruguay). J Sediment Environ 3(4):220–233

Roselli FL (1939) Apuntes de Geología y Paleontología uruguayas. Bol Soc Am Cs Nat "Kraglievich-Fontana" 1(2):29–102

Roselli FL (1976) Contribución al estudio de la Geo Paleontología Departamentos de Colonia y Soriano (República Oriental del Uruguay). Imprenta Cooperativa, Montevideo

Rosenberg G (2009) Malacolog 4.1.1: A Database of Western Atlantic Marine Mollusca. [WWW database (version 4.1.1)]. http://www.malacolog.org/. Accessed 26 Nov 2018

Roy K, Jablonski D, Valentine JW (1995) Thermally anomalous assemblages revisited: patterns in the extraprovincial latitudinal range shifts of Pleistocene marine mollusks. Geology 23(12):1071–1074

Scarabino F (2003) Lista sistemática de los Bivalvia marinos y estuarinos vivientes de Uruguay. Com Soc Malacol Uruguay 8(80–81):229–259

Scarabino F, Zaffaroni JC (2004) Estatus faunístico de veinte especies de moluscos citadas para aguas uruguayas. Com Zool Mus Nac Hist Nat Antr 13(202):1–15

Scarabino V, Maytía S, Cachés M (1975) Carta bionómica litoral del Departamento de Montevideo. I Niveles superiores del Sistema Litoral. Com Soc Malac Uruguay 4(29):117–129

Scarabino F, Zaffaroni JC, Carranza A, Clavijo C, Nin M (2006a) Gasterópodos marinos y estuarinos de la costa uruguaya: faunística, distribución, taxonomía y conservación. In: Menafra R, Rodríguez L, Scarabino F, Conde D (eds) Bases para la conservacion y el manejo de la costa uruguaya. Vida Silvestre, Montevideo

Scarabino F, Zaffaroni JC, Clavijo C, Carranza A, Nin M (2006b) Bivalvos marinos y estuarinos de la costa uruguaya: faunística, distribución, taxonomía y conservación. In: Menafra R, Rodríguez L, Scarabino F, Conde D (eds) Bases para la conservacion y el manejo de la costa uruguaya. Vida Silvestre, Montevideo

Scarabino F, Zelaya DG, Orensanz JM, Ortega L, Defeo O, Schwindt E, Carranza A, Zaffaroni JC, Martínez G, Scarabino V, García-Rodríguez F (2015) Cold, warm, temperate and brackish: Bivalve biodiversity in a complex oceanographic scenario (Uruguay, southwestern Atlantic). Am Malacol Bull 33(2):1–18

Simone LRL (2004) Morphology and phylogeny of the Cypraeoidea (Mollusca, Caenogastropoda). Papel Virtual, Fapesp, São Paulo

Simone LRL, Gonçalves EP (2006) Reanalysis of the southernmost distribution of *Erosaria acicularis* (Caenogastropoda, Cypraeidae) in Brazil. Strombus 13(2):13–14

Sprechmann P (1978) The paleoecology and paleogeography of the Uruguayan coastal area during the Neogene and Quaternary. Zitt 4:3–72

Stuiver M, Reimer PJ, Reimer RW (2018) CALIB 7.1 [WWW program] at http://calib.org. Accessed 18 Oct 2018

Teisseire A (1928) Contribución al estudio de la geología y la paleontología de la República Oriental del Uruguay. Región de Colonia. Anales Univ 122(series 37):319–469

Ubilla M, Martínez S (2016) Geology and paleontology of the quaternary of Uruguay. Springer Brief in Earth Sciences, South America and the Southern Hemisphere, Springer, Basel

Urien CM (1972) Río de la Plate estuary environments. In: Nelson BW (ed) Environmental framework of Coastal Plain Estuaries. Boulder; Geol Soc Am Mem 133, pp 213–234

von Ihering H (1907) Les mollusques fossiles du Tertiaire et du Crétacé Supérieur de l'Argentine. An Mus Nac Buenos Aires 7(3 Série):1–611

von Ihering H (1923) Transgression des Meeres während der Ablagerung der Pampas. Archiv für Molluskenkunde 55(5):183–193

Weber K, Zuschin M (2013) Delta-associated molluscan life and death assemblages in the northern Adriatic Sea: implications for paleoecology, regional diversity and conservation. Palaeogeogr Palaeoclimatol Palaeoecol 370:77–91

Wehmiller JF, York LL, Bart ML (1995) Amino acid racemization geochronology of reworked quaternary mollusks on U.S. Atlantic coast beaches: implications for chronostratigraphy, taphonomy, and coastal sediment transport. Mar Geol 124:303–337

Zuschin M, Oliver PG (2003) Fidelity of molluscan life and death assemblages on sublitoral hard substrata around granitic islands of the Seychelles. Lethaia 36:133–150

Chapter 3
Intertidal Death Assemblages as Proxies of Marine Biodiversity. An Example from Northern Patagonia, Argentina

Fernando M. Archuby and Andrea Roche

Abstract Marine conservation biologists have identified mollusks as one of the appropriate surrogate taxa for characterizing marine benthic diversity. In turn, live/dead comparison studies have overwhelmingly demonstrated that mollusk remains are faithful proxies of the mollusk composition of the living communities from which they come, with positive consequences for the paleoecological evaluation of fossil assemblages. In this contribution, we evaluate the way in which mollusk biodiversity is distributed along the lower intertidal to supratidal (high water mark) dead shell assemblages accumulated on a northern Patagonian rocky shore, in order to explore the usefulness of these assemblages as paleontological proxies and potential surrogates of regional biodiversity. A diversity gradient from the lower intertidal to the supratidal was identified which is probably associated with vertical transport, although the influence of gradients of the living community should be tested to confirm this. The outstanding result of this study is the discovery of high levels of diversity among dead shells (31 bivalves and 39 gastropod species) in a single locality and with a moderate sampling effort. The supratidal death assemblage has higher species richness than expected, possibly caused by stranding of the fauna after storms. Nevertheless, this level shows the lowest level of evenness and a strong bias when samples are not sieved through a fine mesh. The record of marine benthic diversity in death assemblages is a promising area of research that deserves to be explored in depth.

Electronic supplementary material The online version of this chapter (https://doi.org/10.1007/978-3-030-20625-3_3) contains supplementary material, which is available to authorized users.

F. M. Archuby (✉)
Instituto de Investigación en Paleobiología y Geología (IIPG), UNRN-CONICET, General Roca, Río Negro, Argentina
e-mail: farchuby@gmail.com

A. Roche
Escuela Superior de Ciencias Marinas (ESCiMar) and Centro de Investigación Aplicada y Transferencia Tecnológica en Recursos Marinos Almirante Storni (CIMAS), San Antonio Oeste, Río Negro, Argentina
e-mail: andrear.ibmp@gmail.com

S. Martínez et al. (eds.), *Actualistic Taphonomy in South America*, Topics in Geobiology 48, https://doi.org/10.1007/978-3-030-20625-3_3

Keywords Biological surrogates · Conservation paleobiology · Intertidal · Death assemblage · Rocky-bottom · Patagonia · Depth gradient

3.1 Introduction

Dead shells accumulated on the sea floor contain a wealth of information which is useful either for assessing important questions of the genesis of fossil deposits (taphonomy) or for studying living communities. In the search for evidence to determine how representative the fossil record is of communities that lived in the geological past, the discipline of taphonomy has developed tools which provide high-quality information of living ecosystems. This is achieved by allowing their features to be explored on longer timescales, beyond those typically used by ecologists (Kidwell and Tomašových 2017; Olszewski and Kidwell 2007; Tomašových and Kidwell 2009a; Archuby et al. 2015; De Francesco et al. 2013; Yanes et al. 2008; Hassan et al. 2018). Current developments go still further: it is now possible to identify the effect of human impact on ecosystems, by studying the differences between impacted living communities and time-averaged assemblages accumulated over the past decades or centuries (Erthal et al. 2011; Kidwell 2008; Yanes 2012; Dietl et al. 2015). The relevance of this paleobiological information, which offers us an otherwise inaccessible long-term perspective of biodiversity and community change, has given rise to a new discipline: conservation paleobiology (Barnosky et al. 2017; Louys 2012; Rick and Lockwood 2013; Dietl and Flessa 2011; Dietl et al. 2015; Kidwell 2009; Kidwell and Tomašových 2013).

Biodiversity is of fundamental importance to ecology because it is the consequence of how organisms in communities respond to biotic and abiotic factors (Olszewski and Kidwell 2007). The use of biological surrogates (i.e. estimators, such as polychaetes, crustaceans, mollusks, etc.) to evaluate marine biodiversity is a useful practice in conservation biology because it helps overcome the difficulties inherent in surveying benthic communities: time and cost, the occurrence of undescribed species and the problems of species identification (Tyler and Kowalewski 2017; Magierowski and Johnson 2006; Mellin et al. 2011; Warwick and Light 2002). Research focuses on finding appropriate surrogates for the different types of marine communities and their spatial and temporal variations. Mollusks, which are among the groups selected as appropriate surrogates of marine benthic communities, leave abundant mineralized dead remains, which have been proved to be good proxies of the communities from which they derive (Tyler and Kowalewski 2017; Kidwell 2008; Smith 2005).

Assessing how diversity transfers from living communities (life assemblages, LAs) to death assemblages (DAs) is a crucial step towards a better interpretation of diversity in fossil assemblages; this knowledge will also help us to evaluate death assemblages as faithful proxies of living communities. The step from LAs to DAs represents the first filter that modifies diversity measurements, through differential transport and destruction by waves, currents and wind and time-averaging (Archuby et al. 2015; Tomašových and Kidwell 2009a, 2010a).

In the absence of strong reworking of former beds, such as in the case of ravinement, marine beds encompass a short time span and their skeletal content is considered representative of the average composition of successions of communities along hundreds or, at the most, thousands of years (within-habitat time-averaging of Kidwell and Bosence 1991; see also Fürsich and Aberhan 1990). Recently, the quantitative knowledge of the differences between death assemblages and living communities, and the sources of these differences has been greatly improved (Olszewski and Kidwell 2007; Tomašových and Kidwell 2009a, b, 2010a, b, 2011; and many more).

3.2 Death Assemblages, Taxonomic Diversity, and Taphonomic Fidelity

Due to the time-averaged nature of DAs, their species composition is not particularly influenced by the short-term species composition fluctuations of living communities (Fürsich and Aberhan 1990; Tomašových and Kidwell 2010a; Archuby et al. 2015). These short-term fluctuations, such as the local extinction of the surf clam *Mesodesma mactroides* on the Atlantic coasts of Uruguay and Northern Argentina (Fiori and Cazzaniga 1999; Dadon 2005), might give totally different results in samples of living communities separated by only a few weeks. However, in this respect, DAs are highly informative due to their inertia in the face of such fluctuations. Compared with living assemblages, DAs which have accumulated over a few decades to several centuries are expected to have an increase in alpha diversity, a decrease in beta diversity (due to spatial mixing), reduced species dominance and increased frequency of rare species (Tomašových and Kidwell 2010a). Additionally, the ecological information of current ecosystems does not span more than a few decades into the past (Rick and Lockwood 2013). If we consider that human occupation of Patagonia dates from around 17,000–14,000 years BP (Perez et al. 2016), baseline ecological studies might fail to identify the non-impacted conditions when assessing anthropogenic influence, since the impacts were already there.

In turn, death assemblages are used to characterise not only the average species compositions of source communities, but also biotic interactions such as local level predator-prey relationships (e.g., Visaggi and Kelley 2007; Yanes and Tyler 2009; Gordillo and Archuby 2012, 2014; Martinelli et al. 2013; Tyler et al. 2014; Archuby and Gordillo 2018), and to compare these along geographical gradients (e.g., Kelley and Hansen 2007; Visaggi and Kelley 2015; Martinelli et al. 2013). Quantifying predator-prey interactions in living communities implies sampling strategies that are complex and expensive, while the records from death assemblages are a significant source of information.

Studies on taphonomic fidelity (correlation of living and death assemblages) have been developed in marine, freshwater and terrestrial environments (e.g., Fürsich and Flessa 1987; Kidwell and Bosence 1991; Yanes et al. 2008; Tietze and De Francesco 2012; Terry 2010; more references in Archuby et al. 2015). Studies of marine death assemblages are abundant, although they are mostly based on soft-bottom ecosystems (Olszewski and Kidwell 2007; Kidwell 2013), and there are few studies of communities inhabiting rocky bottoms (Zuschin et al. 2000; Zuschin and Oliver 2003; Zuschin and Stachowitsch 2007). Recently, Archuby et al. (2015) assessed the taphonomic fidelity of rocky-bottom communities along 1500 km of the Patagonian Atlantic coast, from death assemblages collected at the high-water mark. These authors found a general agreement between life and death assemblages at the biogeographical province level, working with non-sieved, representative samples (hand-picked along transects). Besides the regional agreement, on smaller geographical scales DAs tended to cluster together and separated from LAs. So far, there are no detailed studies on the nature of DAs on rocky shores. A better understanding of the provenance of the diversity differences between life and death assemblages in modern environments is also crucial for correctly interpreting fossil assemblages and ecosystems (Olszewski and Kidwell 2007).

3.3 Purpose of This Study

In this study, we evaluate the way in which species richness and evenness of mollusk death assemblages is distributed along the depth gradient, from the lower intertidal to accumulations at the high-water mark, in Punta Mejillón, Northern Patagonia, Argentina. Our two goals are to improve the understanding of DAs as paleontological proxies and to evaluate their usefulness as surrogates of shallow benthic living communities. Punta Mejillón has little human impact due to its distance from the nearest city (the town of San Antonio Oeste, 105 km away), the difficulty getting there (sand dunes often cover the route), the need for a four-wheel drive vehicle to reach the beach, and also because it is located in a natural protected area (see The Study Area, below). We aim to determine whether the DAs coming from the same habitat but accumulated at different depths include specimens of different species in different proportions (i.e., there is diversity partitioning of DAs along the gradient). We also test the effect of sieving versus non sieving on species richness and evenness. Specifically, we aim to answer the following questions: (i) How does DA species composition vary along the lower intertidal to supratidal gradient? Is there diversity partitioning along the depth gradient in the rocky intertidal belt of northern Patagonia? (ii) Are death assemblages from rocky shores appropriate surrogates of benthic biodiversity in northern Patagonian shallow marine communities? Is there a horizon along the lower intertidal to supratidal belts that collects most of the information on the death assemblages? In other words, where is it best to sample? (iii) What is the effect of sieving on the biodiversity record?

3.4 Methodology

3.4.1 The Study Area

The study was carried out in Punta Mejillón (PM), located in the Caleta de los Loros natural protected area in Río Negro Province, Argentine Patagonia. The place is difficult to access, which minimizes the impact of tourism and human activities on living communities and death assemblages (Fig. 3.1). Punta Mejillón is on the Atlantic coast (41° 00′ 37″) in the San Matías Gulf. The coastline runs approximately from SE to NW, and the intertidal belt is exposed for more than 300 m during low tides (Fig. 3.2). Biogeographically, PM is in the transition zone between the Argentine and Magellanic Provinces and is characterized by a mixture of species from both biogeographical entities (Balech and Ehrlich 2008). In a recent article, Güller and Zelaya (2017) mention a surprisingly high level of mollusk diversity in the San Matías Gulf, which they describe as a hot-spot of diversity.

The northern part of San Matías Gulf, where Punta Mejillón is located, is subject to high levels of physical disturbance, consisting of strong winds, high tidal amplitudes (up to around 9 m) which leave large areas of the intertidal belt exposed, high-energy flows during high tide and low temperatures (sea surface temperatures 10.1–18.9 °C) (Bertness et al. 2006; Archuby et al. 2015). Due to the high levels of desiccation stress caused by winds, the region is considered an extremely harsh intertidal rocky ecosystem (Bertness et al. 2006), which results in intertidal communities which are strongly organized by physical stress.

3.4.2 Sampling

Sampling was carried out on 29 November 2013 during low tide, between latitudes S 41° 00′ 32″ and 41° 00′ 54″. Samples were collected at four levels: 1. accumulation

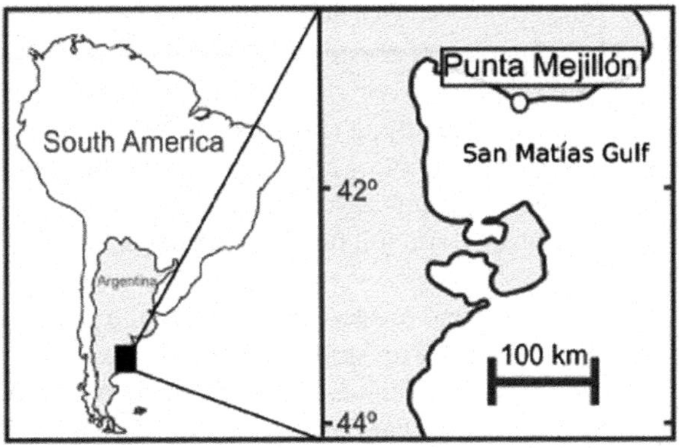

Fig. 3.1 Map of the study area

Fig. 3.2 Pictures of the intertidal belt in Punta Mejillón. **a** Upper intertidal. **b** View from the middle intertidal to the coast. **c** and **d** details of the middle to lower intertidal

of shells at the high-water mark (supratidal or “Supra”); 2. upper intertidal belt (UI); 3. middle intertidal belt (MI); and 4. lower intertidal belt (LI). At each level, several replicates were extracted from the upper 15 cm using a shovel and were pooled together, until completing 15 L of sediment. The replicates were extracted every 10 m along a transect parallel to the coastline. Since the substrate is mostly hard, samples were taken from depressions filled with sediments in the area surrounding the sampling point. In the absence of a suitable place to extract the replicate, the point was skipped, and the sample was taken at the next point. Samples were sieved in the field with a 10 × 10 mm aperture mesh (coarse) above and a 1 mm × 1 mm aperture mesh (fine) below so that large shells were captured separately from small shells (Fig. 3.3). The coarse mesh sieve retains shells that are visible and was considered as a proxy “hand-collecting method”, that was compared with “whole” samples per level (made by the pooling of coarse and fine samples). The 1 mm sieve was used to explore a suitable sampling strategy for rocky-bottom dominated intertidal DAs from the Patagonian Atlantic coast. Kidwell (2002) suggested that sampling with mesh sizes lower than 1 mm might collect a non-representative high amount of larvae and juveniles.

All gastropod and bivalve shells and shell fragments were analyzed and identified to the species level with some exceptions that were unidentifiable due to preservation issues. Other skeletal elements not included in the study were: crab fragments, serpulid tubes, abundant cirriped plates, sea-urchin spines, fragments of bryozoan

Fig. 3.3 Sampling method: sieving samples

colonies, oyster recruits on large valves and polyplacophoran plates. Cirripeds and cirriped plates, although very abundant, were excluded from analysis due to the difficulty in identifying the plates. Gastropod shells and articulated bivalves were assigned one count. Left and right valves of bivalve species were counted separately. The count per species resulted from the sum of articulated specimens plus the most abundant valves (left or right). *Ostrea puelchana* and *Pododesmus rudis* shells that could not be identified were counted together and divided by 2, and then assigned to the respective species. Bivalve fragments were counted if the umbo and at least one-third of the valve were preserved (very small fragments were discarded). Gastropod fragments were counted when they contained the apex and at least half of the shell.

3.4.3 Statistical Methods

Counts were made per level (LI, MI, UI, and Supra). The coarse mesh size fraction of samples was also registered separately for each level. Diversity was estimated using different indices: species richness (S, the raw number of species and by rarefaction), the Shannon-Wiener (H') index, the equitability J index (Hammer and Harper 2006) and the probability of an interspecific encounter (PIE), an evenness index (Hurlbert 1971). Rarefaction to the lowest sample size was calculated in order to evaluate species richness without the effect of sample size. The H' index summarises information on species richness and evenness and correlates with S and sample size, as does the J index. The PIE index was added to obtain an estimation of evenness which was not affected by sample size (Olszewski and Kidwell 2007). Data management and calculation of the PIE index according to Hurlbert's formula was carried out using standard spreadsheet software. Other diversity indices were calculated using PAST v 3.15 (Hammer et al. 2001).

Samples (levels) were plotted using a non-metric multidimensional scaling (NMDS) ordination analysis to evaluate their similarity. The database was first transformed to percentages per sample, then square root transformed, and then a similarity matrix was calculated based on the Bray-Curtis index (Clarke 1993; Clarke and Warwick 2001; Clarke et al. 2006). NMDS was carried out using R software, version 3.4.3 (R Core Team 2017).

To test the effect of using samples sieved with coarse mesh (as proxies for collecting by hand), we compared these with the results obtained for whole samples (coarse + fine mesh) by using diversity indices and an ordination plot (NMDS).

Beta diversity was quantified in order to assess both the existence of a gradient along the coastal profile for the four levels (directional turnover) and non-directional variation for comparing the coarse mesh subsample with the whole sample (whole = coarse plus fine mesh subsamples) (Anderson et al. 2011). To evaluate the gradient in beta diversity, the similarity between the supratidal sample and the samples from all other levels was calculated with the Jaccard similarity index on a presence/absence matrix. The results were plotted in their position on the coastal profile, from Supra to LI. If species turnover along the gradient existed, then a pattern of similarity decrease would be expected from left to right. To determine the differences in species presence/absence in coarse and fine samples, Whittaker's beta diversity index (βw) was calculated between pairs of coarse and whole samples per level, and then plotted in their position on the coastal profile. Higher levels of βw imply a greater mismatch between the coarse mesh samples and the whole samples (Koleff et al. 2003).

3.5 Results

A total of 12,790 mollusk specimens belonging to 31 bivalve and 39 gastropod species were collected (Table 3.1, Supp Appendix A and B). The sample size was uneven between levels due to the variable densities of shells in sediments from the different samples (Fig. 3.4a). Lower and middle intertidal samples contained less than half the specimens of the upper and supratidal samples. The coarse fraction per level fluctuated between 17 and 38% (Supra and MI respectively. Figure 3.4b). MI, with the smallest sample size (1520 specimens), has the highest percentage of the coarse fraction (38%).

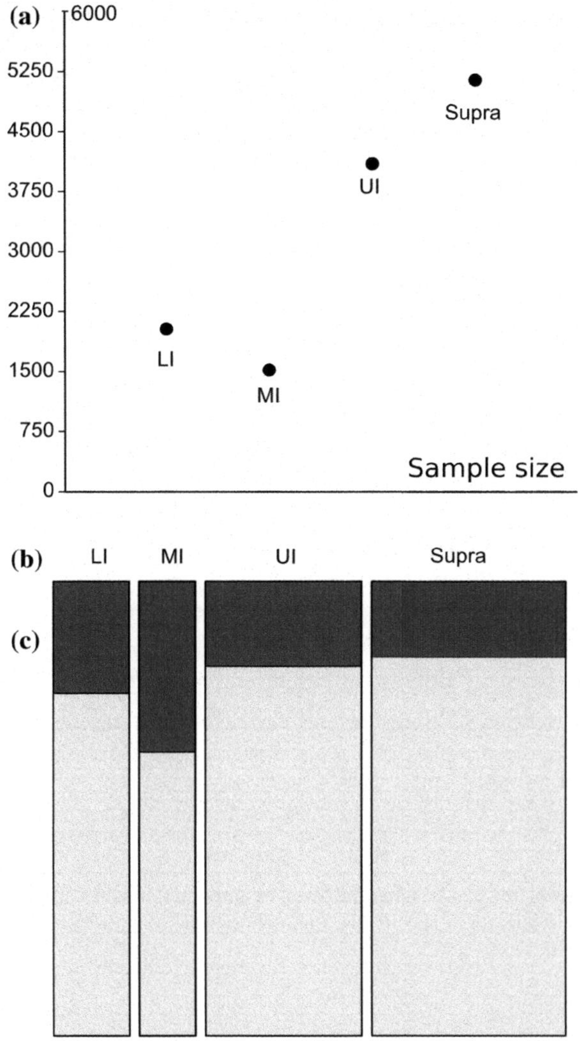

Fig. 3.4 **a** Size of sample per level. Levels, LI: lower intertidal, MI: middle intertidal, UI: upper intertidal, Supra: supratidal. n: number of specimens. **b** Size of samples per level and proportion of specimens captured in the coarse mesh. **c** coarse mesh. Width of bars express sample-size

3.5.1 *Alpha and Beta Diversities Across the Intertidal Gradient*

The 70 species identified in this study are distributed differently across samples (Tables 3.1 and 3.2). The S index is highest in Supra, followed by LI, UI, and MI. However, when standardizing to n = 1520 by rarefaction, the highest diversity is found in LI (46), followed by Supra, MI and UI, which have between 36 and 39 species (Fig. 3.5a, b). The rarefaction curves show that none of the samples have a stabilizing size pattern (Fig. 3.5c), suggesting that larger sample sizes are necessary to accurately document the kind of study.

Evenness differs between levels, and is consistently lowest in Supra, followed by UI, and then LI and MI with higher values (Table 3.2 and Fig. 3.6a, b and c). The H' index is highest for LI, while the J index has MI as the evenest sample. The PIE index, which is more reliable for studies with different sample sizes, is highest for LI, followed by MI, UI, and Supra, coinciding with the H' index.

Multivariate ordination using an NMDS plot indicates a similarity between LI and MI, while UI and Supra remain separate (Fig. 3.7a). The analysis of beta diversity

Table 3.1 Distribution of counts per level and mesh size

	LI		MI		UI		Supra		Total	
Sample type	C	F	C	F	C	F	C	F	C	F
n	2031		1520		4095		5144		12,790	
	510	1521	581	939	773	3322	885	4259	2750	10,040
S	49		37		45		55		70	
	37	32	29	31	36	33	34	46	53	52

LI lower intertidal level, *MI* middle intertidal level, *UI* upper intertidal level, *Supra* supratidal level or high-water mark. *n* number of specimens, *s* number of species (species richness). Sample type: *C* coarse mesh, *F* fine mesh

Table 3.2 Diversity indices calculated per level and for the pooled sample

	LI	MI	UI	Supra	Pooled sample
n	2031	1520	4095	5144	12,790
S	49	37	45	55	70
S-rarefaction	46	37	36	39	
H'	2.412	2.362	2.072	1.893	2.168
J	0.620	0.654	0.544	0.472	0.512
PIE	0.847	0.838	0.791	0.755	0.799

n sample size, *S* species richness, *S-rarefaction* species richness at a sample size of 1520, *H'* Shannon-Wiener index, *J* Pielou's evenness index, *PIE* probability of interspecific encounter diversity index

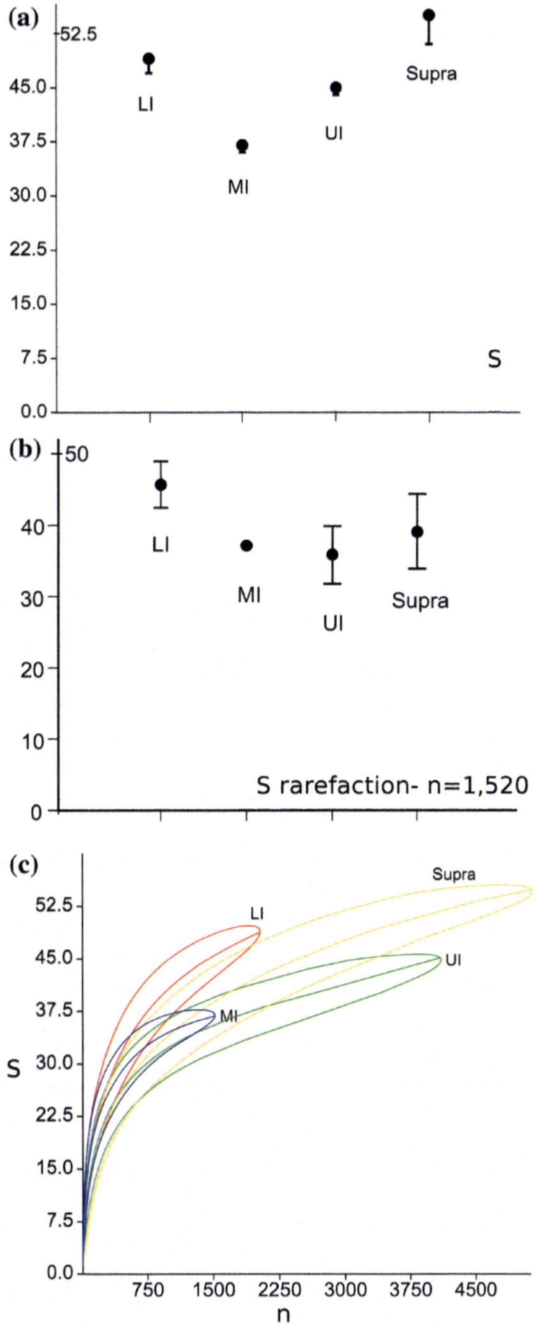

Fig. 3.5 **a** Plot of species richness (S) per level. The bar represents a bootstrap 95% confidence interval. **b** Rarefaction species richness to n = 1520 per sample. The bar includes 2 standard errors. **c** Rarefaction curves per level with 95% bootstrap confidence interval. Species richness on the y axis; sample size on the x axis

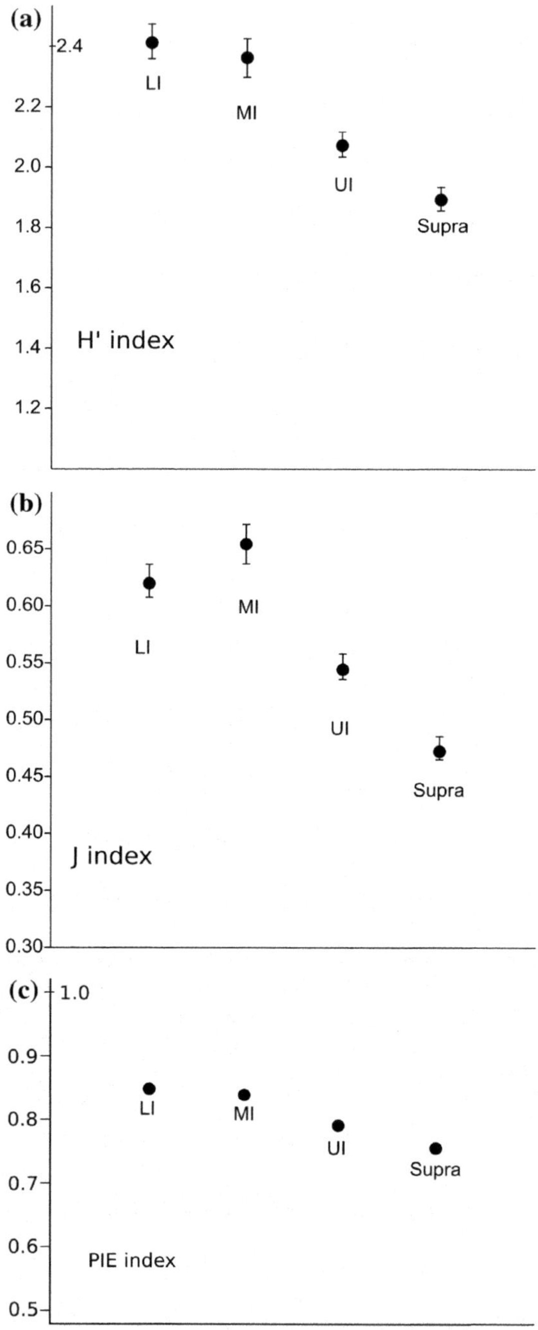

Fig. 3.6 **a** Plot of H' index per level. **b** Plot of J index per level. **C** Plot of PIE index per level

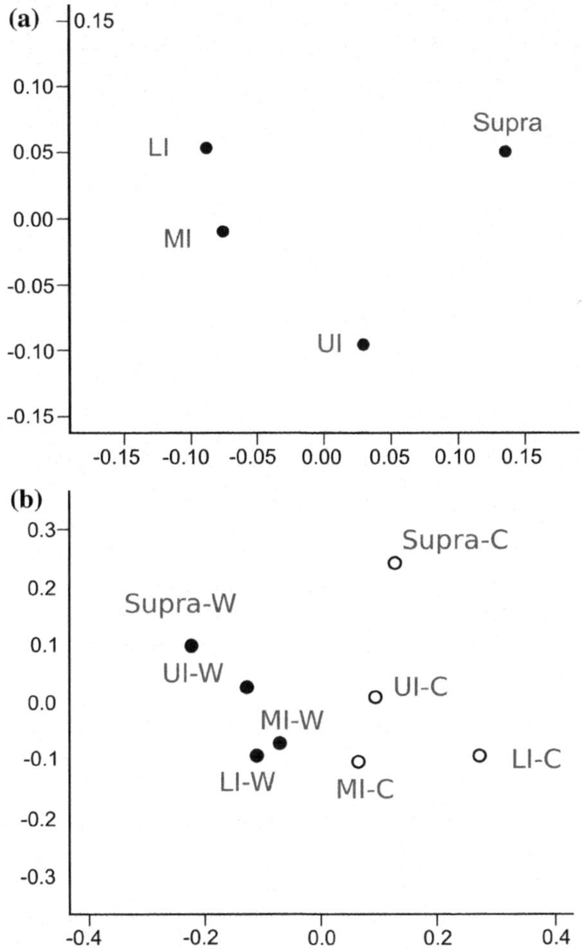

Fig. 3.7 **a** Non-metric multidimensional scaling plot between levels. **b** Non-metric multidimensional scaling plot per level and aperture mesh size. **c** coarse mesh sample; W: whole sample (coarse plus fine mesh sample)

allows the identification of a pattern of decrease along the supratidal to the lower intertidal gradient (Fig. 3.9a).

3.5.2 Effect of Mesh Aperture Size

The samples sieved with coarse mesh have richness and equitability values which are lower than estimations for whole (coarse + fine) samples (Table 3.2 and 3.3). Coarse mesh samples consistently underestimate the species richness of the death

Table 3.3 Diversity indices calculated per level for the coarse aperture mesh sieved samples

Index	LI-C	MI-C	UI-C	Supra-C
n	511	581	773	885
S	37	29	36	34
S-rarefaction	37	28	32	28
H'	2.396	2.279	1.959	1.442
J	0.664	0.677	0.547	0.409
PIE	0.822	0.793	0.684	0.510

n sample size, *S* species richness, *S-rarefaction* species richness at a sample size of 511, *H'* Shannon-Wiener index, *J* Pielou's evenness index, *PIE* probability of interspecific encounter diversity index, *C* coarse mesh

assemblage (Fig. 3.8a), and are less even for all four levels (PIE index, Fig. 3.8b). On the NMDS ordination plot, coarse mesh samples cluster together and are separate from the whole samples (Fig. 3.7b). The comparison of Whittaker's beta diversity indices (βw) shows a large mismatch between the coarse mesh and whole samples at the supratidal level (Fig. 3.9b), suggesting that at this level the species composition of the coarse sample is the least similar to the whole sample.

3.6 Discussion

3.6.1 Alpha and Beta Diversity Trends

There is a general trend in decreasing diversity from the lower intertidal to the supratidal belt, both for species richness and evenness. The pattern is more evident in estimations not dependant on sample size (Figs. 3.5b and 3.6c) than in measurements associated with sample size (Figs. 3.5a and 3.6a, b). In the case of species richness, its estimation via rarefaction to the lowest sample size ($n = 1520$) suggests a trend from LI to UI, although the Supra sample is slightly more diverse than MI and UI (Fig. 3.5b). The PIE index shows a decrease in evenness from LI to Supra, and a similar situation can be observed in the J and H' indices, despite the effect of sample size (Fig. 3.8a, b and c). NMDS ordination does not reflect a clear pattern. However, the values of Bray-Curtis similarity indices between levels follow the LI to Supra gradient (Table 3.4): contiguous samples are more similar to one another than non-contiguous ones. Distance between samples was coded as 1 to 3 (1 for contiguous samples, 2 for LI to UI and Mi to Supra; and 3 for LI to Supra), and the Spearman rank correlation index was calculated between the Bray-Curtis index and distance, thereby obtaining a significant negative correlation of -0.93 ($p = 0.033$). Additional evidence of the influence of the depth gradient in the species composition of samples comes from the evaluation of beta diversity: compared with the Supra level, there is

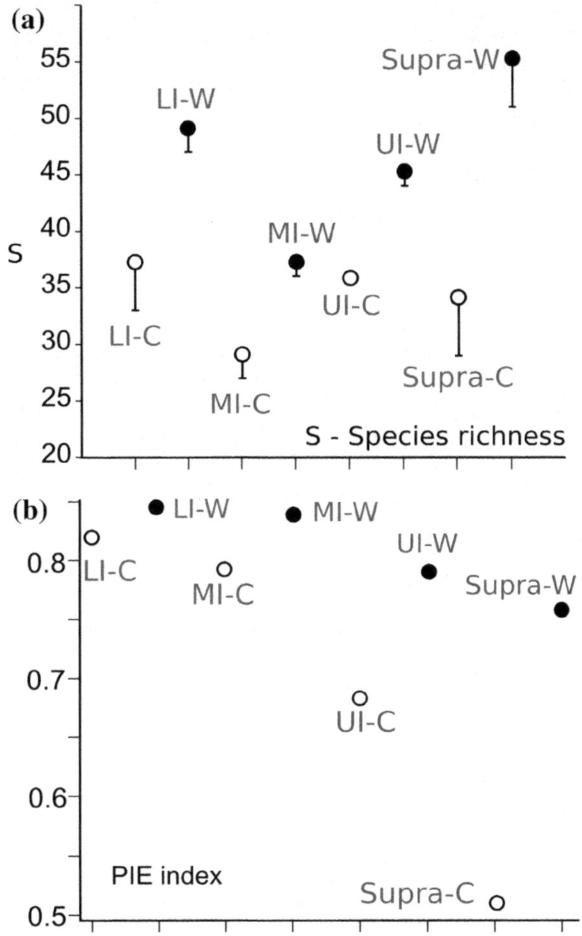

Fig. 3.8 **a** Species richness (S) per level and discriminating coarse mesh sample (C) from whole sample (W). **b** PIE index as estimation of evenness per level and per mesh size. C: coarse mesh sample; W: whole sample (coarse plus fine mesh sample)

a decrease in similarity from UI to LI. This is interpreted as a consequence of the depth gradient, whether due to a taphonomic gradient explained by biostratinomic factors (transport by waves and wind, selective destruction), the species composition gradient of the living community, or both. In turn, harsh environmental conditions (high levels of desiccation, strong winds, and wave energy) suggest that the upper intertidal, and particularly the supratidal, belts should have poorer living community diversity; however, this is not seen in the death assemblages, thus inferring vertical transport from the middle and lower intertidal and shallow subtidal. The higher than expected richness in the Supra sample might be a consequence of the trapping (stranding) of shells above the high water mark during energetic storms; shells and live specimens are stranded above the high water mark, and are no longer reached by usual or normal storm waves (López et al. 2008).

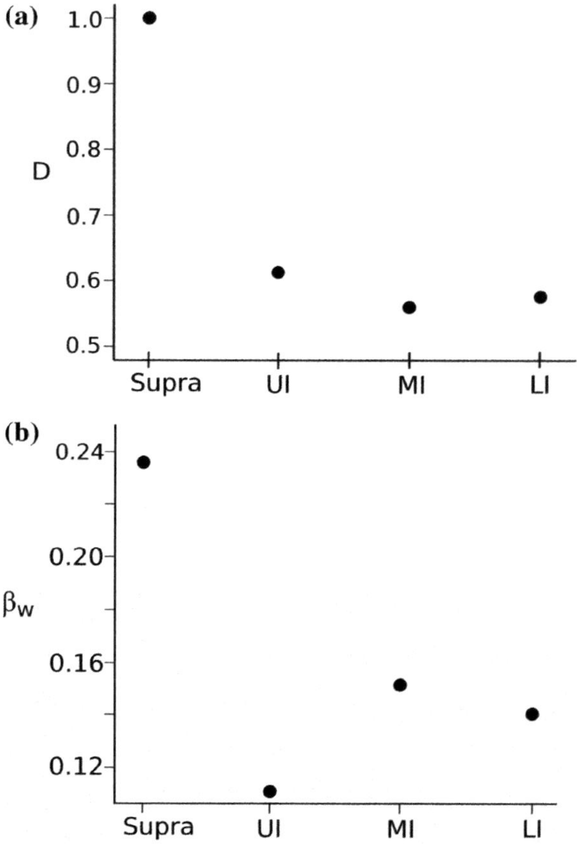

Fig. 3.9 **a** Beta diversity comparison along the supratidal to lower intertidal gradient. D: Jaccard distance index between Supra level and the other three levels. **b** Whittaker beta diversity index between whole sample and coarse mesh sub-sample per level

Table 3.4 Bray-Curtis similarity between levels (same similarity matrix used for NMDS)

Levels compared	Type of comparison	Bray-Curtis index
LI-MI	Contiguous levels	0.869
MI-UI	Contiguous levels	0.834
UI-Supra	Contiguous levels	0.816
LI-UI	One level skipped	0.809
UI-Supra	One level skipped	0.773
LI-Supra	Two levels skipped	0.768

Table 3.5 Bray-Curtis similarity indices of the whole sample (pooled) against samples of every level

Level	Bray-Curtis index
LI	0.858
MI	0.856
UI	0.895
Supra	0.886

Finally, Bray-Curtis similarity was calculated to compare each level with the whole sample (pooled), and it was found that the Supra and UI levels are the most similar to the total sample (Table 3.5). Although in every case abundances were standardized to percentages and square root transformed, the most abundant samples, Supra and UI, might still influence the result and cause this similarity. The best sampling strategy would still be to collect material from every level, but sampling death assemblages accumulated on the high water mark (Supra) level would not lead to important biases. However, it must be considered that the Supra level has the highest bias when fine meshes are not used, as seen below.

Whether diversity along the lower intertidal to supratidal areas of this study follows a gradient of a biological (species composition of the living community), taphonomic (differential transport, destruction and shell production among species) or mixed nature, will hopefully be answered in future studies. As for soft-bottom studies, there is still a need for more actualistic updated research (Olszewski and Kidwell 2007; Kidwell 2015; Tyler and Kowalewski 2017). Our investigation is particularly relevant because it helps to fill the need for studies of this kind on hard-bottom environments (Smith 2008; Archuby et al. 2015).

3.6.2 Mesh Size Matters: The Effect of Sieving on Biodiversity Assessment

There are relevant differences between the coarse fraction and the whole sample (whole samples are composed of coarse and fine mesh samples. For an explanation, see Sampling in Methodology). Coarse samples are less even (Fig. 3.8b) and have lower species richness (as expected, since "coarse" samples are part of the "whole" samples of each level; Fig. 3.8a). Species richness in coarse samples ranges from 80 to 62% of the values obtained for the whole samples, with equivalent sampling efforts. The differences observed are also reflected in the ordination plot, in which all coarse samples are clustered together and separated from whole samples (NMDS, Fig. 3.7b). Discordance in terms of beta diversity is more marked in Supra than in the rest of the samples. This means that in Supra, the mismatch between coarse and whole samples is highest. This is still more relevant if we consider that many minute species, such as *Phlyctiderma semiaspera*, *Turbonilla macaensis* or *Anachis isabellei*, are present in the coarse samples because they were stuck to mytilid byssus, which would not otherwise have been sampled by hand. Without these queue-jumper species, the biases of coarse samples would have been even larger. Olszewski and Kidwell (2007) detected that the evenness of death assemblages is more similar to live assemblages when samples are sieved with a mesh size finer than 2 mm, which was also the case for species richness. A positive bias in evenness and species richness in coarse mesh samples with respect to fine mesh samples (which are in turn more similar to live assemblages), can be caused by the greater durability of large mollusk shells, lower temporal volatility of adult specimens in living communities, or both (Kidwell 2002; Olszewski and Kidwell 2007).

The positive effect of sieving with fine meshes in diversity studies based on DAs is supported by evidence from living communities. In a compilation of mollusk diversity (including bivalves, gastropods, polyplacophorans, scaphopods and cephalopods) in the San Matías and San José gulfs, Güller and Zelaya (2017) noted that out of the total 196 species described for the whole area, 61 (31.1%) have a maximum size smaller than 10 mm. As a consequence, almost one-third of the species in the assemblage have a lower probability of being collected, and would perhaps be neglected if fine mesh size had not been used. The most abundant gastropod and bivalve species found by these authors, respectively *Parvanachis isabellei* and *Crenella divaricata*, fit this condition.

3.6.3 *Sampling Issues: Features of Death Assemblages Along the Intertidal Belt*

Dead shell assemblages differ from living communities due to time-averaging and differential transport and destruction (Kidwell 2001). The effect of tidal regime together with the action of waves on the bottom differ in intensity along the intertidal belt. As a consequence of these differences in the intensity of transport, destruction, and sorting of shells, some variation might be expected in shell density, bioclast size, and vertical transport, which would affect the results of sampling. Shell density varies along the intertidal belt, as evidenced in the individuals counted per sample (Fig. 3.4a and Table 3.1). The rarefaction curves indicated that none of the samples were large enough to be representative of species diversity, so more sampling is therefore needed, especially from the lower and middle intertidal levels. The proportion of coarse to fine mesh shells varies little; the exception is MI, but the fact that this level has the smallest sample size could explain the difference (Fig. 3.4b). Strong vertical transport of shells in DAs is also evident since the species composition of all the samples includes at least some taxa that are characteristic of subtidal to lower intertidal belts (e.g. *Aulacomya atra*, *Venus antiqua*, some *Buccinanops* species, etc.). On the other hand, these rocky Patagonian shores have little diversity of living fauna in the upper intertidal belt, and almost no marine life higher up, in the supratidal fringe (Bertness et al. 2006 and personal observations).

3.6.4 Mollusk Shell Dead Shell Assemblages as Samples of Living Marine Biodiversity

Two theoretical frameworks have come together in this study: 1. conservation biology has supplied information on the use of biological surrogates of marine benthic communities (Magierowski and Johnson 2006; Smith 2005, 2008; Mellin et al. 2011; Tyler and Kowalewski 2017); and 2. taphonomy, and the new, related discipline conservation paleobiology, have contributed with the assessment of the processes that operate between living communities and accumulations of their remains, for those taxa that bear mineralized or highly durable tissues (Kidwell 2001; Tomašových and Kidwell 2009a; Dietl and Flessa 2011). Put simply: if a particular taxon is an appropriate surrogate for a living community, and the accumulation of its durable remains or death assemblages are good proxies of the living counterpart of the taxon, then the death assemblages are highly valuable tools as rapid and faithful proxies of the living communities.

Different studies coincide on the point that exhaustively sampling living marine diversity is almost impossible, very expensive and particularly time-consuming, mainly due to the difficulties involved in accessing study sites, poor taxonomic knowledge and the high diversity of marine communities (Warwick and Light 2002; Magierowski and Johnson 2006; Smith 2008; Mellin et al. 2011; Tyler and Kowalewski 2017). Besides, results show that at least in some cases (depending on habitat type and spatial scale), mollusks are appropriate surrogates of marine communities (e.g., Smith 2008; Tyler and Kowalewski 2017). Dead shell assemblages represent time-averaged relics of the communities they come from. Their differences from living assemblages are explained mainly by their time-averaged nature: they are composed of a mixture of successive communities that lived in the same area and are modified by vertical and lateral transport and other biostratinomic agents (Kidwell 2001, 2013; Archuby et al. 2015). One of the expectations with respect to the features of death assemblages is an increase in species richness and evenness (Tomašových and Kidwell 2010a). Olszewski and Kidwell (2007) demonstrated that on average death assemblages surpass living communities in species richness and evenness, although particular examples might have a different pattern (with little frequency). The only study of live/dead comparisons along the Patagonian Atlantic coast detected a systematic increase in both diversity measurements in every single comparison (Archuby et al. 2015), even when sediments were not sieved (just hand collected) and live and dead samples did not coincide in time or extent (living communities were only sampled from the middle intertidal a few years before the collection of death assemblages).

In order to evaluate to what degree our death assemblages can provide relevant information on regional biodiversity, we compared them with data from surveys of benthic communities. Relevant studies on living mollusk diversity in the San Matías and San José gulfs are summarised in Table 3.6, including species richness (discriminating between bivalves and gastropods), the nature of the sample (life or death assemblage), the extent of the sampling area, the sampling effort (in terms of

Table 3.6 Living and death assemblages studied in San Matías and San José Gulfs

Locality	**Biv**	Gast	Moll	DA/LA	Extent	K	SE	n
1. PM (this study)	31	39	70	DA	Small	1	Moderate	12,681
2. PM (Archuby et al. 2015) (2)	4	2	6	LA	Small	1	Low	5573
3. SMG (Avaca et al. 2008. NOR)	12	9	21	LA	Medium	122	High	np
4. LG (Archuby et al. 2015) (2)	3	1	4	LA	Small	1	Low	3903
5. PD (Archuby et al. 2015) (2)	4	2	6	LA	Small	1	Low	3564
6. PD (Archuby et al. 2015) (1)	10	14	24	DA	Small	1	Low	686
7. PL (Archuby et al. 2015) (1)	8	13	21	DA	Small	1	Low	1826
8. SMG (Avaca et al. 2008 NOE)	19	4	23	LA	Medium	32	High	np
9. SJG, SMG. (Zaixo et al. 1998)	25	36	61	LA	Large	120	Very high	np
10. SAB (Güller and Zelaya 2017) (3)	71	48	119	LA	Large	np	Very high	16,479 (4)
11. PL (Güller and Zelaya 2017) (3)	54	43	97	LA	Large	np	Very high	2861 (4)
12. SJG (Güller and Zelaya 2017) (3)	26	39	65	LA	Large	np	Very high	np
13. PD (Güller and Zelaya 2017) (3)	58	44	102	LA	Large	np	Very high	np
14.SMG (Güller and Zelaya 2017) (3)	60	81	141	LA	Large	85	Very high	30,481

(1) Data from Archuby et al. (2015) plus additional information that completes the death assemblage. (2) Samples taken only at the middle intertidal horizon. (3) Only bivalves and gastropods, restricted to species found alive. (4) Authors provide self sampled sizes but some species were not found but posteriorly added from literature, with no indication of additional sample sizes. *np* not provided in the source literature. Acronyms: *PM* Punta Mejillón, *SJG* San José Gulf, *SMG* San Matías Gulf, *PD* Playas Doradas, *PL* Puerto Lobos, *LG* Las Grutas, *SAB* San Antonio Bay. *Biv*, *Gast*, *Moll*bivalves, gastropods and mollusks respectively, *DA* death assemblage, *LA* living assemblage, *Extent* area comprised in the sampling, *K* number of samples, *SE* sampling effort, *n* number of specimens

cost and time of the sampling process) and, when available, sample size. Archuby et al. (2015) studied a series of live and dead mollusk assemblages, spanning 1500 km from Punta Mejillón to Puerto Deseado in the South of Patagonia. Their case studies from the San Matías Gulf were numbered 2, 4, 5 (LAs), 6 and 7 (DAs). Avaca et al. (2008) provided two databases of living communities, sampled in the North and Northwest margins of the San Matías Gulf. Case study 9 corresponds to an extensive study in the San José Gulf and the closest part of the San Matías Gulf (Zaixso et al. 1998). Recently, Güller and Zelaya (2017) published a highly-qualified study with information on mollusk species from the San Matías and San José gulfs, which includes results from their own samples and from an exhaustive bibliographic compilation. The authors used 85 sampling points from the intertidal to a depth of 25 m in the subtidal, and also took some (not detailed) samples from deeper bottoms, obtaining a total of 30,481 mollusk specimens, including empty shells and valves. Their database, and most of their study focused on four main areas which together account for 119 species of bivalves (49) and gastropods (70) for both gulfs (when considering only live species found in their samples). When other studies are added, the species count for the whole area reaches 141 species (60 bivalves and 81 gastropods).

Death assemblages offer relatively high levels of diversity compared to life assemblages (Fig. 3.10). Individual (single place) samples of living communities such as 2, 4 and 5 represent low effort sampling but with very little diversity (up to 6 species). Several-point samples of living communities demand high levels of effort (availability of vessels, complex sampling devices, diving). Case studies 3 and 8 are samples of living communities with 122 and 32 sampling points respectively, where only 21 (case study 3) and 23 (case study 8) species were collected. In these two cases, the sampling area was considered of medium size (narrow fringes parallel to the coast). Case study 9 (Zaixso et al. 1998) is a 120 sample point survey in which a species richness of 61 was collected. Case study 12, taken from Güller and Zelaya (2017), is based on Zaxso's data plus additional information. Case studies 10, 11 and 13 are based on multi-point sampling of living communities across wide areas of the San Matías Gulf. Case study 14 represents a synthesis of multipoint sampling and all available published information on mollusk diversity in the San Matías and San José gulfs. Güller and Zelaya (2017) and Zaixso et al. (1998) sieved their samples with less than 2 mm aperture meshes, while Avaca et al. (2008) used 40 mm aperture nets. According to Güller and Zelaya (2017), sampling without fine meshes drastically reduces diversity, since they detected that more than 45% of species have shells with a maximum size smaller than 15 mm. This difference in sampling strategy might have caused the reduced diversity record in Avaca's samples. Güller and Zelaya (2017) consider that the lower than expected diversity found in the San José Gulf compared with their results in other areas is due to the different sampling methods used (they did not actually sample this gulf, but instead summarised information from other studies).

The three death assemblage case studies, 1 (this study), 6 and 7 (Archuby et al. 2015), are single point samples that represent moderate or low effort but offer a relatively high number of specimens and high diversity. Case studies 6 and 7 sampled

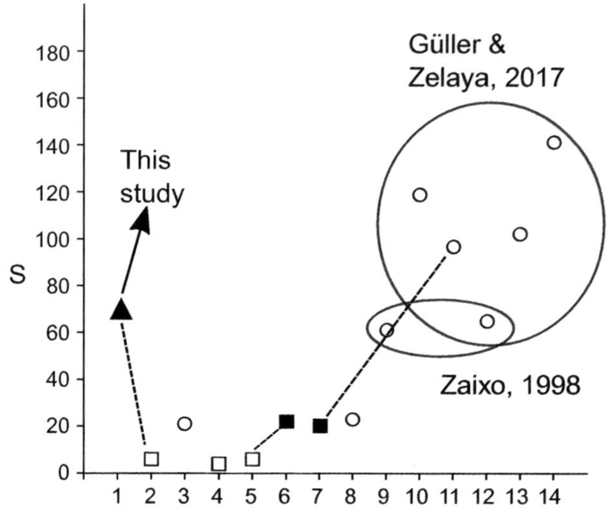

Fig. 3.10 Plot of calculated species richness in different case studies in the San Matías and San José gulfs. Order of case studies and acronyms of location sampling areas, as in Table 3.6. (1: PM (this study). 2: PM (Archuby et al. 2015) (2). 3: SMG (Avaca et al. 2008. NOR). 4: LG (Archuby et al. 2015) (2). 5: PD (Archuby et al. 2015) (2). 6: PD (Archuby et al. 2015). (1). 7: PL (Archuby et al. 2015) (1). 8: SMG (Avaca et al. 2008. NOE). 9: SJG, SMG. (Zaixo et al. 1998). 10: SAB Güller and Zelaya 2017 (3). 11: PL Güller and Zelaya 2017 (3). 12: SJG Güller and Zelaya 2017 (3). 13: PD Güller and Zelaya 2017 (3). 14: SMG Güller and Zelaya 2017 (3). Empty symbols: life assemblages. Filled symbols: death assemblages. Squares: low sampling effort. Triangles: moderate sampling effort. Circles: high sampling effort. Slashed lines join samples from comparable localities

24 and 21 species respectively, which is similar to the 21 and 23 species in Avaca et al. (2008) which required 122 and 32 samples for a similar result. However, those samples were taken without sieving. In our study, a single point in Punta Mejillón, sampled using 1 × 1 mm aperture mesh, detected higher species richness than 122 samples throughout the entire San José Gulf.

This study has confirmed that sampling species richness requires sieving with fine mesh sieves for both living communities (Güller and Zelaya 2017) and their associated death assemblages (Kidwell 2002). In turn, death assemblages give excellent results at equivalent levels of sampling effort, if compared with the sampling of life assemblages. This can be explained by the time-averaging and spatial homogenization of successive communities which accumulated to constitute the death assemblage (Tomašových and Kidwell 2010a).

A word of caution is needed here: the case studies of living communities used for comparisons are mainly based on samples taken from subtidal soft bottoms, while our study was carried out in a rocky-bottom intertidal belt. More quantitative research is needed in order to compare life and death assemblages from equivalent habitats and to evaluate the partitioning of biodiversity along depth gradients and in different types of seafloor in the San Matías and San José gulfs.

3.7 Conclusions

One of the outstanding conclusions of this work is that death assemblages accumulated in rocky-bottom coastal environments are highly informative of the regional biodiversity. In this study, we showed that a single point sample contains more species than almost every study of living communities based on dozens of samples for the same region.

The lower intertidal to supratidal (high-water mark) depth gradient contains dead shell accumulations that reflect a gradient in diversity (a decrease in species richness and evenness), as well as a pattern of species turnover. The gradient in the death assemblages is mainly explained by differential transport upwards, and also by a gradient in species turnover in the living community.

Representative samples of death assemblages must be obtained with the use of fine (up to 2 mm aperture) meshes. Supratidal death assemblages, i.e., shells accumulated in the high water mark, are an acceptable proxy of the whole intertidal to subtidal assemblage if a horizon is to be chosen for sampling. The supratidal is more diverse than expected in terms of species richness, probably due to the supply of shells by strong storms, which were then trapped, out of the reach of normal waves or normal storm waves. In turn, supratidal samples show the highest bias in coase-mesh sieved samples (considered as equivalent to hand-collecting). Results from this study suggest that none of the levels is fully representative, and a sample which pools the different levels is recommended. A live/dead comparison is needed to assess this question in more depth. Exploring the live/dead mismatch in rocky-bottom intertidal environments is useful for improving our knowledge of benthic marine life, and for filling a void in the studies of rocky bottoms, as well as for evaluating the novel idea of conservation paleobiology with respect to the detection of diversity-altered, human-impacted ecosystems.

Acknowledgements This work was supported by the research project PI 40-A-383 of Universidad Nacional de Río Negro.

References

Anderson MJ, Crist TO, Chase JM, Vellend M, Inouye BD, Freestone AL, Sanders NJ, Cornell HV, Comita LS, Davies KF, Harrison SP, Kraft NJ, Stegen JC, Swenson NG (2011) Navigating the multiple meanings of b diversity: a roadmap for the practicing ecologist. Ecol Lett 14(19):19–28

Archuby FM, Adami M, Martinelli J, Gordillo S, Boretto GM, Malve ME (2015) Regional-scale compositional and size fidelity of rocky intertidal communities from the Patagonian Atlantic coast. Palaios 30:627–643

Archuby FM, Gordillo S (2018, September) Drilling predation traces on recent limpets from northern Patagonia, Argentina. Pal Elect 21.3.36A:1–23

Avaca MS, Narvarte MA, Gonzaléz R (2008) Asociaciones macrobentónicas en la zona norte del Golfo San Matías (Río Negro, Argentina). IBMP—Serie Publicaciones, VII:39–58

Balech E, Ehrlich MD (2008) Esquema biogeográfico del mar Argentino. Rev INIDEP 19:45–75

Barnosky AD, Hadly EA, Gonzalez P, Head J, Polly PD, Lawing AM, Eronen JT, Ackerly DD, Alex K, Biber E, Blois J, Brashares J, Ceballos G, Davis E, Dietl GP, Dirzo R, Doremus H, Fortelius M, Greene HW, Hellmann J, Hickler T, Jackson ST, Kemp M, Koch PL, Kremen C, Lindsey EL, Looy C, Marshall CR, Mendenhall C, Mulch A, Mychajliw AM, Nowak C, Ramakrishnan U, Schnitzler J, Das Shrestha K, Solari K, Stegner L, Stegner MA, Stenseth NC, Wake MH, Zhang Z (2017) Merging paleobiology with conservation biology to guide the future of terrestrial ecosystems. Science, 355(6325):eaah4787

Bertness MD, Crain CM, Silliman BR, Bazterrica MC, Reyna MV, Hildago F, Farina JK (2006) The community structure of Western Atlantic Patagonian rocky shores. Ecol Mon 76(3):439–460

Clarke KR (1993) Non-parametric multivariate analyses of changes in community structure. Austral J E 18(1):117–143

Clarke KR, Somerfield PJ, Chapman MG (2006) On resemblance measures for ecological studies, including taxonomic dissimilarities and a zero-adjusted Bray-Curtis coefficient for denuded assemblages. J Exp Mar Biol Ecol 330(1):55–80

Clarke KR, Warwick RM (2001) Change in marine communities: an approach to statistical analysis and interpretation, 2nd edn. PRIMER-E, Plymouth

Dadon J (2005) Changes in the intertidal community structure after a mass mortality event in sandy beaches of Argentina. Cont Zool 74(1–2):27–39

De Francesco CG, Tietze E, Cristini PA (2013) Mollusk Successions of Holocene Shallow-Lake Deposits from the Southeastern Pampa Plain. Argentina Palaios 28(12):851–862

Dietl GP, Kidwell SM, Brenner M, Burney DA, Flessa KW, Jackson ST, Koch PL (2015) Conservation paleobiology: leveraging knowledge of the past to inform conservation and restoration. Ann Rev Earth Planet Sci 43:79–103

Dietl GP, Flessa KW (2011) Conservation paleobiology: putting the dead to work. Trends Ecol Evol 26(1):30–37

Erthal F, Kotzian CB, Simoes MG (2011) Fidelity of Molluscan Assemblages from the Touro Passo formation (Pleistocene-Holocene), Southern Brazil: Taphonomy as a tool for discovering natural baselines for freshwater communities. Palaios 26(7):433–446

Fiori SM, Cazzaniga NJ (1999) Mass mortality of the yellow clam, Mesodesma mactroides (Bivalvia: Mactracea) in Monte Hermoso beach. Argentina Biol Cons 89(3):305–309

Fürsich FT, Aberhan M (1990) Significance of time averaging for palaeocommunity analysis. Lethaia 23(2):143–152

Fürsich FT, Flessa K (1987) Taphonomy of tidal flat molluscs in the northern Gulf of California: paleoenvironmental analysis despite the perils of preservation. Palaios 2(6):543–559

Gordillo S, Archuby FM (2012) Predation by drilling gastropods and asteroids upon mussels in rocky shallow shores of Southernmost South America: paleontological implications. Act Palaeont Polonica 57(3):633–646

Gordillo S, Archuby FM (2014) Live-live and live-dead interactions in marine death assemblages: The case of the patagonian clam venus antiqua. Act Palaeont Polonica 59(2):429–442

Güller M, Zelaya DG (2017) A hot-spot of biodiversity in Northern Patagonia, Argentina. Biodiv Cons 14:1–14

Hammer Ø, Harper DAT (2006) Paleontological data analysis. Blackwell Publishing, Oxford

Hammer Ø, Harper DAT, Ryan PD (2001) Paleontological statistics software package for education and data analysis. Palaeont Elect 4(1):9–18

Hassan GS, Rojas LA, De Francesco CG (2018) Incorporating taphonomy into community-based paleoenvironmental reconstructions: can diatom preservation discriminate among shallow lake sub-environments? Palaios 33:376–392

Hurlbert SHS (1971) The nonconcept of species diversity: a critique and alternative parameters. Ecology 52(4):577–586

Kelley PH, Hansen TA (2007) Latitudinal patterns in naticid gastropod predation along the east coast of the United States: a modern baseline for interpreting temporal patterns in the fossil record. In Sediment-organism interactions: a multifaceted ichnology. SPEM Special Publication. SEPM, pp 287–299

Kidwell SM (2001) Preservation of species abundance in marine death assemblages. Science 294(November):1091–1094

Kidwell SM (2002) Mesh-size effects on the ecological fidelity of death assemblages: a meta-analysis of molluscan live-dead studies. Geobios 35(suppl):107–119

Kidwell SM (2008) Ecological fidelity of open marine molluscan death assemblages: Effects of post-mortem transportation, shelf health, and taphonomic inertia. Lethaia 41(3):199–217

Kidwell SM (2009) Evaluating human modification of shallow marine ecosystems: mismatch in composition of molluscan living and time-averaged death assemblages. In: Dietl GP, Flessa KW (eds) Conservation paleobiology: using the past to manage for the future. pp 113–139

Kidwell SM (2013) Time-averaging and fidelity of modern death assemblages: building a taphonomic foundation for conservation palaeobiology. Palaeontology 56(3):487–522

Kidwell SM (2015) Biology in the Anthropocene: challenges and insights from young fossil records. Proc Nal Acad Sci 112(16):4922–4929

Kidwell SM, Bosence DWJ (1991) Taphonomy and time-averaging of marine shelly faunas. In: Allison PA, Briggs DEG (eds) Taphonomy: releasing the data locked in the fossil record. pp 115–209

Kidwell SM, Tomašových A (2013) Implications of time-averaged death assemblages for ecology and conservation biology. Ann Rev Ecol Evol Syst 44(1):539–563

Kidwell SM, Tomašových A (2017) Nineteenth-century collapse of a benthic marine ecosystem on the open continental shelf. Proc Roy Soc B 284:1–9

Koleff P, Gaston KJ, Lennon JJ (2003) Measuring beta diversity for presence—absence data. JAnimal Ecol 72:367–382

López RA, Penchaszadeh PE, Marcomini SC (2008) Storm-related strandings of mollusks on the Northeast Coast of Buenos Aires, Argentina. J Coast Res 244:925–935

Louys J (2012) Paleontology in ecology and conservation: an introduction. In: Louys J (ed) Paleontology in ecology and conservation. Springer, Berlin-Heidelberg, pp 1–7

Magierowski RH, Johnson CR (2006) Robustness of surrogates of biodiversity in marine benthic communities. Ecol Appl 16(6):2264–2275

Martinelli JC, Gordillo S, Archuby FM (2013) Muricid drilling predation at high latitudes: insights from the Southernmost Atlantic. Palaios 28(1):33–41

Mellin C, Delean S, Caley J, Edgar G, Meekan M, Pitcher R, Przeslawski R, Williams A, Bradshaw C (2011) Effectiveness of biological surrogates for predicting patterns of marine biodiversity: a global meta-analysis. PLoS ONE 6(6):e20141

Olszewski TD, Kidwell SM (2007) The preservational fidelity of evenness in molluscan death assemblages. Paleobiology 33(1):1–23

Perez SI, Postillone MB, Rindel D, Gobbo D, Gonzalez PN, Bernalet V (2016) Peopling time, spatial occupation and demography of Late Pleistocene-Holocene human population from Patagonia. Quat Int 425:214–223

R Core Team (2017) R: a language and environment for statistical computing. Available at: https://www.r-project.org

Rick TC, Lockwood R (2013) Integrating paleobiology, archeology, and history to inform biological conservation. Cons Biol 27(1):45–54

Smith SDA (2005) Rapid assessment of invertebrate biodiversity on rocky shores: where there's a whelk there's a way. Biodivers Conserv 14:3565–3576

Smith SDA (2008) Interpreting molluscan death assemblages on rocky shores: are they representative of the regional fauna? J Exp Mar Biol Ecol 366(1–2):151–159

Terry RC (2010) The dead do not lie: using skeletal remains for rapid assessment of historical small-mammal community baselines. Proceedings R Soc2 B 77(1685):1193–1201

Tietze E, De Francesco CG (2012) Compositional fidelity of subfossil mollusk assemblages in streams and lakes of the Southeastern Pampas. Argentina Palaios 27(6):401–413

Tomašových A, Kidwell SM (2009a) Fidelity of variation in species composition and diversity partitioning by death assemblages: time-averaging transfers diversity from beta to alpha levels. Paleobiology 35(1):94–118

Tomašových A, Kidwell SM (2009b) Preservation of spatial and environmental gradients by death assemblages. Paleobiology 35(1):119–145

Tomašových A, Kidwell SM (2010a) Predicting the effects of increasing temporal scale on species composition, diversity, and rank-abundance distributions. Paleobiology 36(4):672–695

Tomašových A, Kidwell SM (2010b) The effects of temporal resolution on species turnover and on testing metacommunity models. Am Nat 175(5):587–606

Tomašových A, Kidwell SM (2011) Accounting for the effects of biological variability and temporal autocorrelation in assessing the preservation of species abundance. Paleobiology 37(2):332–354

Tyler CL, Kowalewski M (2017) Surrogate taxa and fossils as reliable proxies of spatial biodiversity patterns in marine benthic communities. Proc R Society B 284(1850):20162839

Tyler CL, Leighton LR, Kowalewski M (2014) The effects of limpet morphology on predation by adult cancrid crabs. J Exp Mar Biol Ecol 451:9–15

Visaggi CC, Kelley PH (2007) Relationship between drilling predation and shell morphology of patellid limpets from Southwestern England. Denver, Colorado, USA, GSA Anual Meeting, p 19813

Visaggi CC, Kelley PH (2015) Equatorward increase in naticid gastropod drilling predation on infaunal bivalves from Brazil with paleontological implications. Palaeogeogr Palaeoclimatol Palaeoecol 438:285–299

Warwick RM, Light J (2002) Death assemblages of molluscs on St Martin's Flats, Isles of Scilly: a surrogate for regional biodiversity? Biodiv Cons 11:99–112

Yanes Y (2012) Anthropogenic effect recorded in the live-dead compositional fidelity of land snail assemblages from San Salvador Island. Bahamas. Biodiv Cons 21(13):3445–3466

Yanes Y, Tyler CL (2009) Drilling predation intensity and feeding preferences by *Nucella* (Muricidae) on limpets inferred from a dead-shell assemblage. Palaios 24(5):280–289

Yanes Y, Tomašových A, Kowalewski M, Castillo C, Aguirre J, Alonso MR, Ibáñez M (2008) Taphonomy and compositional fidelity of Quaternary fossil assemblages of terrestrial gastropods from carbonate-rich environments of the Canary Islands. Lethaia 41(3):235–256

Zaixso HE, Lizarralde Z, Pastor C, Gomes-Simes E, Romanello E, Pagnoni G (1998) Distribución espacial del macrozoobentos submareal del golfo San José (Chubut, Argentina). Rev Biol Mar Oceanogr 33(1):43–72

Zuschin M, Hohenegger J, Steininger FF (2000) A comparison of living and dead molluscs on coral reef associated hard substrata in the northern Red Sea—implications for the fossil record. Palaeogeogr Palaeoclimat Palaeoecol 159:167–190

Zuschin M, Oliver PG (2003) Fidelity of molluscan life and death assemblages on sublittoral hard substrata around granitic islands of the Seychelles. Lethaia 36(2):133–149

Zuschin M, Stachowitsch M (2007) The distribution of molluscan assemblages and their postmortem fate on coral reefs in the Gulf of Aqaba (Northern Red Sea). Mar Biol 151(6):2217–2230

Chapter 4
Alien Species, a Natural Experiment in Actualistic Taphonomy

Sergio Martínez, Alejandra Rojas, Fernanda Cabrera and Diego Antuña

Abstract The use of alien invasive species is applied for the first time to evaluate the taphonomic damage during a period of up to two decades, in natural conditions. A preliminary study using shells of the gastropod *Rapana venosa* collected in a microtidal beach showed that the specimens suffered important damage in the considered time span, comparable with the one registered in presumably older shells.

Keywords Alien species · Mollusks · *Rapana venosa* · Rio de la Plata · Uruguay

4.1 Introduction

One of the main difficulties facing Actualistic Taphonomy is time averaging, present even at small time scales. For example, hundreds or thousands of years are represented in the frequently studied nearshore "death assemblages" (e.g. Kidwell and Bosence 1991; Flessa 1993; Meldahl et al. 1997; Kosnik et al. 2009). As a consequence, although fine-tuned compared with the regular fossil record, the precise temporal sequence of the taphonomic processes that lead to the final product (the collected specimen) is still not gathered from the fossil record.

On the other hand, the actualistic approach has a time limitation in the usually short-term laboratory experiments (Experimental Taphonomy), which span only weeks or months (e.g. Flessa and Brown 1983; Briggs 1995; Chattopadhyay et al. 2013). In the field, the involved time span can be years with respect to these studies; SSETI (Shelf and Slope Experimental Taphonomy Initiative, e.g., Powell et al. 2002, 2008, 2010, 2011a, b; Parsons-Hubbard et al. 2011) is an important project, which has been under development for around two decades (and counting). In this project, skeletons are disposed on the sea bottom with an experimental design that permits the periodic evaluation of taphonomic modifications. The samples are placed in mesh bags, with some shells dispersed in the substrate in order to register its burial through

S. Martínez (✉) · A. Rojas · F. Cabrera · D. Antuña
Departamento de Paleontología, Facultad de Ciencias, Universidad de la República, Montevideo, Uruguay
e-mail: smart@fcien.edu.uy

S. Martínez et al. (eds.), *Actualistic Taphonomy in South America*,
Topics in Geobiology 48, https://doi.org/10.1007/978-3-030-20625-3_4

photography. Nevertheless, even in the relatively realistic conditions in which these studies are performed, they are not fully natural.

Unfortunately for the more or less pristine ecosystems, but fortunately for Actualistic Taphonomy, there is presently a natural experiment occurring during the time lapse required to fill the aforementioned gap: the presence of alien species.

4.2 Alien Species and Taphonomy

According to the Convention of Biological Diversity (CBD 2002), an alien species is "a species, subspecies or lower taxon, introduced outside its natural past or present distribution; includes any part, gametes, seeds, eggs, or propagules of such species that might survive and subsequently reproduce", and an alien invasive species is "An alien species whose introduction and/or spread threaten biological diversity".

Although not a new phenomenon, the presence of alien and alien invasive species has increased exponentially during the last decades, and they are a very relevant target for research, including the monitoring of ecosystems. One of the consequences of this close tracking of alien species is that we have a fairly good record of their primary detection and the posterior expansion of their populations.

In the Río de la Plata area, the arrival and establishment of several alien species is well documented in the last decades. Therefore, the variable "time" is well constrained and comprises a time-lapse that is adequate for research into taphonomical patterns and processes that have not yet been studied in fully natural conditions. In other words, investigation of the taphonomical aspects of alien species allows the time gap between research on present natural phenomena *stricto* sensu (e.g. those related to putrefaction or disarticulation), and those that comprise hundreds or thousands of years ago (e.g. coastal shell death assemblages) to be closed.

The goal of this chapter is to preliminarily characterize the taphonomic attributes of an alien invader species arrived at the Río de la Plata about two decades ago, which is now well established. This is the first study on the taphonomy of an alien invader species with a known date of arrival and of the presence of stable populations.

4.3 Materials and Methods

The Asian gastropod *Rapana venosa* (Valenciennes 1846), an estuarine species that is capable of reaching oceanic waters, was selected for this preliminary study. It is a successful invader gastropod, originally from Asia, but presently widely distributed along several coasts of the world (Mann et al. 2004; Brugnoli et al. 2014).

Rapana venosa (Fig. 4.1) is currently found in the middle and external sector of the Río de la Plata (Fig. 4.2), being initially detected at the turn of the XX and XXI centuries in the middle part of the estuary (Scarabino et al. 1999; Pastorino et al. 2000). Later, Lanfranconi et al. (2009) and Carranza et al. (2010) confirmed

Fig. 4.1 *Rapana venosa* in different stages of shell damage

the existence of stable populations and its presence in the exterior limit, very close to the Atlantic Ocean (Fig. 4.2). According to Carranza et al. (2010), specimens are 85 mm in height on average, while Lanfranconi et al. (2009) stated that this dimension reaches 81 mm. According to the records, they live at depths of up to 20 m, and perhaps up to 50 m (Giberto and Bruno 2014).

Sampling was performed in a sandy beach of Gorriti island, Punta del Este Uruguay, collecting all shells and fragments of *R. venosa* larger than 2 cm, in an area of about 150 m^2, with 100 specimens recovered in total. As happens in other localities (e.g., Giberto and Bruno 2014), shells are left in the beach by storms. The Uruguayan coast is microtidal, and differences in tide levels are about 40 cm, with the sea level changes induced by southern winds being far more important than those induced by tides (Chebataroff 1972; Nagy et al. 1997; Gautreau 2006).

Fig. 4.2 Location of the sampling site

Taphonomic signatures were observed in this instance by one person (D.A.) in view of the results about the multiple operators shown by Rothfus (2004). Attributes were taken into account (i) for all specimens, (ii) for the last whorl only (except fragmentation), and (iii) for the spire only (except fragmentation), and were classified in the following categories:

1. Fragmentation. 0: >80% of the shell, 1: between 80 and 30%, 2: <30%.
2. Color. 0: minimum alteration, 1: intermediate, 2: no coloration.
3. Corrasion. 0: minimum alteration, 1: intermediate, 2: intense.
4. Bioerosion. 0: minimum, 1: intermediate, 2: intense.
5. Bioencrustation. 0: minimum, 1: intermediate, 2: intense.

4.4 Results

Of the 100 specimens, 83 represent the last whorl, eight the spire, and only nine are complete shells (i.e. last whorl + spire). From the 92 specimens with the last whorl, 35 have lost the external lip. The results of the remaining taphonomic observations are summarized in Tables 4.1, 4.2, 4.3 and 4.4.

In general, damage to the shells is intense; very few shells remain pristine or nearly pristine, with the spire being broken or destroyed. The last whorl is better preserved, and since it represents a great part of the shell, it is well represented in our samples. Nevertheless, when considered as individual specimens, last whorls reveal serious damage too. The proportion of color loss is equal in the three categories, with

Table 4.1 Taphonomic alteration, all specimens, n = 100. X^2 was calculated for an expected proportion of 1/3 each case

	0	1	2	X^2	*p*
Fragmentation	30	23	47	9.1492	0.01031
Color	22	39	39	5.7859	0.055413
Bioerosion	22	49	29	11.792	0.0027506
Bioencrustation	82	13	5	107.65	4.21E-24
Corrasion	26	36	38	2.4826	0.28901

Table 4.2 Taphonomic alteration, spire, n = 10. X^2 was calculated for an expected proportion of 1/3 each case

	0	1	2	X^2	*p*
Color	4	2	4	0.80909	0.66728
Bioerosion	2	4	4	0.80909	0.80909
Bioencrustation	9	1	0	14.748	0.0006272
Corrasion	2	5	3	1.4152	0.49284

Table 4.3 Taphonomic alteration, last whorl, external view, n = 92. X^2 was calculated for an expected proportion of 1/3 each case

	0	1	2	X^2	*p*
Color	24	38	30	3.2181	0.20008
Bioerosion	22	43	27	7.8495	0.019747
Bioencrustation	82	6	4	128.98	9.80E-29
Corrasion	19	43	30	9.4151	0.0090269

Table 4.4 Taphonomic alteration, last whorl, internal view, n = 92. X^2 was calculated for an expected proportion of 1/3 each case

	0	1	2	X^2	*p*
Color	32	25	35	1.7178	0.42363
Bioerosion	34	24	34	2.1744	0.33716
Bioencrustation	90	1	1	172.23	3.98E-38

bioerosion having a medium incidence, and bioencrustation being low. Corrasion is equally distributed among categories.

When considering only the spire, all factors have similar frequencies, except bioerosion, which has a very low incidence.

In the external part of the last whorl, corrasion is important; on the contrary, bioencrustation is very low.

The internal part of the last whorl shows equal proportions among the categories of color loss and bioerosion. Bioencrustation is almost absent.

4.5 Discussion

Fragmentation, bioerosion, and color loss are significant, being important taphonomic issues in the studied period. On the contrary, bioencrustation is poor. These features can be related to the stressful, high-energy shallow waters and hard or at least consolidated substrates present in the studied locality. Additionally, fragmentation was surely favored by the intense bioerosion. This taphonomic condition is not very different from that found in other mollusks from the beach death assemblage, most of which are presumably older (up to thousands of years) than our specimens. This will be the matter of further research, but pilot studies suggest that this hypothesis may be true.

As stated earlier, the goal of this study is to show taphonomic changes on a decadal scale, using an alien species as a model for the first time. Although the use of an alien species in a coastal environment make differences with other analyses like SSETI significant, some broad comparisons can be made.

Taylor and Wilson (2002) coined the term sclerobiont to make reference to both bioencruster and bioeroder organisms. Using this concept and terminology, Brett et al. (2011) presented the results of SSETI experiments after 1–13 years, comprising several environments (deeper than coastal), in bivalves. They found a shell coverage of ca. 70% after 12/13 years. In our case, and after a similar time lapse, the incidence of sclerobionts is far less, and can be attributed to the environmental differences. The stressful coast on which *Rapana venosa* lives is much less productive and biodiverse than the locations where the SSETI seeds were implanted, and in our case, we are dealing only with preservable evidence. The environmental factor was recognized within SSETI locations indeed; Brett et al. (2011) found drastic differences between stressed environments and carbonate ones in relation to the presence of covering organisms, along the 12–13 year samples. Besides, in our case, episodes of burial could have contributed to the absence of bioencrusters.

Powell et al. (2008) signaled in their experiment that dissolution was more intense in the outer surface of the shell than in the inner one, as expected by Cutler and Flessa (1995) and Kidwell et al. (2001). This is true in our specimens as well. Although we used the category "corrasion" (abrasion + dissolution), Powell et al. (2008) reported that there was no inverse relation between abrasion and dissolution in the same paper, which validates our comparison.

4.6 Conclusions

This first, preliminary evaluation of taphonomic conditions in a gastropod alien species on a decadal scale, in a coastal environment, showed that important damage is reached early, both by biotic and abiotic factors. Bioencrustation, on the contrary, is not significant, at least compared with that found in less stressful environments.

Acknowledgements This work was funded by ANII, Project FCE_1_2017_1_135699.

References

Brett CE, Parsons-Hubbard KM, Walker SE, Ferguson C, Powell EN, Staff GM, Aston-Alcox KA, Raymond A (2011) Gradients and patterns of sclerobionts on experimentally deployed bivalve shells: Synopsis of bathymetric and temporal trends on a decadal time scale. Palaeogeogr Palaeoclimatol Palaeoecol 312:278–304

Briggs DEG (1995) Experimental taphonomy. Palaios 10:539–550

Brugnoli E, Giberto DA, Lafranconi A, Schiariti A, Aguilera F, Bremec CS, Barrero G, Muniz P (2014) El gasterópodo invasor *Rapana venosa* (Valenciennes, 1846) y sus posibles efectos en el ecosistema costero estuarial del Río de la Plata. In: Goso C (ed) Nuevas miradas a la problemática de los ambientes costeros. DIRAC, Montevideo, pp 1–18

Carranza A, de Mello C, Ligrone A, González S, Píriz P, Scarabino F (2010) Observations on the invading gastropod *Rapana venosa* in Punta del Este, Maldonado Bay, Uruguay. Biol Inv 24:995–998

CBD (2002) Conference of the parties to the convention on biological diversity, decision VI/23, Annex, Introduction, footnotes i and ii

Chattopadhyay D, Rathie A, Das A (2013) The effect of morphology on postmortem transportation of bivalves and its taphonomic implications. Palaios 28:203–209

Chebataroff J (1972) Costas platenses y atlánticas del Uruguay. Talleres Gráficos Bouzot, Montevideo

Cutler AH, Flessa KW (1995) Bioerosion, dissolution and precipitation as taphonomic agents at high and low latitudes. Senck mar 25:115–121

Flessa KW (1993) Time-averaging and temporal resolution in recent marine shelly faunas. In: Kidwell SM, Behrensmeyer AK (eds) Taphonomic approaches to time resolution in fossil assemblages. Paleontological Society Short Courses in Paleontology vol 6, pp 9–33

Flessa KW, Brown TJ (1983) Selective solubility of macroinvertebrate calcareous hard parts: a laboratory study. Lethaia 16:193–205

Gautreau P (2006) La Bahía de Montevideo: 150 años de modificación de un paisaje costero y subacuático. In: Menafra R, Rodríguez-Gallego L, Scarabino F, Conde D (eds) Bases para la conservación y manejo de la costa Uruguaya. Vida Silvestre, Montevideo, pp 401–441

Giberto DA, Bruno LI (2014) Recent records of the exotic gastropod *Rapana venosa* (Valenciennes, 1846) along the Argentine coastline: is the invasion progressing southwards? Pan-Am J Aquat Sci 9:324–330

Kidwell SM, Bosence DWJ (1991) Taphonomy and time-averaging of marine shelly faunas. In: Allison PA, Briggs DEG (eds) Taphonomy: releasing the data locked in the fossil record. Plenum Press, New York, pp 115–209

Kidwell SM, Rothfus TA, Best MMR (2001) Sensitivity of taphonomic signatures to sample size, sieve size, damage scoring system, and target taxa. Palaios 16:26–52

Kosnik MA, Hua Q, Kaufman DS, Wüst RA (2009) Taphonomic bias and time-averaging in tropical molluscan death assemblages: differential shell half-lives in great barrier reef sediment. Paleobiology 35:565–586

Lanfranconi A, Hutton M, Brugnoli E, Muniz P (2009) New record of the alien mollusc *Rapana venosa* (Valenciennes 1846) in the Uruguayan coastal zone of Río de la Plata. Pan-Am J Aq Scienc 4:216–221

Mann R, Occhipinti A, Harding JM (2004) Alien species alert: *Rapana venosa* (veined whelk). ICES Cooperative Research Report, 264, 14 p

Meldahl KH, Flessa KW, Cutler AH (1997) Time averaging and postmortem skeletal survival in benthic fossil assemblages: quantitative comparisons among Holocene environments. Paleobiology 23:207–229

Nagy GJ, Martínez CM, Caffera RM, Pedrosa G, Forbes EA, Perdomo AC, López Laborde J (1997) The hydrological and climatic setting of the Río de la Plata. In: Wells PG, Daborn GR (eds) The Río de la Plata. An environmental overview. An EcoPlata Project Background Report. Dalhousie University, Halifax, pp 17–70

Parsons-Hubbard KM, Brett CE, Walker SE (2011) Taphonomic field experiments and the role of the shelf and slope experimental taphonomy initiative (SSETI). Palaeogeogr Palaeoclimatol Palaeoecol 312:195–208

Pastorino G, Penchaszadeh PE, Schejter L, Bremec C (2000) *Rapana venosa* (Valenciennes, 1846) (Mollusca: Muricidae): a new gastropod in south Atlantic waters. J Shellfish Res 19:897–899

Powell EN, Parsons-Hubbard KM, Callender WR, Staff GM, Rowe GT, Brett CE, Walke SE, Raymond A, Carlson DD, White S, Heise EA (2002) Taphonomy on the continental shelf and slope: two-year trends: Gulf of Mexico and Bahamas. Palaeogeogr Palaeoclimatol Palaeoecol 184:1–35

Powell EN, Callender WR, Staff GM, Parsons-Hubbard KM, Brett CE, Walke SE, Raymond A, Ashton-Alcox KA (2008) Molluscan shell condition after eight years on the sea floor—taphonomy in the Gulf of Mexico and Bahamas. J Shellfish Res 27:191–225

Powell EN, Brett CE, Parsons-Hubbard KM, Callender WR, Staff GM, Walker SE, Raymond A, Ashton-Alcox KA (2010) The relationship of bionts and taphonomic processes in molluscan taphofacies formation on the continental shelf and slope: eight-year trends: Gulf of Mexico and Bahamas. Facies 57:15–37

Powell EN, Staff GM, Callender WR, Ashton-Alcox KA, Brett CE, Parsons-Hubbard KM, Walker SE, Raymond A (2011a) Taphonomic degradation of molluscan remains during thirteen years on the continental shelf and slope of the northwestern Gulf of Mexico. Palaeogeogr Palaeoclimatol Palaeoecol 312:209–232

Powell EN, Staff GM, Callender WR, Ashton-Alcox KA, Brett CE, Parsons-Hubbard KM, Walker SE, Raymond A (2011b) The influence of molluscan taxon on taphofacies development over a broad range of environments of preservation: the SSETI experience. Palaeogeogr Palaeoclimatol Palaeoecol 312:233–264

Rothfus TA (2004) How many taphonomists spoil the data? Multiple operators in taphofacies studies. Palaios 19:514–519

Scarabino F, Menafra R, Etchegaray P (1999) Presencia de *Rapana venosa* (Valenciennes, 1846) (Gastropoda: Muricidae) en el Río de la Plata. Bol Soc Zool Uruguay (2a Época) 11:40

Taylor PD, Wilson MA (2002) A new terminology for marine organisms encrusting hard substrates. Palaios 17:522–525

Chapter 5
Actualistic Taphonomy of Freshwater Mollusks from the Argentine Pampas: An Overview of Recent Research Progress

Claudio G. De Francesco, Eleonor Tietze, Paula A. Cristini and Gabriela S. Hassan

Abstract Historically, the development of taphonomic studies in freshwater environments has been scarce and mainly restricted to the Northern Hemisphere, with almost no information on how taphonomic processes work (and eventually differ) in the Southern Hemisphere. We present here an overview of the main results obtained in actualistic taphonomic studies on freshwater mollusks conducted in the Pampa Plain of Argentina, since 2008. The research included patterns of distribution of live and dead mollusks, live-dead fidelity at local and regional scales and dead and fossil shell preservation. Studies were conducted above and below the sediment-water interface, including both descriptive and experimental approaches.

Keywords Molluscan preservation · Taphofacies · Live-dead fidelity · Experimental Taphonomy · Taphonomically active zone

5.1 Introduction

Actualistic taphonomy comprises the study of death, decay, and burial of organisms in contemporary settings to aid interpretations of paleontological patterns provided by the fossil record (Kowalewski and LaBarbera 2004). The discipline has signifi-

C. G. De Francesco (✉) · E. Tietze · G. S. Hassan
Instituto de Investigaciones Marinas y Costeras (IIMyC), CONICET, UNMdP, CC, 1260, Correo Central, B7600WAG Mar del Plata, Argentina
e-mail: cgdefra@mdp.edu.ar

P. A. Cristini
Instituto de Investigaciones en Ciencia y Tecnología de Materiales (INTEMA), CONICET, UNMdP, Av. Colón 10850, B7606BWV Mar del Plata, Argentina

S. Martínez et al. (eds.), *Actualistic Taphonomy in South America*,
Topics in Geobiology 48, https://doi.org/10.1007/978-3-030-20625-3_5

cantly grown in the last 40 years, mainly in the context of the emergence of a series of experimental studies in modern environments aimed at establishing taphonomic signatures associated with local physical, chemical, and biological conditions (i.e., taphofacies, sensu Speyer and Brett 1986). Much of the work on actualistic taphonomy was done with mollusks, and conducted mainly in marine and estuarine settings (see special issues of Palaios edited by Kidwell and LaBarbera 1993; Kowalewski and LaBarbera 2004; Kowalewski and Rothfus 2012, as examples). During these 40 years, articles on actualistic taphonomy in shallow-marine settings experienced a sharply increase, covering a broad spectrum of subdisciplines, such as experimental taphonomy, comparative taphonomy, fidelity, and time averaging.

The development of actualistic taphonomy in freshwater environments, on the other hand, has been scarce and restricted to specific regions of the globe (mainly the Northern Hemisphere). Nevertheless, the preliminary studies brought some interesting results. Among them, the pioneer study by Cohen (1989) in the large Lake Tanganyika, which assessed the morphological fidelity of gastropods of the genus *Paramelania* between life and death assemblages, allowed recognizing that taphonomic mixture obscures the interpretation of evolutionary sequences in the stratigraphic record. Briggs et al. (1990), analyzing life and death assemblages of mollusks in the braided River Lech (northern Austria), found that variations in the numbers of species and individuals, as well as in the patterns of breakage reflected variations related to the local depositional environment and the post-mortem history of the shells, rather than changes in general climatic conditions. Cummins (1994) studied life and death assemblages of unionids (Bivalvia) in several streams and reservoirs of east-central Ohio (USA), finding that death assemblages preserved the rank orders of abundance and biomass of life assemblages only in those environments that were not environmentally disturbed. More recently, Brown et al. (2005) correlated the degree of alteration of dead mussel shells (Bivalvia) in the North Fork Holston River (Virginia, USA), to different extirpation histories related to mercury contamination. Newell et al. (2007) determined that shells of the bivalve *Unio* can suffer in situ abrasion within a channel of the Sakmara River (southern Urals, Russia). Nielsen et al. (2008) compared shell preservation of the river pearl mussel *Margaritifera margaritifera* in rivers from the northeastern Finnish Lapland, finding differential preservation related to acidic conditions of soils. All these preliminary results have provided interesting clues for understanding specific freshwater environments and for comparing with results obtained in marine environments.

Kotzian and Simões (2006) and Martello et al. (2006) carried out the first actualistic taphonomic studies of freshwater mollusks in the Southern Hemisphere. They explored compositional fidelity and shell alteration in mollusks from the Touro Passo River (Brazil). These exploratory works provided an incipient suggestion (from one limited case study) that a good rank order correlation for life-death assemblages occurred and that dissolution might be the main taphonomic process affecting shells in freshwater.

Despite these specific advancements, the knowledge on the taphonomy of other freshwater environments in the Southern Hemisphere remained largely ignored. Framed within this theoretical background, we initiated in 2008 a research project aimed to understand the taphonomic processes affecting molluscan preservation in freshwater environments of the Pampa plain of Argentina. The Pampa plain is a vast region characterized by a flat geomorphology that promotes a high abundance of lentic water bodies in a poor drainage net. These lakes have proven valuable for reconstructing Holocene paleoenvironments and paleoclimates (Stutz et al. 2012; De Francesco et al. 2013; Hassan 2013; Hassan et al. 2014).

As the construction of a complete taphonomic framework for modern freshwater systems requires knowledge of the degree of variation to be expected between and within habitats, two important goals were to assess how the taphonomic signal varied among different freshwater environments and to determine if the variability in taphonomic signatures at large scales (among sites) reflected the variability occurring at small scales (within sites). In the present chapter, we present a summary of our main results after ten years of actualistic taphonomic research. Our research project included studies of live-dead fidelity and taphofacies above and below the sediment-water interface.

5.2 Study Area

Our actualistic taphonomic studies were carried out in shallow lakes and streams of the southeastern Pampa plain (Buenos Aires Province, Argentina, Figs. 5.1 and 5.2). The Pampa region, one of the largest flatlands in the world, includes the center-east part of Argentina, most of Uruguay and the southernmost Brazilian state, Rio Grande do Sul. The Buenos Aires Province has an area of 307,000 km^2 at the core of the Pampa Plain (Castro Berman et al. 2018). The gentle slope is only interrupted by two ranges of low mountains, Tandilia (500 m) and Ventania (1,000 m) situated toward the southeast. From a geological and geomorphological point of view, the study area is a flat plain crossed by a few streams and rivers and dominated by shallow lakes (Diovisalvi et al. 2015). The flat geomorphology is responsible for the disparity between the abundance of lentic water bodies and the low density of the drainage net. The soils of the area are generally fertile with a high nutrient content, composed mainly of loess and with a marked capacity for cationic interchange, predominantly involving calcium (Rodrigues Capítulo et al. 2010). The climate is temperate humid or sub-humid with a mean annual temperature of 15 °C and a mean annual precipitation of 1,100 mm (Feijoó and Lombardo 2007). Precipitation patterns also display large variability, both geographically and inter-annually. This large interannual variability in combination with poorly developed drainage systems results in recurrent and extensive floods, alternating with drought periods. During flood periods, lakes can exceed their capacity, resulting in very slow surface flows that result in lake interconnections. All these processes affect the lake water residence time, the water content of soils, and the depth of the water table (Diovisalvi et al. 2015).

Lakes are very shallow (average depths between 1 and 1.5 m), polimyctic, slightly alkaline (pH 8–10) and are situated in nutrient-rich soil drainages, which makes their

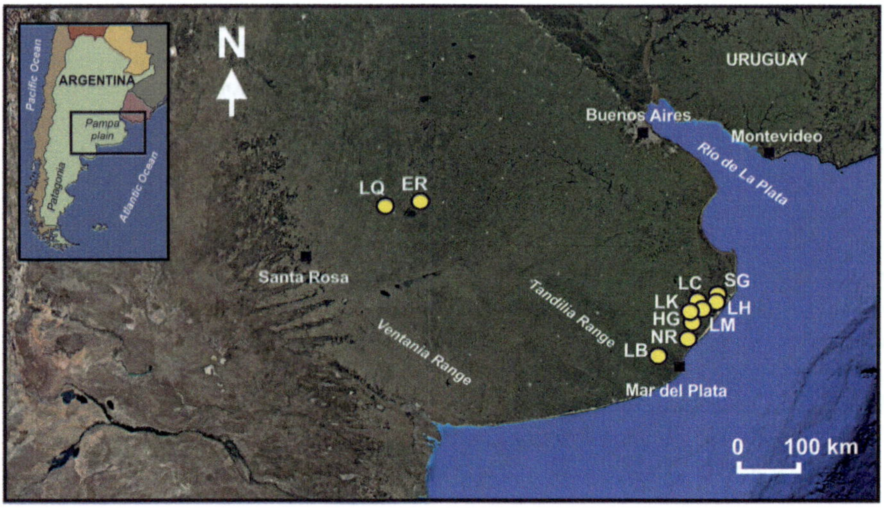

Fig. 5.1 Location of the study area showing the lakes sampled in our research project (yellow circles). LQ: Los Quilmes, ER: El Recado, LB: La Brava, NR: Nahuel Rucá, HG: Hinojales San Leoncio, LK: Lonkoy, LM: Las Mostazas, LC: Los Carpinchos, LH: Los Horcones, SG: Salada Grande

trophic states vary from meso-eutrophic to highly hypertrophic depending on local land use intensity. In the Pampa region, the intensity of land use for agriculture has been increasing gradually since the last decades of the 19th century, but it has been heavily intensified during the last 35 years. As expected, high nutrient loads usually enter the already naturally eutrophic shallow lakes (Quirós et al. 2006). In addition, since 1996 the massive adoption of glyphosate for agricultural production in Argentina led most shallow Pampean lakes to be susceptible to pesticide contamination through surface run-off, leaching, direct overspray and/or spray drift (Castro Berman et al. 2018). A highly vegetated littoral zone occurs in the most freshwater lakes, while scarce or nil vegetation (lower productivity) is represented in more saline lakes (Fig. 5.2).

5.3 The Freshwater Malacofauna of the Pampas

The freshwater malacofauna of the Pampas is characterized for being quite poor, i.e. represented by a small number of families and genera as well as low richness, attributes shared with the Neotropical Region, and differing from the higher richness of other world regions (see Tietze et al. 2011). In our study area, the molluscan richness was represented by 13 species of gastropods and only one species of bivalve (Cristini and De Francesco 2012; Tietze and De Francesco 2010, 2012, 2017; Tietze et al. 2011). Most species corresponded to minute-sized mollusks, and displayed

Fig. 5.2 Some of the Pampean streams and lakes sampled during our actualistic taphonomy project. **a** Arroyo El Peligro, located in the mountainous area close to La Brava lake, **b** Arroyo Grande, a typical sinuous Pampean stream, located in the area close to Nahuel Rucá lake, **c** Nahuel Rucá lake (mean conductivity = 1.9 mS/cm) showing the typical littoral emergent macrophytes that surrounds the lake, which are characteristic of most freshwater lakes, **d** Los Horcones lake (mean conductivity = 5 mS/cm), is a more brackish lake with a lower development of vegetation, **e** Los Quilmes lake (mean conductivity = 32 mS/cm) is a large saline lake located towards the west of Buenos Aires Province, with no emergent vegetation, **f** El Recado lake (mean conductivity = 80 mS/cm) is a hypersaline lake located towards the west of Buenos Aires Province, close to Los Quilmes lake. A vegetation decrease is observed from freshwater to saline lakes

tiny aragonitic shells. These characteristics are quite different from those of previous taphonomic studies conducted in other regions of the world, which were mostly characterized by larger and more robust species as well as by faunas dominated by bivalves. It becomes noticeable that extrapolating such previous results to these circumstances may lead to biases and erroneous interpretations of the fossilization process in freshwater systems of the Pampas. Therefore, we have adapted previous methodologies and erected new taphonomic variables and methodological strategies to deal with Pampean mollusk assemblages.

5.4 Compositional Fidelity of Death Assemblages

Live-dead comparisons are the most common actualistic methods of quantifying the fidelity (faithfulness) of the ecological information preserved in the fossil record (Kidwell 2013). The aim of these studies is to assess the extent to which death assemblages (DAs) reflect the composition, structure, and scale of the original life assemblages (LAs). Compositional fidelity usually focuses on presence-absence and proportional abundance data of species.

As mentioned in the previous section, the Pampas is a flat region dominated by shallow lakes and crossed by few streams and rivers, which make water flow rates to be relatively low. At the beginning of our research project, Tietze and De Francesco (2012) asked if the subtle differences between these two main types of freshwater systems (streams and lakes) might be able to promote differences in the compositional fidelity of DAs. In other words, they asked whether DAs from shallow lakes and streams faithfully reflected the richness and abundances of LAs or if any fidelity bias might occur in any of them. Also, they wondered how reliably differences between areas were captured by DAs.

5.4.1 Differences Between Shallow Lakes and Streams

Tietze and De Francesco (2012) conducted fidelity studies in two areas separated by approximately 55 km. One area was located close to the Tandilia mountain, where most streams originate, in the surroundings of La Brava lake, while the other was situated close to the marine coast, where most streams debouch into the sea, in the proximities of Nahuel Rucá lake (Fig. 5.1). In each area, five lotic (streams) and five lentic (lakes) sites were selected and sampled seasonally during 2008. Both LAs and DAs were analyzed.

The presence-absence fidelity was similar in lotic and lentic environments but differed between the two study areas, being higher in the area close to Nahuel Rucá lake. Regarding abundances, all sites from both areas showed good live-dead agreement, as expected from within-habitat time averaging. This suggested that fidelity was not being affected significantly by postmortem bias and/or between-habitat time averag-

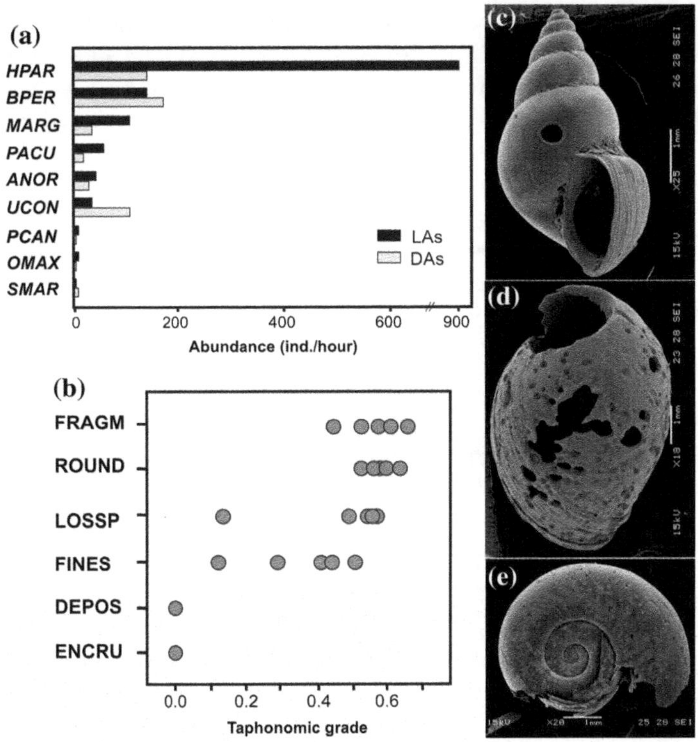

Fig. 5.3 Summary of taphonomic analyses conducted in Nahuel Rucá lake. **a** Mollusk abundance of life assemblages (LAs) and death assemblages (DAs). HPAR: *Heleobia parchappii*, BPER: *Biomphalaria peregrina*, MARG: *Musculium argentinum*, PACU: *Physa acuta*, ANOR: *Antillorbis nordestensis*, UCON: *Uncancylus concentricus*, PCAN: *Pomacea canaliculata*, OMAX: *Omalonyx* sp., SMAR: *Stenophysa marmorata* (modified after Tietze and De Francesco 2017), **b** plot showing the taphonomic grades of the different variables measured on death assemblages. FRAGM: Fragmentation, ROUND: Edge rounding of commissure and hinge line (for bivalves) and aperture (for gastropods), DEPOS: Presence of deposits of ferruginous material, ENCRU: Encrustations of periphyton and/or egg capsules, FINES: Fine-scale surface alteration (damage to shell surface resulting from some combination of dissolution, abrasion and microboerosion), LOSSP: Loss of proteinaceous parts (i.e. periostracum, ligament, and operculum) (modified after Tietze and De Francesco 2017), **c**–**e** SEM photographs (modified after Tietze and De Francesco 2014) of **c** *Heleobia parchappii* shell exhibiting a hole tentatively attributed to dissolution produced by microbial activity, **d** highly-fragmented *Chilina parchappii* shell exhibiting a very high degree of fine-scale surface alteration and **e** *Biomphalaria peregrina* shell showing moderate surface alteration and fragmentation

ing. Nahuel Rucá lake exhibited lower abundances in DAs than in LAs (Fig. 5.3a), while the opposite pattern was recorded in La Brava lake. The lower abundances in DAs may be related to lower shell input rates, which may be related either to a recent decrease in dead–shell production rates or to a recent increase in shell destruction rates.

When comparing fidelity between lentic and lotic environments, Tietze and De Francesco (2012) found that in La Brava area, lentic and lotic sites were different in composition, but these differences were equally captured by LAs and DAs. In contrast, in the Nahuel Rucá area, lentic and lotic environments had comparable variation in species composition in both LAs and DAs. The bivalve *Musculium argentinum* and the pulmonated snail *Chilina parchappii* were more abundant in lotic communities. The coexistence of these two species may be explained by the spatial heterogeneity of the sites, where the accumulation of soft sediments, the presence of hard substrates, and patchy vegetation are common (Tietze and De Francesco 2010). The species that were more abundant in lentic communities (i.e., *Biomphalaria peregrina*, *Heleobia parchappii*) were also highly abundant in lotic communities.

5.4.2 Differences Among Shallow Lakes

As detailed above, Tietze and De Francesco (2012) noticed some environmental differences in fidelity between La Brava and Nahuel Rucá lakes, suggesting a possible difference in the magnitude of time averaging affecting these two areas. To explore more in detail this hypothesis, Tietze and De Francesco (2017) compared fidelity among three shallow lakes of similar size and depth (Lakes Las Mostazas, Los Carpinchos, and Nahuel Rucá, Fig. 5.1), located in the area close to Nahuel Rucá lake, which exhibited subtle differences in water quality (mainly conductivity) and substrate. In fact, lakes showed subtle differences in conductivity, with relatively low values that ranged between 0.3 and 4.3 mS/cm. The results indicated that the lake exhibiting more brackish conditions (Las Mostazas, 4.3 mS/cm) had lower richness in LAs than in DAs, while lakes with lower conductivity showed the opposite pattern (higher richness in LAs than in DAs). Such pattern is suggesting a higher residence time in the taphonomically active zone (TAZ), i.e., the zone located at the sediment-water interface and the bioturbated layer just beneath it, where dissolution proceeds at highest rates (Davies et al. 1989) of the most brackish lakes. DAs formed in such lakes would be more time-averaged than those deposited in more freshwater conditions.

De Francesco and Hassan (2017) started an actualistic field-based study of live-dead fidelity of mollusks in six selected shallow lakes (La Brava, Nahuel Rucá, La Salada Grande, Los Horcones, Los Quilmes and El Recado, Fig. 5.1) representing a wider gradient of conductivity (from 1 mS/cm in La Brava lake to 80 mS/cm in El Recado lake) than that studied by Tietze and De Francesco (2017). All selected lakes were dominated by the same snail species, the euryhaline mud snail *Heleobia parchappii*. The study focused only on *H. parchappii* (selected as target species) to avoid possible biases related to different intrinsic factors (e.g., skeletal mineralogy, robustness, size, mode of life, behavior) in different species. Live-dead fidelity included the comparison of the abundance and size of shells occurring alive (LAs) with those occurring dead (DAs). The preliminary results indicated that dead shells were more abundant in saline lakes (where, paradoxically, live snails were less abundant) suggesting a gradual cumulative pattern of time-averaged deposits. Shell accumulation would be favored in brackish-saline conditions because of less postmortem destruction. This might be a consequence of their lower productivity that would promote less microbial activity on deposited shells (maceration of shell organic matrix), which is one of the most important taphonomic factors responsible for shell dissolution in Pampean lakes (see below).

5.5 Taphofacies

The success of actualistic taphonomy as a discipline has been to a large extent due to the worldwide spread of studies related to the comparison and quantification of taphonomic damage among environments. The primary aim of such modern taphofacies analyses has been to improve the quality of paleoenvironmental interpretations by considering the state of preservation of organic remains, including patterns of damage to skeletal hard parts and styles of skeletal concentration (Kidwell et al. 2001).

The analysis of modern taphofacies in the Pampa plain was conducted with the aim of identifying the main taphonomic processes that affected mollusk shells in the same freshwater environments where compositional fidelity had been previously assessed (Tietze and De Francesco 2012, 2017; De Francesco and Hassan 2017). For instance, taphofacies would help to elucidate the processes responsible for the differences in fidelity observed between Nahuel Rucá and La Brava lake (the former exhibited lower abundances in DAs than in LAs, while the opposite pattern was recorded in the latter) as well as among lakes differing in salinity. Several questions were addressed: Did lower abundances in DAs relate to a recent decrease in dead–shell production rates? Or did they relate to a recent increase in shell destruction rates? Were shells deposited in saline lakes in a better taphonomic condition than shells deposited in freshwater lakes? Can taphonomic signatures of death assemblages be useful to discriminate environments?

5.5.1 *Differences Between Shallow Lakes and Streams*

Framed within the objective of comparing preservation between streams and lakes, Tietze and De Francesco (2014) analyzed the main taphonomic variables responsible for shell preservation in the areas of La Brava and Nahuel Rucá lakes, in the same sites where compositional fidelity had been previously assessed (Tietze and De Francesco 2012). Seven taphonomic variables were analyzed for each sample: (1) loss of proteinaceous parts (periostracum, ligament and operculum); (2) fine-scale surface alteration (damage to shell surface from some combination of dissolution, abrasion, and microbioerosion); (3) disarticulation (only for bivalves); (4) encrustation; (5) presence of deposits of ferruginous material; (6) edge rounding; and (7) fragmentation.

In both areas, shells were mainly affected by the loss of proteinaceous parts, fine-scale surface alteration, fragmentation, and edge rounding (Fig. 5.3b–e). Loss of proteinaceous parts together with fine-scale surface alteration has been mostly attributed to biological processes (such as microorganism decay). Tietze and De Francesco (2014) concluded that the main process affecting shell surface was dissolution mainly due to biological factors, such as productivity, organic matter decay, or respiration (these factors can favor dissolution through the generation of enriched-CO_2 microenvironments). Previous taphonomic studies in freshwater (Cummins 1994; Nielsen et al. 2008; Kotzian and Simões 2006) also recognized dissolution as the main taphonomic agent, but in those cases, it was chemical rather than biological in origin, promoted by the natural acidic conditions (pH values of 5.5–6.5) and low concentration of calcium carbonate existent in the water. The high frequency of articulated bivalves together with the absence of abrasion indicated, possibly, little transportation, with in situ death assemblage formation.

Although the general taphonomic condition of shells was similar in La Brava and Nahuel Rucá, some variables exhibited small-scale differences between streams and lakes.

Contrary to expectations, streams (high-energy environments) showed better preservation than lakes (low-energy environments). This pattern was more evident in Nahuel Rucá lake and suggested that shell preservation was more influenced by organic matter availability than by environmental energy (waves, currents). Nahuel Rucá lake is a highly productive lake, and the decomposition of large amounts of organic matter generates more material available for microorganisms, some of which (euendolithic species) are responsible for microboring on mollusk shells (Fig. 5.3c). In addition, carbon dioxide, methane, organic acids, and alcohols released during decomposition of organic matter, may be also favoring dissolution of carbonate remains. A rapid dissolution of remains through organic decomposition may also prevent encrustation and the presence of ferruginous deposits. The hypothesis of a high shell destruction rate in Nahuel Rucá lake due to microbial decomposition of organic matter present in the environment may also explain why shell abundances are lower in DAs than in LAs.

5.5.2 Differences Among Shallow Lakes

Tietze and De Francesco (2017) noticed differences in fidelity among shallow lakes of similar size and depth, mainly related to differences in salinity. In fact, the most brackish lake (Las Mostazas) displayed lower richness in LAs than in DAs, while lakes with lower conductivity showed the opposite pattern, suggesting a higher residence time in the TAZ and, consequently, higher time-averaging in the most brackish lakes.

When analyzing taphonomic damage in DAs from the same lakes, Tietze and De Francesco (2017) found that between-lake variability was higher than within-lake variability, as consequence of the different species dominating DAs at each lake. In fact, the best state of preservation (low rates of loss of proteinaceous parts) was recorded in Los Carpinchos lake, a lake dominated by the freshwater limpet *Uncancylus concentricus*, a species that is more prone to be destroyed due to its thin shell. The presence of well-preserved *U. concentricus* shells in DAs suggested that they constituted recently dead remains; i.e., shells that spanned very low residence time in the TAZ. Such explanation is consistent with previous findings regarding fidelity that suggest that this lake (and Nahuel Rucá) is subject to high destruction rate and might thus be marked by a lower within-habitat time averaging than that of the most brackish lake (Las Mostazas). In the latter lake, shells were well-preserved when compared with shells from the two previous lakes. Indeed, the fact that only the shells collected from LM exhibited incrustations, a taphonomic variable associated with the time of exposure at the sediment-water interface, is consistent with the previous interpretations based on LA-DA comparisons that indicate a higher time of permanence in the TAZ in this lake and, thus, a higher within-habitat time averaging. De Francesco and Hassan (2017) found that the abundant concentrations of *H. parchappii* present in the DAs of slightly-productive brackish-saline lakes (Fig. 5.4) were, in average, less fragmented than those recorded in DAs from highly-productive freshwater lakes, suggesting less destructive conditions for postmortem preservation in the former.

5.6 Actualistic Taphonomy Below the Sediment-Water Interface

All previous results have been obtained with studies conducted above the sediment-water interface (SWI). This has been the general rule for most taphonomic studies worldwide, being at present only limited information on the processes occurring below the SWI (see Walker and Goldstein 1999; Powell et al. 2011 for a review on this subject). This clearly biased knowledge of preservation towards the processes acting above the SWI, raising questions about what occurs below: For instance, are hard parts differentially affected above and below the SWI? Is there a taphonomic tiering with depth?

Cristini and De Francesco (2012, 2017, 2019) studied taphonomic processes below the SWI in the same shallow lakes (Nahuel Rucá, Los Carpinchos, and Las Mostazas) where taphofacies were analyzed above the SWI. Both observational and experimental approaches were carried out. The former consisted of the taphonomic analysis of sedimentary cores extracted from the sediments while the latter involved long-term field experiments.

Fig. 5.4 Dense concentrations of *Heleobia parchappii* in the littoral ridges of a saline lake (El Recado). **a** General view of the concentrations (scale bar = 1 m), **b** detail showing the abundance of *H. parchappii* shells within ridges (scale bar = 1 cm), **c** individual shell of *Heleobia parchappii* from the ridges exhibiting excellent preservation (scale bar = 1 mm)

5.6.1 Taphonomic Patterns Below the Sediment-Water Interface

In a preliminary study, Cristini and De Francesco (2012) analyzed the preservation of mollusks below the SWI in Nahuel Rucá lake by extracting short sedimentary cores (~50 cm) from the littoral zone of the lake. In all cores the recovered material was dominated by fragments, suggesting a high destruction rate (Fig. 5.5). Of the eight species recorded above the SWI in Nahuel Rucá lake (Tietze and De Francesco 2012), five were also recorded below (only *P. acuta*, *A. nordestensis*, and *Omalonix* sp. were not present). In addition, the estuarine mud snail *Heleobia australis* and the pulmonate *Drepanotrema kermatoides* were also recorded. Species richness decreased with depth, being only the dominant *H. parchappii* and unidentified frag-

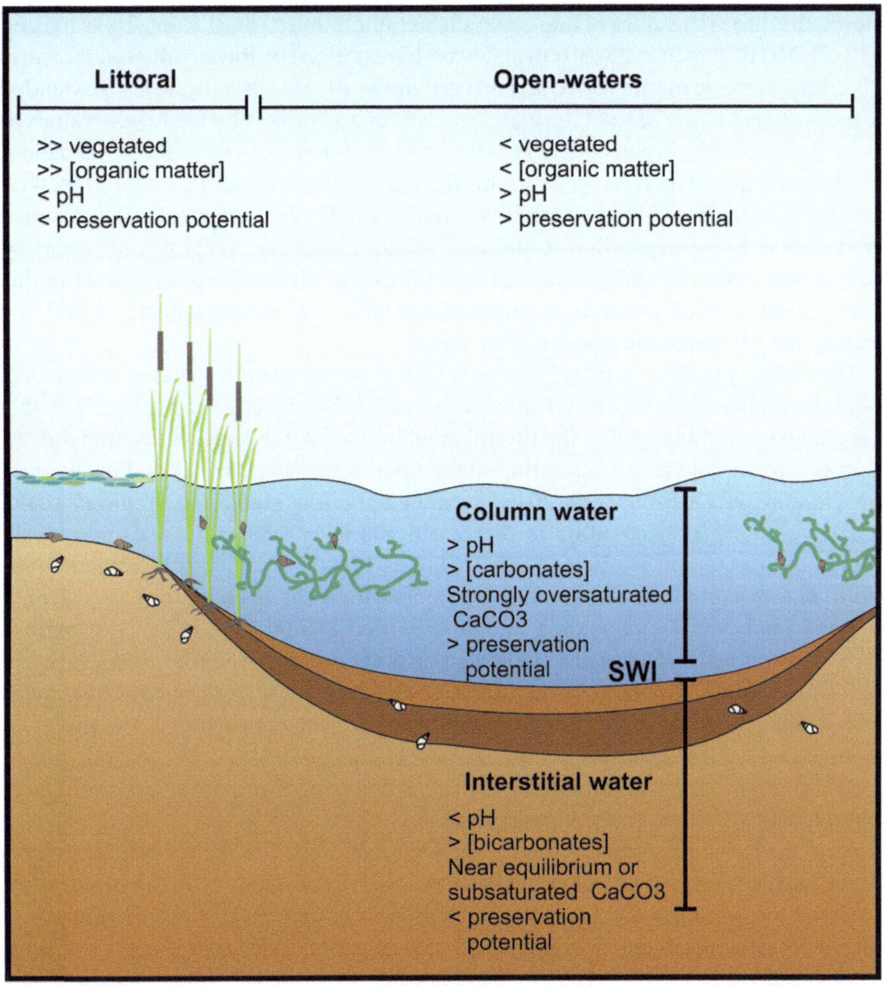

Fig. 5.5 Schematic representation of the main variables affecting mollusk shell preservation in Pampean shallow lakes. Differences in vegetation cover, organic matter concentration and pH are compared between littoral and open waters sub-environments. Additionally, the concentrations of carbonates, bicarbonates, pH and the saturation of $CaCO_3$ are compared between the column and interstitial waters. The relative impact of environmental differences on the preservation potential of mollusks is indicated (modified after Cristini et al. 2017). SWI: Sediment-water interface

ments recorded near the bottom of the cores. Shells were in general highly altered, showing a high degree of breakage at all depths.

Three different zones, characterized by differences in physico-chemical variables, were recognized along the depth. The first zone extended from the SWI to the depth of 10 cm and was characterized by higher values of moisture (55–77%), organic matter content (7–21%) and carbonates (3–6%). Shells recovered from this zone showed intermediate to high values of fine-scale surface alteration (0.8–2). Zone 2 was located between 10 cm and circa 25–30 cm and was characterized by lower values of moisture (25–57%), organic matter (0–10%) and carbonates (0–5%). Finally, Zone 3 extended from 25–30 cm to the base of the cores and was characterized by the lowest values of moisture (30–45%), organic matter (3–8%) and carbonates (1–3%). Shells from these two deeper zones displayed very variable degrees of fine-scale surface alteration, with no clear pattern. Based on these results, Cristini and De Francesco (2012) concluded that in zone 1 (corresponding to the first 10 cm below the SWI) would occur the highest dissolution of shell remains. Higher concentrations of organic matter in this zone generates great amounts of material available for microorganisms, which are responsible for dissolution on mollusk shells.

The study by Cristini and De Francesco (2012) provided the first baseline to understand the taphonomic pathway followed by mollusks in freshwater lakes. Yet, this was exclusively conducted in the littoral zone of one lake (Nahuel Rucá) and did not address taphonomic processes acting in the open water zones of lakes. For that reason, Cristini and De Francesco (2017) expanded previous studies on a regional scale, by analyzing shell preservation in two additional lakes (the same lakes previously studied by Tietze and De Francesco 2017: Las Mostazas and Los Carpinchos). In addition, they compared preservation between littoral and open-water zones of each lake and analyzed if taphonomic alteration varied locally (within-lake) or regionally (between-lake). The results indicated an important between-lake variability in shell preservation. In Nahuel Rucá lake, shells recovered from open-water zones were better preserved than those obtained from littoral zones (Fig. 5.5), while Las Mostazas lake did not show differences between these two areas. The littoral zone of Los Carpinchos lake was completely sterile, suggesting that dissolution might have destroyed all shells from death assemblages.

As in the previous study by Cristini and De Francesco (2012), the highest percentage of moisture, organic matter, and carbonate were recorded at the top of the cores (~10 cm) and diminished gradually downward. Furthermore, pH showed the opposite trend increasing when the other physicochemical variables decreased and vice versa. Cristini and De Francesco (2017) concluded that the sediment layer located within the first 10 cm below the SWI in the littoral zone, had less favorable conditions for shell preservation. More acidic conditions and higher concentrations of organic matter in the first 10 cm below the SWI would explain the lower abundance of mollusks in Nahuel Rucá lake as well as the sterile level found in Los Carpinchos lake. In Las Mostazas lake, on the other hand, pH was rather alkaline and constant through depth (Cristini et al. 2017). According to the results obtained here, Cristini and De Francesco (2017) suggested that the layer located within the first 10 cm below the SWI represented the TAZ of these lakes.

5.6.2 Field Experiments

Cristini and De Francesco (2012) also conducted a long-term field experiment (5 months) in Nahuel Rucá lake to determine if shell dissolution differed at different depths below the SWI and if periostracum had an effect on preservation. Experimental approaches are needed to identify the variables responsible for the observed patterns. Experimental taphonomy has been recognized as one of the most productive fields to understand the fossilization process and guide interpretations on assemblage formation (see Parsons-Hubbard et al. 2011 and references therein). The experimental array consisted of a series of 0.5-cm-mesh bags containing empty shells (with and without periostracum) of the gastropods *Heleobia parchappii* and *Biomphalaria peregrina*, attached to 1.8-m wood poles buried below the sediment-water interface at five different depths (5, 20, 35, 50, and 65 cm). In both species, dissolution was higher in shells without periostracum. Among them, *B. peregrina* exhibited significantly higher values than *H. parchappii*, probably because *B. peregrina* shells are larger and, thus, exhibit a higher exposed surface vulnerable to dissolution. No significant differences among depths were obtained for any species. This was interpreted as a consequence of the short duration of the experiment (5 months), which was probably not enough to evidence dissolution in a quantifiable way.

For that reason, Cristini and De Francesco (2019) conducted a new taphonomic field experiment of longer duration (30 months) to evaluate the effect of depth below the SWI and time of exposure on shell dissolution. The experiment was deployed at three different depths below the SWI (5, 20, and 35 cm) in the littoral zone of Nahuel Rucá lake. In that experiment, surface alteration of *H. parchappii* and *B. peregrina* exhibited significant differences in depth, with shells placed at 5 cm being more altered than those located below. These experimental results agreed with previous interpretations, based on patterns of shell dissolution and physicochemical variables extracted from short sedimentary cores (Cristini and De Francesco 2017) that suggested less favorable conditions for shell preservation within the first 10 cm below the SWI.

5.7 Discussion: Research Contributions and Future Directions

Our results contributed significantly to understanding patterns of molluscan preservation in freshwater systems of southern South America. Prior to 2008, no information on molluscan actualistic taphonomy existed for the Pampa plain of Argentina even though mollusks constituted one of the most common bioindicators recorded in Quaternary sediments. Despite their ubiquity, mollusks usually exhibited very low diversity, which limited their potential use as bioindicators to reconstruct the nature and dynamics of ancient water bodies. As is generally known, paleontologists use taxonomic uniformitarianism ("the ecology of modern organisms is the key to the ecology of past organisms" sensu Dodd and Stanton 1990) to interpret ancient environments. Due to the very low diversity of life and death mollusk assemblages in the Pampas plain (most species are found in any water body), this methodological procedure brought very limited paleoenvironmental information for the area. In contrast, our taphonomic studies (mainly modern taphofacies) allowed recognizing that the taphonomic signature of Pampean mollusks may provide additional paleoenvironmental information.

As an example, De Francesco et al. (2013) analyzed the molluscan fauna preserved in three fossil (Holocene) successions recovered from lacustrine sediments. Both paleoecology and taphonomy of fossil assemblages were assessed to infer ancient environmental conditions for Pampean lakes during the last 6,000 years B.P. Authors recognized a shift from brackish to freshwater euryhaline conditions, which was reflected both in changes in richness and abundance of species (paleoecology) and in shell preservation (taphonomy). Even though the sedimentary succession was always dominated by the euryhaline *Heleobia parchappii*, shell preservation varied during the lapse, with shells exhibiting better taphonomic conditions when water bodies developed brackish conditions at circa 6,000 years B.P. On the other hand, the preservation was poorer in times of freshwater input. Taphonomic signatures, thus, provided a tool for recognizing changes in relation to the balance between brackish and freshwater dominance. These hydrological changes constitute a natural feature of Pampean lakes in response to alternations between drought and flood periods and, therefore, their recognition in the fossil record brings interesting perspectives as a natural baseline for conservation paleobiology purposes. Similar studies conducted in other Quaternary Pampean successions located in the Salado River Basin (Pisano and Fucks 2016) also reached similar conclusions, supporting our interpretations.

Even more interesting is the fact that a differential preservation exists between mollusks and diatoms (one of the commonest bioindicators used in Quaternary studies for reconstructing environments). While diatoms exhibited a higher alteration in sedimentary levels interpreted as brackish and less-productive, mollusks were more altered at the topmost levels, characterized as freshwater highly productive lakes (Hassan et al. 2014). This contrasting response was interpreted because of the differential reaction of carbonate and silica to dissolution agents acting in fresh and saline water, as indicated by studies on actualistic taphonomy of diatoms (see Hassan et al. this volume). This highlights the relevance of including taphonomic traits in Quaternary paleoenvironmental or paleoclimatic studies and pose questions about the biases that may occur if such aspects were not considered.

To sum up, our main contributions for the last 10 years were related to the knowledge of the taphonomic processes suffered by dead mollusks during their permanence in the taphonomically active zone. Considering the studies carried out to date, we can highlight three main contributions to the field of taphonomy in freshwater environments:

(1) *Most freshwater death assemblages from the Pampa plain are within-habitat time-averaged.* Within-habitat time-averaged assemblages are accumulated over a period of relative environmental stability, within a single environment

(Kidwell and Bosence 1991). In other words, death assemblages are mostly composed of indigenous shells that lived in the local environment, exhibiting a high live-dead agreement. Transported remains (from other environments) are rare. This is very good news for the use of death assemblages as indicative of past local communities, suggesting that ecological information can be extracted from fossil assemblages with a high degree of confidence.

(2) *Dissolution is the main taphonomic process affecting shell preservation.* Biological rather than chemical factors are probably responsible for such patterns. Dissolution can be generated by algal and fungal microboring as well as by microbial maceration of the organic matrix (Best and Kidwell 2000). The high productivity of the numerous nutrient-enriched shallow lakes occurring in the Pampa plain may favor dissolution through organic matter decomposition, by generating material available for microorganisms to decompose (mainly cyanobacteria, which are enhanced by the higher availability of phosphorus). Some of them (endolithic cyanobacteria) are responsible for microboring on mollusk shells. In fact, Tietze and Esquius (2018) recognized cyanobacteria of the family Pseudanabaenaceae as the producers of holes and tunnels on shells of *Heleobia parchappii*.

(3) *Most dissolution occurs within the first 10 cm from the sediment-water interface.* As mentioned above, dissolution (of biological origin) was the most important variable affecting shell alteration. The effect of dissolution, characterized by minor pitting, was more marked at the sediment-water interface and the first 10 cm below it in the littoral zone of shallow lakes (Cristini and De Francesco 2012, 2017, 2019). This coincides with the highest concentrations of organic matter and less alkaline pH, which are associated to aragonite and calcite saturation indices near equilibrium or slightly subsaturated within the first 10 cm of sediments (Cristini et al. 2017). In some cases, where macrophyte root abundance was higher, the effect extended to nearly 25 cm. Based on these evidences, Cristini and De Francesco (2017) hypothesized that the layer located within the first 10 cm below the SWI represented the TAZ of these lakes. This layer was lower than that recorded for the TAZ in marine settings (e.g., 50 cm, Walker 2001). The presence of deep burrowers in marine environments (in freshwater only shallow burrowers occur) may be responsible for the deeper extent of the TAZ recorded there. No available data on shell age-frequency distributions in death assemblages exists for Pampean lakes, which avoids estimating scales of time-averaging. Therefore, we cannot at present discriminate loss rates as to model the dynamics of shell loss below the SWI. Tomašových et al. (2014) demonstrated the existence of a two-phase dynamic model, where the timing of a drop in disintegration rate is controlled by a sequestration rate that operates at millennial scales. In this model, shell loss rate is partitioned within the mixed layer, between the TAZ (characterized by high disintegration rates) and a sequestration zone (SZ, sensu Olszewski 2004) with low disintegration rates. These two zones can exist as discrete, surficial and deeper layers or, alternatively, the SZ can comprise patchy microenvironments within the TAZ. Further

studies aimed to recognize the existence of a TAZ and a SZ in these lakes are needed to allow modeling shell loss rate in these environments.

Our research has important implications for a variety of paleontological studies, as well as related disciplines such as ecology and geology. The scarce studies conducted at present in freshwater poses our results as a preliminary baseline for comparing with most taphonomic information coming from marine or estuarine settings (most intensely studied). In addition, the results of our taphonomic research program form part of a larger corpus of actualistic data intended to develop a comprehensive taphonomic model for South America in a regional project called the South American School of Taphonomy (SST) that includes multiple depositional settings, bioindicators, and spatial scales (see Ritter et al. 2016). One of the greatest challenges of the SST is to integrate the information from the variety of environments we study.

Acknowledgements Financial support for this research was granted by the Agencia Nacional de Promoción Científica y Tecnológica (PICT 16-20706/04 and PICT 2727/13), the Consejo Nacional de Investigaciones Científicas y Técnicas (PIP 1328/08 and PIP 0063/10), and the Universidad Nacional de Mar del Plata (EXA 369/07, EXA587/12). Special thanks to Pedro Urrutia (Estancia Nahuel Rucá), Santiago González Aguilar (Las Mostazas), Nicolás Bustamante (Camping Ruca Lauquen), Héctor Sanabria (Estancia La Confianza), Josefina Pueyrredón (Lonkoy), Víctor Albarenga (Estancia San Leoncio), Cacho Soraire (Laguna Salada Grande), Luis Romero (Club de Pescadores San Nicolás) and Enrique Bidondo (Estancia Vidania) for permission to access private land where the lakes are located. We also thank Gustavo Bernava for performing the chemical analyses. Alejandro Canepuccia, Sebastián Grondona, Christian Ituarte, Darío Lazo, Diego Navarro, Silvina Stutz and Marcela Tonello aided at some stage of the research. Special thanks to the editors Sergio Martínez, Alejandra Rojas and Fernanda Cabrera for inviting us to participate with this chapter and Adam Tomašových for the helpful review.

References

Best MMR, Kidwell SM (2000) Bivalve taphonomy in tropical mixed siliciclastic-carbonate settings. I. Environmental variation in shell condition. Paleobiology 26(1):80–102

Briggs DJ, Gilbertson DD, Harris AL (1990) Molluscan taphonomy in a braided river environment and its implications for studies of quaternary cold-stage river deposits. J Biogeogr 17:623–637

Brown ME, Kowalewski M, Neves RJ, Cherry DS, Schreiber ME (2005) Freshwater mussel shells as environmental chronicles: geochemical and taphonomic signatures of mercury-related extirpations in the North Fork Holston River, Virginia. Environ Sci Technol 39:1455–1462

Castro Berman M, Marino DJG, Quiroga MV, Zagarese H (2018) Occurrence and levels of glyphosate and AMPA in shallow lakes from the Pampean and Patagonian regions of Argentina. Chemosphere 200:513–522

Cohen AS (1989) The taphonomy of gastropod shell accumulations in large lakes: an example from Lake Tanganyika, Africa. Paleobiology 15(1):26–45

Cristini PA, De Francesco CG (2012) Análisis tafonómico de moluscos por debajo de la interfase agua-sedimento en la laguna Nahuel Rucá (provincia de Buenos Aires, Argentina). Ameghiniana 49(4):594–605

Cristini PA, De Francesco CG (2017) Molluscan taphonomic patterns below the sediment-water interface in freshwater shallow lakes from the southeastern Pampa plain, Argentina. Palaios 32:528–542

Cristini PA, De Francesco CG (2019) Taphonomic field experiment in a freshwater shallow lake: molluscan shell alteration below the sediment-water interface. J Molluscan Stud (accepted)

Cristini PA, Tietze E, De Francesco CG, Martínez DE (2017) Water geochemistry of shallow lakes from the southeastern Pampa plain, Argentina and their implications on mollusk shells preservation. Sci Total Environ 603–604:155–166

Cummins RH (1994) Taphonomic processes in modern freshwater molluscan death assemblages: implications for the freshwater fossil record. Palaeogeogr Palaeocl Palaeoecol 108:55–73

Davies DJ, Powell EN, Stanton RJ (1989) Relative rates of shell dissolution and net accumulation—a commentary: can shell beds form by the gradual accumulation of biogenic debris on the sea floor? Lethaia 22:207–212

De Francesco CG, Hassan GS (2017) Differential preservation of the snail *Heleobia parchappii* (d'Orbigny) in freshwater and saline shallow lakes of the Argentine pampas. In: Abstracts of the workshop actualistic taphonomy in South America, Facultad de Ciencias, Montevideo, 9–11 Oct 2017

De Francesco CG, Tietze E, Cristini PA (2013) Mollusk successions of Holocene shallow-lake deposits from the southeastern Pampa plain, Argentina. Palaios 28:851–862

Diovisalvi N, Bohn VY, Piccolo MC, Perillo GME, Baigún C, Zagarese HE (2015) Shallow lakes from the Central Plains of Argentina: an overview and worldwide comparative analysis of their basic limnological features. Hydrobiologia 752:5–20

Dodd JR, Stanton RJ (1990) Paleoecology: concepts and applications. Wiley, New York, p 512

Feijoó CS, Lombardo RJ (2007) Baseline water quality and macrophyte assemblages in Pampean streams: a regional approach. Water Res 41:1399–1410

Hassan GS (2013) Diatom-based reconstruction of middle to late Holocene paleoenvironments in Lake Lonkoy, southern Pampas, Argentina. Diatom Res 28:473–486

Hassan GS, Tietze E, Cristini PA, De Francesco CG (2014) Differential preservation of freshwater diatoms and mollusks in late Holocene sediments: paleoenvironmental implications. Palaios 29:612–623

Kidwell SK (2013) Time-averaging and fidelity of modern death assemblages: building a taphonomic foundation for conservation paleobiology. Palaeontology 56:487–522

Kidwell SK, Bosence DWJ (1991) Taphonomy and time-averaging of marine shelly faunas. In: Allison PA, Briggs DEG (eds) Taphonomy: releasing data locked in the fossil record: topics in geobiology, vol 9. Plenum, New York, pp 115–209

Kidwell SK, LaBarbera M (1993) Experimental taphonomy. Palaios 8(3):1–2

Kidwell SK, Rothfus TA, Best MMR (2001) Sensitivity of taphonomic signatures to sample size, sieve size, damage scoring system, and target taxa. Palaios 16:26–52

Kotzian CB, Simões MG (2006) Taphonomy of recent freshwater molluscan death assemblages, Touro Passo Stream, southern Brazil. Rev Bras Paleontolog 9(2):243–260

Kowalewski M, LaBarbera M (2004) Actualistic taphonomy: death, decay, and disintegration in contemporary settings. Palaios 19:423–427

Kowalewski M, Rothfus TA (2012) Death and disintegration in Bahamas: taphonomic patterns and processes in tropical island settings. Palaios 27:123–126

Martello AR, Kotzian CB, Simões MG (2006) Quantitative fidelity of Recent freshwater mollusk assemblages from the Touro Passo River, Rio Grande do Sul, Brazil. Iheringia, Sér Zool 96(4):453–465

Newell AJ, Gower DJ, Benton MJ, Tverdokhlebov VP (2007) Bedload abrasion and the in situ fragmentation of bivalve shells. Sedimentology 54:835–845

Nielsen JK, Helama S, Nielsen JK (2008) Taphonomy of freshwater molluscs in carbonate-poor deposits: a case study of the river pearl mussel in northeastern Finnish Lapland. Norw J Geol 88:103–116

Olszewski T (2004) Modeling the influence of taphonomic destruction, reworking, and burial on time-averaging in fossil accumulations. Palaios 19:39–50

Parsons-Hubbard KM, Brett CE, Walker SE (2011) Taphonomic field experiments and the role of the shelf and slope experimental taphonomy initiative. Palaeogeogr Palaeocl 312:195–208

Pisano MF, Fucks EE (2016) Quaternary mollusc assemblages from the lower basin of Salado River, Buenos Aires Province: their use as paleoenvironmental indicators. Quat Int 391:100–111

Powell EN, Staff GM, Callender WR, Ashton-Alcox KA, Brett CE, Parsons-Hubbard KM, Walker SE, Raymond A (2011) Taphonomic degradation of molluscan remains during thirteen years on the continental shelf and slope of the northwestern Gulf of Mexico. Palaeogeogr Palaeocl Palaeoecol 312:209–232

Quirós R, Boveri MB, Petracchi CA, Rennella AM, Rosso JJ, Sosnovsky A, von Bernard HT (2006) Los efectos de la agriculturización del humedal pampeano sobre la eutrofización de sus lagunas. In: Galizia Tundisi J, Matsumura-Tundisi T, Galli CS (eds) Eutrofização na América do Sul: Causas, conseqüências e tecnologias de gerenciamento e controle. Rede EUTROSUL, PROSUL, São Carlos, Brazil, pp 1–16

Ritter MN, De Francesco CG, Erthal F, Hassan GS, Tietze E, Martínez SA (2016) Manifesto of the South American School of (actualistic) taphonomy. Palaios 31:20–24

Rodrigues Capítulo A, Gómez N, Giorgi A, Feijoó CS (2010) Global changes in pampean lowland streams (Argentina): implications for biodiversity and functioning. Hydrobiologia 657:53–70

Speyer SE, Brett CE (1986) Trilobite taphonomy and middle Devonian taphofacies. Palaios 1:312–327

Stutz SM, Borel M, Fontana SL, Tonello MS (2012) Holocene changes in trophic states of shallow lakes from the Pampa plain of Argentina. Holocene 22:1263–1270

Tietze E, De Francesco CG (2010) Environmental significance of freshwater mollusks in the southern Pampas, Argentina: to what detail can local environments be inferred from mollusk composition? Hydrobiologia 641:133–143

Tietze E, De Francesco CG (2012) Compositional fidelity of subfossil mollusk assemblages in streams and lakes of the southeastern Pampas, Argentina. Palaios 27:401–413

Tietze E, De Francesco CG (2014) Taphonomic differences in molluscan shell preservation in freshwater environments from southeastern Pampas, Argentina. Palaios 29:501–511

Tietze E, De Francesco CG (2017) Compositional fidelity and taphonomy of freshwater mollusks in three pampean shallow lakes of Argentina. Ameghiniana 54(2):208–223

Tietze E, Esquius KS (2018) First record of cyanobacteria microboring activity in Pampean shallow lakes of Argentina. Rev Bras Paleontolog 21(2):187–193

Tietze E, De Francesco CG, Núñez MV (2011) What can gastropod assemblages tell us about freshwater environments? In: Bianchi A, Fields J (eds) Gastropods: diversity, habitat and genetics. Nova Science Publishers, New York, pp 1–34

Tomašových A, Kidwell SM, Barber RF, Kaufman DS (2014) Long-term accumulation of carbonate shells reflects a 100-fold drop in loss rate. Geology 42(9):819–822

Walker SE (2001) Below the sediment-water-interface: a new frontier in taphonomic research. Palaios 16:113–114

Walker SE, Goldstein ST (1999) Taphonomic tiering: experimental field taphonomy of molluscs and foraminifera above and below the sediment–water interface. Palaeogeogr Palaeocl Palaeoecol 149:227–244

Chapter 6
Actualistic Taphonomy of Freshwater Diatoms: Implications for the Interpretation of the Holocene Record in Pampean Shallow Lakes

Gabriela S. Hassan, Claudio G. De Francesco and Marisel C. Díaz

Abstract Recent findings of noticeable changes in diatom preservation in the Holocene record of Pampean shallow lakes evidenced the need for carrying out actualistic studies to decipher the environmental significance of taphonomic signatures. In this chapter, we review a series of field and experimental studies recently conducted, focusing on the effect of contemporary environmental gradients on the dissolution and fragmentation of diatom valves. Field studies signaled salinity, carbonates, and bicarbonates as the main drivers of dissolution of the target taxon *Cyclotella meneghiniana* in shallow lakes covering wide environmental gradients. Laboratory controlled experiments demonstrated a negative effect of NaCl and HCO_3Na on valve preservation, evidenced by an increase in dissolved silica and by the occurrence of valves showing advanced stages of dissolution. Detailed analysis of taphonomic attributes and their relationship with live/dead agreement at a freshwater lake showed that within-lake taphofacies are useful to discriminate between diatom subenvironments. The joint analysis of compositional and taphonomic variations in the recent sedimentary record of the lake demonstrated the usefulness of taphonomic analyses to uncover subtle paleoenvironmental variations, which could be overlooked if only traditional compositional analyses were performed. The strong link between compositional and taphonomic patterns at different working scales highlighted the usefulness of including taphonomic analyses when conducting diatom-based paleoenvironmental studies in environmentally heterogeneous shallow lakes.

Keywords Diatom preservation · Taphonomic signatures · Taphofacies · Experimental taphonomy · Live/dead fidelity

G. S. Hassan (✉) · C. G. De Francesco · M. C. Díaz
Instituto de Investigaciones Marinas y Costeras (IIMyC), CONICET, UNMdP, CC 1260 Correo Central, Mar del Plata B7600WAG, Argentina
e-mail: ghassan@mdp.edu.ar

S. Martínez et al. (eds.), *Actualistic Taphonomy in South America*, Topics in Geobiology 48, https://doi.org/10.1007/978-3-030-20625-3_6

6.1 Introduction

Diatoms (Bacillariophyceae) are among the most ubiquitous bioindicators used to reconstruct paleoenvironmental changes in freshwater environments. These eukaryotic algae are small (2–500 μm) and diverse (>24,000 known species) and can occupy a variety of planktonic, benthic, and subaerial habitats (Julius and Theriot 2010). Since the recognition of diatoms as potential indicators of water pollution and paleoenvironmental changes in the late nineteenth century, diatom analyses of lake sediments became increasingly used as a tool to understand different aspects of lake dynamics, water acidification, eutrophication, and climate change (see Battarbee et al. 2001 and references therein). Although the inner contents of diatom cells are similar to other eukaryotic algae, the uniqueness of diatoms relies on the composition and morphology of their cell walls, which constitute their preservable parts and can thus be incorporated to the fossil record (Battarbee et al. 2001). The cell wall, known as the frustule, is composed of silica (SiO_2) arranged in two halves (i.e. valves) and several bands or belts known as the griddle bands. The frustule has many openings allowing the contact between the cytoplasm and the surrounding environment, which can vary considerably in shape and architecture among species, from simple pores to specialized structures of extremely complex micro-architecture (Julius and Theriot 2010). Overall, given its particular chemical composition, small size, and variability in shape and structure, the taphonomic constraints to the preservation of the diatom frustule can differ strongly from those of carbonatic macro-remains (such as mollusks), which have been the main focus of taphonomic research since the establishment of taphonomy as a discipline.

The recognition of taphonomy as a key factor in structuring the diatom sedimentary record has a long history (e.g. Lewin 1961). Removal via outflows, transport from upstream sources, resuspension and reworking of sediments within lake basins, sediment bioturbation, and silica dissolution have been advocated as potential causes of taphonomic biases in continental settings (Battarbee 1986). Among them, dissolution has received most of the attention, given the relevance of this process in saline and alkaline environments (Ryves et al. 2001). Dissolution is enhanced in solutions with low silica concentrations (e.g. Flower and Ryves 2009), elevated temperature, pH (e.g. Lewin 1961; Barker 1992; Flower 1993), and salinity (e.g. Mikkelsen 1977; Barker et al. 1994). These processes can alter not only total diatom abundances in sediments but also assemblages composition (Warnock and Scherer 2015), as species differ in their preservation potential, being delicate and less silicified taxa highly sensitive to cell wall dissolution (Fritz 2007). Therefore, dissolution can lead to an underestimation of the relative abundances of these delicate taxa, biasing the diatom record towards robust species (Ryves et al. 2009). Contrastingly, the causes of fragmentation of diatom frustules have been relatively poorly studied, (e.g. Haberyan 1985; Ryves et al. 2006). Higher fragmentation has been associated with high energy environments, such as littoral and shallow lacustrine zones (Round 1964; Flower and Nicholson 1987). The effect of grazing has also proven to be significant in causing frustule breakage (Haberyan 1985). Therefore, biological and physical environmental variables, such as fetch, depth, wind speed and grazers abundance may play a key role in explaining diatom breakage. Overall, the joint effect of biological, chemical and physical factors in promoting diatom breakage and dissolution can be significant in continental settings, particularly in saline, alkaline and very shallow lakes with high nutrient content, being these changes highly dependent on local taxonomy and variable among environments and regions.

The study of diatoms as a tool to understand Holocene environmental changes in the Pampean Region has increased significantly during the last decade (García-Rodríguez et al. 2009; Stutz et al. 2010; Hassan 2013; Hassan et al. 2011, 2012, 2014). In these studies, the shifts in the composition of sedimentary diatom assemblages during the last *ca.* 5000 cal. yrs BP were addressed and interpreted as indicative of past changes in salinity and nutrient status of the lakes. Diatom records showed a shift from brackish/saline and low trophic conditions at the mid-Holocene towards freshwater/eutrophic conditions developed since *ca.* 1500–1200 cal. yrs BP (Hassan et al. 2011, 2014; Hassan 2013). Hassan et al. (2014) observed that these shifts in diatom assemblages composition were also accompanied by noticeable changes in preservation, evidenced by marks on frustule breakage and dissolution. These changes were explained as a consequence of past variations in lake salinity, suggesting the potential use of taphonomic signatures as sources of paleoenvironmental information. These results pointed out the significance of incorporating taphonomic analyses when conducting paleoenvironmental studies and highlighted the need for conducting actualistic studies in Pampean shallow lakes, in order to elucidate the environmental significance of diatom taphonomic signatures.

Under this scenario, we have recently started a research project on actualistic taphonomy of diatoms in Pampean shallow lakes, in order to assess the effect that physicochemical lake characteristics exert on valve preservation. The project included controlled laboratory experiments and field research covering regional and local scales. By conducting this project, we intend to combine the information provided by diatom taphonomy with the classic inferences based on community composition, in order to develop new tools to improve the quality of future paleoenvironmental studies in the area. In this chapter, we review the advances of a series of studies conducted in the context of this project during the last four years and discuss the expected future developments in this field of research.

6.2 Environmental Significance of Modern Diatom Assemblages

The quality of diatom-based paleoenvironmental reconstructions relies on a complete knowledge of the environmental requirements and taphonomic biases of the species found as fossils. Reconstructions assume that environmental requirements of the fossil diatom taxa used as bioindicators have remained constant during the period considered and, consequently, are similar to those of their closest living representatives (taxonomic uniformitarianism, Dodd and Stanton 1990). In this way, the environmental information obtained from living organisms can be used as modern analogs and extrapolated to the fossil record, particularly in Quaternary research (Hassan et al. 2011). Hence, the study of modern diatom-environment relationships constitutes an essential first step to gather ecological information, although the application of this data to paleoenvironmental reconstruction should also consider the biases suffered by the information contained in assemblages during their fossilization.

Pampean diatoms in modern shallow lakes have demonstrated a great potential as proxies for paleoenvironmental reconstructions, as their distribution is strongly related to key environmental variables (Hassan et al. 2009, 2011). Assemblages are mostly dominated by benthic and epiphytic taxa, because of the very shallow depth and high vegetation cover that characterize these environments. Tychoplanktonic taxa (i.e. taxa that are closely related to benthic habitats, occurring intermittently in the water column after being swept up from the bottom; Smol and Stoermer 2010) are also abundant in these wind-stressed shallow lakes, being *Cyclotella meneghiniana* one of the most widely distributed diatom species in the region (Hassan and De Francesco 2018). In a study covering Pampean shallow lakes and streams, the composition of assemblages was strongly linked to conductivity, pH and nutrient gradients (Hassan et al. 2009). When integrated with datasets from the adjacent Monte and Espinal regions (i.e. semiarid regions located at the west of the Pampean Region, see Fig. 2 in Hassan et al. 2011), the longer gradient of temperature was also significantly linked to diatom composition (Hassan et al. 2011). Overall, the strong relationship between diatom assemblages composition and environmental variables directly (temperature) or indirectly (conductivity and pH) linked to climate enhance their potential application to paleoclimatic reconstruction, provided they were faithfully preserved in the sedimentary record. Moreover, the significant role of nutrients in structuring modern communities also emphasizes their use as biomonitors of anthropogenic impacts in the region (Hassan et al. 2009).

Internal within-lake variability in diatom assemblages composition is high in these lakes, as a consequence of their characteristic environmental heterogeneity (Hassan 2018). The extensive development of the littoral zone relative to the pelagic zone promotes the development of complex lateral environmental gradients, which support diverse and productive periphytic and benthic communities (Wetzel 2001). Consequently, diatom assemblages in these lakes allow to distinguish between littoral and open-waters habitats, as they differ in species composition. Tychoplanktonic taxa, such as *Cyclotella meneghiniana*, *Aulacoseira granulata* and *A granulata* var. *angustissima* tend to be more abundant in open-waters, whereas epiphytic and benthic taxa usually dominate littoral sediments (Hassan 2018). Moreover, the analysis of epiphytic assemblages inhabiting five macrophyte taxa from Nahuel Rucá lake demonstrated that diatom species can also be used to discriminate between free-floating and submersed/emergent vegetation (Rojas and Hassan 2017). Hence, besides its potential use to reconstruct regional environmental gradients, these assemblages can also be used to discriminate among within-lake subenvironments, providing useful tools for the recognition of subtle environmental changes in the fossil record (Hassan et al. 2018).

This good performance of Pampean diatom assemblages as modern analogs for past environmental conditions opened interrogations about the impact of taphonomic processes in structuring the sedimentary record of these bioindicators, which laid the foundations of our project on actualistic taphonomy. Particularly, the main questions to be addressed were: (1) Is the compositional and environmental information stored by living communities faithfully reflected by death assemblages? (2) Can taphofacies analyses be used to discriminate between the observed local and regional environmental gradients? (3) Which are the environmental causes of diatom taphonomic signatures? and (4) Can diatom-based paleoenvironmental inferences be improved by combining the use of taphonomic and compositional lines of evidence? In the following sections, we will discuss our main advances directed to answer these questions.

6.3 Ecological Fidelity of Death Assemblages

Understanding the faithfulness with which sedimentary diatom assemblages represent the composition of the source communities from which they were derived constitutes a key point in diatom analysis (Battarbee et al. 2001). Diatom-based paleoenvironmental reconstructions are based on the uniformitarian transference of ecological information from modern taxa to the past and assume that death assemblages integrate small-scale temporal and spatial variability of living communities into time-averaged assemblages (Hassan et al. 2008). However, as previously mentioned, diatom assemblages are subjected to a wide array of taphonomic biases in shallow lakes, which may lead to a low compositional agreement between living communities and death assemblages. In fact, if the variability introduced by taphonomic processes exceeds the original biological variability, then the capability of death assemblages to preserve the original ecological information might be completely obliterated. Likewise, these taphonomic processes might shift the record of the responses of diatom assemblages to environmental gradients, thus testing the assumptions of paleoecological and paleoenvironmental analyses (Tomašových and Kidwell 2009b).

Live-dead comparisons are the most common actualistic methods of evaluating the faithfulness of the preservation of ecological information (Kidwell 2013). In live-dead studies, the ecological properties of the life assemblages (LAs) are compared with those of death assemblages (DAs, Kidwell 2013). They have mostly focused on compositional fidelity (i.e., preservation of composition, richness and abundance, Tomašových and Kidwell 2009a, 2011), whereas works on environmental fidelity (i.e., the ability of death assemblages to detect environmental gradients) have been particularly scarce (Tomašových and Kidwell 2009b). In order to assess the degree of ecological fidelity exhibited by death assemblages in Pampean shallow lakes, Hassan (2015) analyzed three live-dead diatom datasets. Two main points were evaluated:

(1) the ability of death assemblages to preserve the composition of the original living community (compositional fidelity); and (2) the ability of death assemblages to preserve the original responses of living communities to environmental gradients (environmental fidelity).

6.3.1 Compositional Fidelity

Compositional fidelity of sedimentary diatom assemblages was addressed in three shallow lakes located in the southeastern Pampa Plain: Las Mostazas, Los Carpinchos and Nahuel Rucá (Fig. 6.1, see De Francesco et al. this volume for a description of the Study Area). These are typical Pampean shallow lakes, as they are small (~250 ha), very shallow (maximum depths of 1.5–2 m), hypereutrophic, and surrounded by littoral zones densely populated by emergent and submersed macrophytes, forming a ring that surrounds entirely the lakes and clearly differentiates littoral from open waters areas. The three lakes are located along a north-south gradient covering approximately 60 km, and range from freshwater to slightly brackish salinity (0.3–4.3 mS/cm). A total of 132 surface sediment samples were collected seasonally from each lake's littoral and open waters subenvironments, each one representing a live-dead pair.

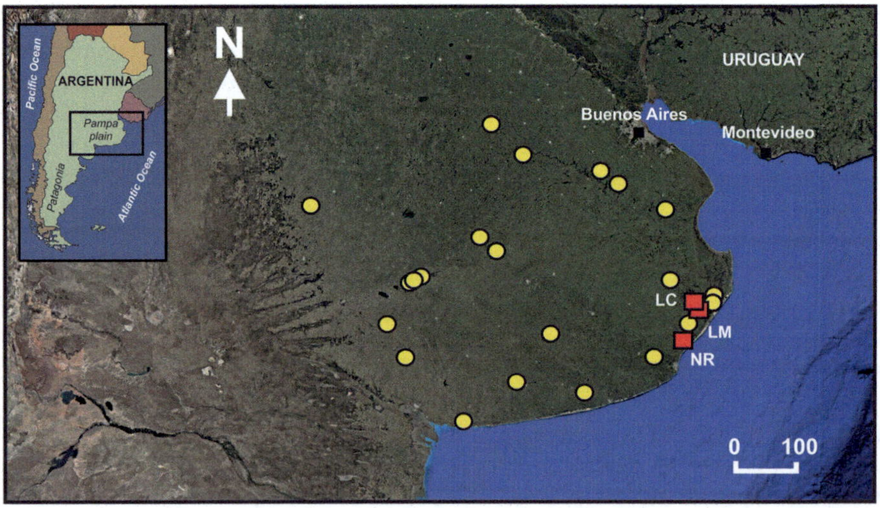

Fig. 6.1 Location map showing all the lakes sampled (yellow circles) in the regional taphofacies analysis conducted by Hassan and De Francesco (2018). The situation of the three lakes used in fidelity, within-lake taphofacies, and modern-fossil comparisons are highlighted (red squares, modified after Hassan and De Francesco 2018). NR: Nahuel Rucá Lake; LM: Las Mostazas Lake; LC: Los Carpinchos Lake

Compositional fidelity was assessed by comparing richness, diversity, abundances and multivariate dispersions among life (i.e. cells containing intact chloroplasts) and death assemblages. These estimates differed significantly between LAs and DAs, indicating a broadly unfaithful representation of diatom communities by death assemblages. The observed differences were interpreted mainly as consequences of (1) differences in the temporal resolution between time-averaged DAs and non-averaged LAs, and (2) differential preservation of diatom taxa related to the intrinsic properties of their valves, specifically the over-representation of robust species and under-representation of fragile ones. The influence of time averaging was expected, however, as DAs constitute time-averaged representations of living communities and integrate dead-valves inputs over long periods of time (Kidwell 2002; Tomašových and Kidwell 2011). Even when diatom living communities are sampled in a seasonal basis, the samples collected in each lake would represent only snapshots of the full seasonal variability, which can be high in diatoms given their very short generation times (from hours to a few days, Baars 1981). Enrichment of death assemblages can also be related to the input of valves from lake subenvironments not included in the sampling strategy, such as epiphytic taxa living upon macrophytes (e.g., *Lemnicola hungarica* and *Cocconeis placentula*, Rojas and Hassan 2017). Hence, it was expected that, given longer sampling periods and more intensive sampling of all lake subenvironments, pooled community richness and diversity values would more closely approximate those of the DAs.

Further examination of the differences in the proportional abundances of taxa between LAs and DAs allowed identifying species which were significantly under or over-represented in death assemblages. Most of the taxa over-represented in DAs (i.e., *Navicula peregrina, Surirella striatula, Aulacoseira granulata, Cyclotella meneghiniana*) were characterized by highly silicified and robust valves, whereas under-represented taxa (i.e., *Nitzschia amphibia, Pseudostaurosira americana, Aulacoseira granulata* var. *angustissima*) were weakly silicified, and more prone to suffer fragmentation given their elongated shape and denser areolation patterns (Fig. 6.2). Given the very shallow nature of these lakes, their bottom sediments are likely subjected to significant reworking by wind action, being the resulting turbulence a possible cause of breakage and subsequent dissolution for delicate taxa (Ryves et al. 2006). As an example, the under-representation of *Aulacoseira granulata* var. *angustissima* and the over-representation of *A. granulata* in DAs of Nahuel Rucá lake were accompanied by differences in the preservation of both species. Valves of the delicate *Aulacoseira granulata* var. *angustissima* showed less fragmentation than *A. granulata*, as they constitute recently dead remains and tend to remain a shorter time in surface sediments before being more easily destroyed (Fig. 6.2). Hence, as the probability of breakage of diatom valves is strongly related to their length/width ratio and pore row frequency (Haberyan 1985), differential fragmentation of taxa in these wind-stressed shallow lakes is more likely responsible for the observed deviations in proportional abundances.

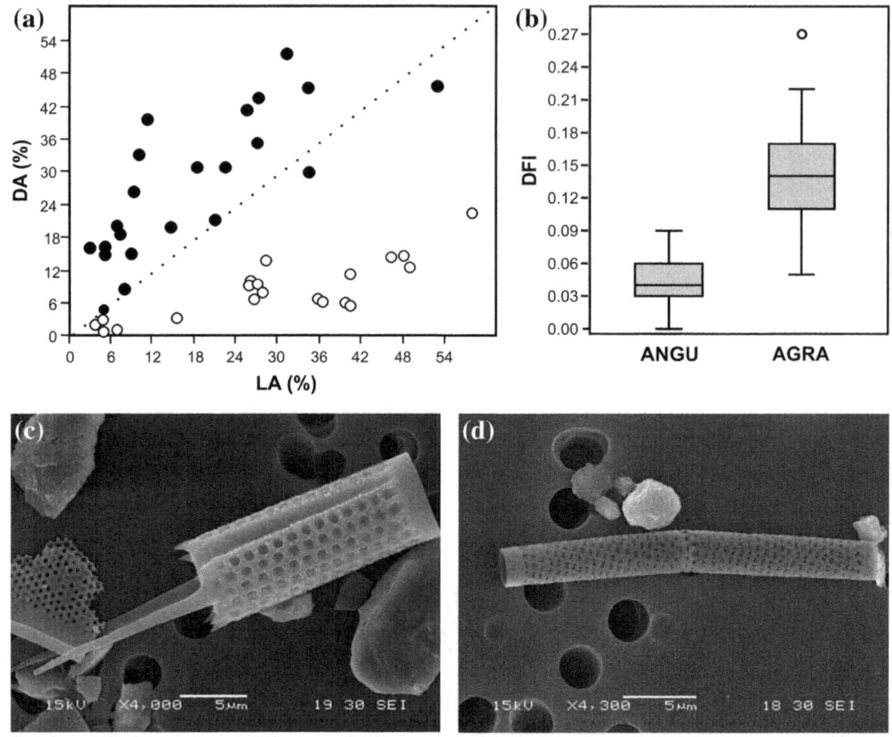

Fig. 6.2 Compositional fidelity analyses of Nahuel Rucá open waters samples showing **a** the proportional abundances in life assemblages (LA) versus the proportional abundances in death assemblages (DA) for *Aulacoseira granulata* var. *angustissima* (ANGU, white circles) and *Aulacoseira granulata* (AGRA, black circles). The under-representation of ANGU and the over-representation of AGRA in death assemblages are evidenced by the displacement of samples from the 1:1 expected relationship (dotted line, modified after Hassan 2015). Examination of the Diatom Fragmentation Index (DFI) in the same samples **b** showed a significantly higher fragmentation of AGRA (Kruskal-Wallis $p < 0.001$), which tend to remain as identifiable fragments in the sediments long before the destruction of ANGU valves (modified after Hassan et al. 2018). This difference in the preservation of both taxa can be explained by the larger length/width ratio and pore row frequency exhibited by ANGU (**d**) when compared to AGRAN (**c**)

6.3.2 Environmental Fidelity

Given the low compositional fidelity exhibited by DAs, the question of whether this unfaithfulness affects the environmental information stored by sedimentary assemblages became a significant issue for diatom-based paleoenvironmental reconstructions in these lakes. If the environmental significance of DAs became biased during its formation, then the inference of environmental parameters from sedimentary diatom composition should incorporate these biases to avoid interpretative errors. Hence, the degree to which DAs capture the original environmental patterns affecting the structure of LAs was evaluated through comparisons at local (within-lake) and regional (between-lake) scales from the same three lake dataset (Hassan 2015).

Results of gradient analyses demonstrated that, despite this low compositional fidelity, DAs were able to capture gradients of conductivity, pH, depth, and vegetation even better than LAs, as the percentage of variance in diatom composition explained by these environmental variables was higher in DAs than in LAs, both at regional and local scales. This higher environmental sensitivity of DAs over LAs has been also mentioned in mollusk-based environmental fidelity studies and related to the capability of DAs to capture community composition (especially rare species) over longer temporal durations (see Tomašových and Kidwell 2009b and references therein). In the case of the studied datasets, it was also related to the structural redundancy exhibited by diatom communities (i.e., more than one mutually exclusive subset of species significantly captures community structure based on the full set of species, Clarke and Warwick 1998). For example, the under-representation of *Aulacoseira granulata* var. *angustissima* in DAs of Nahuel Rucá lake was compensated by the over-representation of *A. granulata* in the same assemblages, as both taxa responded similarly to the studied environmental gradients (Fig. 6.2). In this case, redundancy in environmental information implies that the between-sample relationships of life assemblages can be significantly preserved by DAs even if some taxa are randomly removed by taphonomic processes or are missed due to incomplete sampling (Tomašových and Kidwell 2009b). Overall, this good live-dead agreement implies that past environmental conditions inferred from fossil diatom assemblages would be even more accurate than those based on a single sampling of the original living community in these lakes, supporting one of the main assumptions of diatom-based paleoenvironmental reconstructions. However, the good preservation of environmental gradients observed in diatom assemblages did not imply good preservation of environmental characteristics not considered in that study, such as nutrient status, or longer conductivity and pH gradients, highlighting the importance of conducting more studies to cover new environmental gradients of paleoenvironmental interest.

6.4 Taphofacies Analyses

The low live-dead compositional agreement exhibited by diatom assemblages in Pampean shallow lakes suggested a possible role of taphonomic processes in the structuration of the diatom record. Moreover, the variability in dissolution and fragmentation profiles observed in the fossil record of these lakes pointed valve dissolution and breakage as probable causes of biases in assemblages composition (Hassan et al. 2014), highlighting the importance of analyzing the causes of diatom taphonomic signatures in modern settings. Accordingly, a series of studies intending to assess the environmental significance of diatom dissolution and breakage marks were conducted in a number of Pampean shallow lakes. The purpose was to identify potential taphonomic biases (i.e. postmortem information loss), but also to assess the positive contributions of taphonomic signatures to the interpretation of past environmental changes. Taphofacies analysis (i.e., facies defined on the basis of diagnostic taphonomic traits; Speyer and Brett 1986) constitute a powerful methodology in that direction, as it allows to underscore the non-random relationship between taphonomic signatures and environmental conditions in modern settings, highlighting the feasibility of reconstructing paleoenvironments from taphonomic analyses (Parsons and Brett 1991). Accordingly, the potential use of diatom taphofacies to infer environmental conditions was assessed at (1) regional (i.e. between-lake comparisons; Hassan and De Francesco 2018), and (2) local (i.e. within-lake comparisons; Hassan et al. 2018) scales in modern shallow lakes.

6.4.1 Regional Scales: Between-Lake Taphofacies Comparisons

The impact of long environmental gradients on diatom preservation was assessed in twenty-five lakes of the Pampean Region (Fig. 6.1) representing a strong salinity gradient (0.4–13.2 ppt, Hassan and De Francesco 2018). The preservation was evaluated through the analysis of frustule fragmentation and dissolution on a target taxon, *Cyclotella meneghiniana*. Given the strong dependency of preservation on valve morphology, the use of a target taxon such as *C. meneghiniana* allowed to assess changes in preservation independently of the assemblage composition. This taxon had already been used as an indicator of diatom dissolution (e.g. Hassan et al. 2014), not only because of its abundance in modern and past environments but also for its moderate sensitivity to dissolution and its easily recognizable valves, which can be identified even under very poor preservation conditions (Barker 1992). The study of its preservation in Pampean lakes indicated a significant relationship between valve dissolution, salinity, pH, carbonate, bicarbonate and silica concentrations, whereas environmental constraints on valve fragmentation were less clearly identified (Fig. 6.3). According to Generalized Linear Model (GLM) analyses, dissolution was negatively correlated to silica concentration, and positively to the rest of these variables, whereas fragmentation was only marginally correlated to salinity (Fig. 6.3).

The application of further direct gradient analyses (Redundancy and Canonical Correspondence Analyses, ter Braak 1994) allowed to assess and compare the percentage of variance in compositional and taphonomical data explained by environmental variables. Salinity explained significant portions of variance both in taphonomic and compositional analyses, although the effect on assemblage composition was only marginal and lower: while it explained a 33.3% of the taphonomic variation ($p < 0.001$), only an 11.7% of the compositional variance was explained by this environmental variable ($p < 0.01$). These results were comparable to previous regional studies covering a wider region (Central Argentina), in which log-Conductivity accounted for 4.8% of the variance in diatom assemblage composition (Hassan et al. 2011). These findings indicated that although this environmental fac-

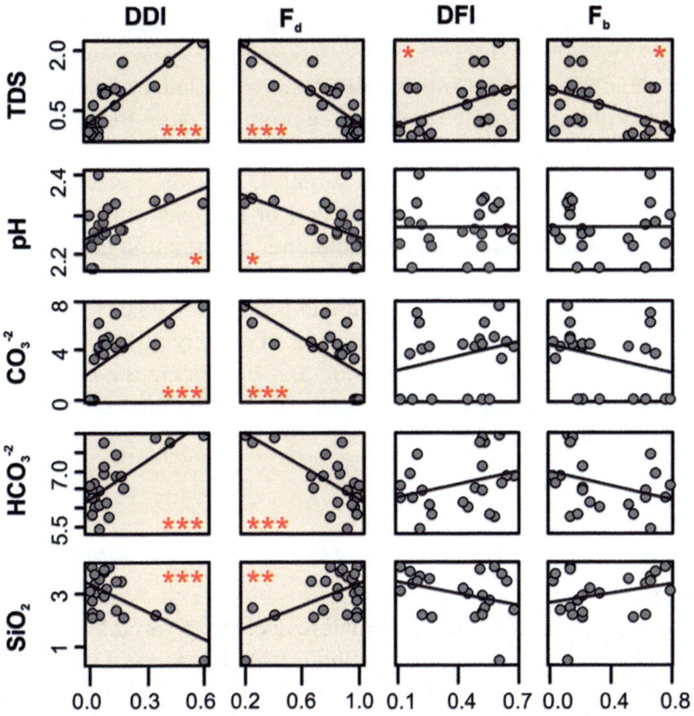

Fig. 6.3 Bivariate plots of selected environmental variables (log-transformed) and taphonomic indices along a regional gradient covering 21 lakes (modified after Hassan and De Francesco 2018). Significant relations, according to GLM analyses are highlighted, and significances at $p < 0.001$ (***), $p < 0.01$ (**), and $p < 0.05$ (*) are indicated by asterisks. DDI: Diatom Dissolution Index (average dissolution grade of the assemblage, Ryves et al. 2006), DFI: Diatom Fragmentation Index (average breakage grade of the assemblage, Hassan et al. 2014), Fd: F index of dissolution (ratio of undissolved valves to all valves counted, Ryves et al. 2009), Fb: F index of fragmentation (ratio of unbroken valves to all valves counted, Ryves et al. 2009), TDS: Total Dissolved Solids (ppt)

tor has an influence in structuring the composition of diatom living communities in Pampean shallow lakes, it plays a stronger and more significant role in the preservation of their frustules after their deposition in death assemblages. Moreover, examination of the multivariate distances among sites showed a significant correlation between taphonomic and compositional distances (based on Manhattan and Bray-Curtis distance metrics, respectively), demonstrating the existence of common causal environmental stressors in the structuration of both patterns. Two causes were suggested to explain this relationship: (1) inter-specific variation in preservation, related to intrinsic properties of frustules, that can cause differential preservation and the consequent changes in species composition along environmental gradients; and (2) the same environmental factors are affecting both the distribution and preservation of the taxa, i.e. by altering the taphonomic signature of frustules and also shifting the species composition of the living communities according to their optima and tolerances.

Whatever the case, these results highlight the potentiality of combining diatom taphonomy and community paleoecology as a promising tool for paleoenvironmental reconstruction. If diatom taphonomic signatures can be used to trace past changes in key environmental variables (such as salinity), then the potential applicability of preservational profiles to the reconstruction of past environmental fluctuations increases significantly. In such a case, taphonomic information can be used to gain paleoenvironmental information, and not simply to assess the loss of information from the fossil record. This was the case of Pampean shallow lakes, where valve dissolution proved to be a more sensitive indicator of salinity differences among lakes than shifts in relative abundances of diatom taxa, enhancing the paleoenvironmental significance of diatom taphonomic signatures at long environmental gradients (Hassan and De Francesco 2018).

6.4.2 Local Scales: Within-Lake Taphofacies Comparisons

The scale of analysis of taphonomic signatures is highly variable, as variability in death assemblages structure and preservation is usually observed at different spatial scales, varying from local changes over centimeters to regional variation over kilometers or more (Staff and Powell 1990). The study of diatom taphofacies at regional scale demonstrated the potential use of this variability in addressing long environmental gradients. At local scales, variability can be promoted by subtle gradients in environmental or sedimentological parameters, but also by taphonomic factors acting during and after deposition (see Hassan et al. 2018 and references therein). In shallow lakes, the extensive development of the littoral zone relative to the pelagic zone provides a wide suite of available microhabitats that promote the growth of periphytic, benthic and pelagic communities (Wetzel 2001), being highly probable for these within-lake subenvironments to imprint distinctive taphonomic signatures on diatom valves. In such cases, recognizing within-lake taphofacies would become a useful tool for identifying subtle environmental differences from fossil diatoms in shallow lakes.

Accordingly, Hassan et al. (2018) conducted a study intending to determine whether diatom taphofacies are useful to discriminate subtle within-lake environmental differences in five sub-environments (free-floating and attached macrophytes, water column, open waters, and littoral bottom sediments) from Nahuel Rucá lake. In that study, compositional (relative abundances, diversity, richness, evenness and relative abundances) and taphonomic (dissolution and fragmentation) variables were obtained from a set of 129 samples (48 of surface sediments, 24 of water and 57 of macrophytes) covering both seasonal and spatial variability in diatom composition. The studied sub-environments differed in depth, macrophyte coverage, and water chemistry, being these differences reflected both by compositional and taphonomic data (Fig. 6.4).

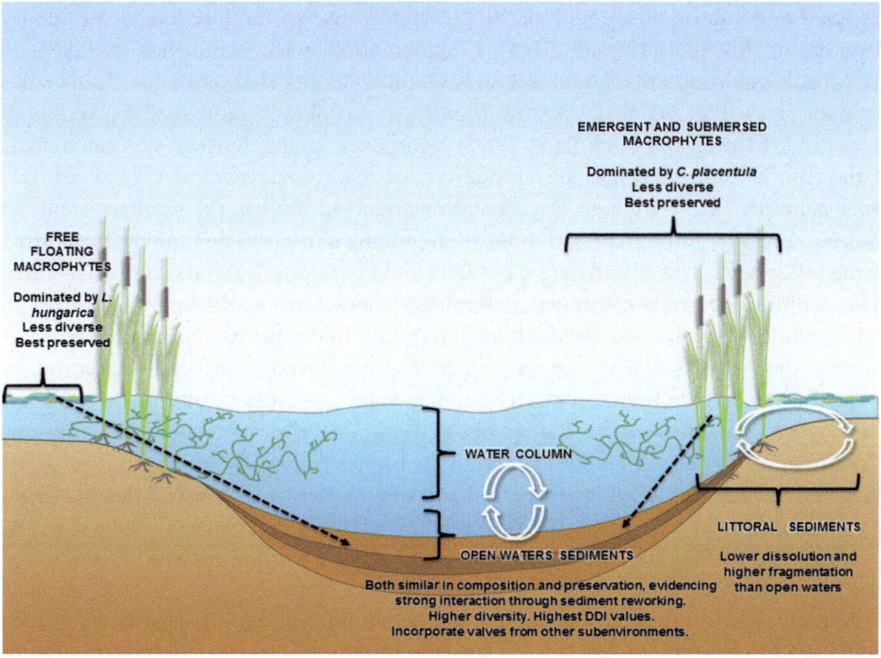

Fig. 6.4 Summary of the taphonomic processes taking place in the different within-lake sub-environments of Nahuel Rucá lake. Reproduced from Hassan et al. (2018), with permission

Assemblages from littoral, open-waters, and water column were dominated by the tychoplanktonic species *C. meneghiniana, A. granulata* and *A. granulata* var. *angustissima*, and differed from epiphytic assemblages, which were dominated by the epiphytic taxa *Cocconeis placentula* and *Lemnicola hungarica*. Moreover, diversity and richness were higher in planktonic and sedimentary assemblages, whereas dominance showed higher values in epiphytic samples. This difference was also reflected by taphonomic variables: epiphytic assemblages showed the best preservation, evidenced both by the lowest fragmentation and dissolution indices (Fig. 6.4). This good preservation is related to the conditions of their habitat, as they are protected by the epiphytic biofilm structure and the sheltered areas among macrophytes (Rojas and Hassan 2017), and therefore tend to be less subjected to the taphonomic biases generated by physical and chemical environmental factors than planktonic and sedimentary assemblages. Benthic and planktonic sub-environments, on the other hand, constitute less stable habitats in which sediment reworking pose physical constraints to diatom preservation. Although the compositional analysis prevented the discrimination among planktonic, littoral and open waters sub-environments, the taphonomic analysis was useful to detect preservational differences among them (Fig. 6.4).

Diatom breakage was the main taphonomic attribute identified in Nahuel Rucá assemblages, being marks of dissolution less frequent (Fig. 6.4). As Nahuel Rucá is a very shallow freshwater lake, the prevalence of fragmentation over dissolution was expected and can be advocated as the probable cause of the low live-dead fidelity observed in this lake (Hassan 2015). Fragmentation marks were more frequent in littoral sub-environments. Given the very shallow depths characteristic of this environment (less than 60 cm), frustule breakage is probably enhanced by sediment reworking (Flower and Nicholson 1987). Moreover, as the densely-vegetated areas of the littoral support abundant populations of macroinvertebrates (Tietze and De Francesco 2017), the highest fragmentation found in the littoral sediments can be also explained by biological action through grazing activity. The higher dissolution, on the other hand, was found in open-waters and planktonic habitats, and was related to the within-lake gradients in water chemistry characterizing the lake (Cristini et al. 2017), while in the littoral zone, characterized by lower pH and carbonate concentrations, diatom dissolution was prevented and preservation favored (Barker et al. 1994; Hassan and De Francesco 2018). Open waters and water column, on the other hand, were characterized by higher pH and carbonate/bicarbonate concentrations (Cristini et al. 2017), which could explain the higher frequency of dissolution marks exhibited by diatom assemblages in these sub-environments. Therefore, despite being low, dissolution was significantly higher in open waters and plankton, being probably enhanced by the continuous reworking and mixing exerted by wind action in the lake.

Overall, diatom taphofacies analysis demonstrated to be useful to capture environmental differences also at local scales, being taphonomic signatures related to within-lake differences in depth, macrophyte coverage, and water composition. Under such circumstances, combining taphonomic and compositional analyses would allow to uncover subtle within-lake environmental gradients not identifiable when only compositional analyses are performed. In this case, compositional analyses would allow to discriminate among different microhabitats, leading to paleoenvironmental inferences that would not be possible from taphonomic analyses only.

6.5 Experimental Taphonomy

The patterns of preservation of diatom assemblages observed at both local and regional scales clearly signaled the significant impact that environmental variables exert over the dissolution and breakage of the frustules. Particularly, the influence of salinity and pH can be relevant in Pampean shallow lakes, where the precipitation/evaporation balance plays a significant role in controlling water depth and chemistry. Changes in the hydrological cycle caused by climatic conditions or by anthropogenic activities can cause significant fluctuations in the pH and brine concentration of waters, which can be very alkaline and range from oligo to hypersaline (Fernández Cirelli and Miretzky 2004). Previous hydrochemistry studies in the Southern Pampas indicated that NaCl and HCO_3Na are the dominant salts in surface and interstitial waters (Fernández Cirelli and Miretzky 2004; Hassan et al. 2012; Cristini et al. 2017). Hence, diatom silica dissolution caused by high concentrations of these salts, as well as elevated pH, are among the most probable processes explaining the observed patterns of preservation in the Pampean Region. In order to test this hypothesis, a series of short-term experiments on the dissolution of silica from cell walls under controlled laboratory conditions were recently conducted (Diaz and Hassan 2016, 2017).

Experiments were run on artificial assemblages constructed by mixing samples from different subenvironments (epiphytes, plankton, and sediments) from Nahuel Rucá lake. Experimental sets were designed to test for the effect of three concentrations of the two salts, NaCl (0.6, 1.2 and 3 M) and HCO_3Na (0.6, 0.9 and 1.2 M), and two pH values (7 and 10) on diatom dissolution. An aliquot of each experimental assemblage (M1, M2, and M3) was thoroughly mixed with each solution to a final volume of 100 ml and stored polyethylene flasks. To avoid biological contamination a drop of mercury (ll) chloride ($HgCl_2$) was added to each flask (Ryves 1994). Each one of the treatments and controls was made by triplicate and kept in an immersion bath under a controlled temperature of 20 °C. Aliquots of the experimental solutions were removed once each 5 days during 20 days and analyzed for changes in dissolved silica concentration (SiDi), relative and absolute abundances of diatoms, and dissolution indices based on the target taxon *Cyclotella meneghiniana*. All the experimental solutions increased significantly the rate of dissolution of silica from diatom walls, particularly since day 10. The two salts, however, differed in their relative effect, being the final SiDi concentrations obtained with HCO_3Na solutions almost twice than those obtained with NaCl (Diaz and Hassan 2016). The effect of high pH was also evident: whereas the pH 7 solution did not produce increased dissolution, the difference in SiDi concentrations between initial and final experimental times was significant at pH 10 (Diaz and Hassan 2017). These increased SiDi values were accompanied by significant changes in the dissolution indices calculated from *C. meneghiniana* valves, which reached maximum values at pH 10, whereas no significant changes in relative or absolute abundances of diatoms were registered during the experiments. Evidences of dissolution were recorded as increases in the perimeter of individual areolae and decreases in the distances among the conjoining of neighboring areolae, as were observed in many taxa (Fig. 6.5). This effect of early stages of the dissolution process was also signaled by previous articles, both in marine and continental settings (Ryves et al. 2001; Warnock and Scherer 2015), in some cases accompanied by a lack of delicate structures and striae enlargement (Mikkelsen 1977; Ryves et al. 2001).

Overall, these experimental results highlighted the impact that water chemistry exerts on diatom dissolution in Pampean shallow lakes. Given the naturally high pH, NaCl and HCO_3Na concentrations characteristic of many of these lakes, these experimental findings can be confidently extrapolated to the interpretation of the dissolution trends found in modern surface sediments. Even as no significant changes in relative or absolute abundances were detected, this can be a consequence of the short exposure times covered by the experiments, instead of a definitive result. Previous studies over longer experimental times found significant changes in diatom abundances (Barker et al. 1994; Ryves et al. 2001), although the results were dependent of temperature and pre-treatments (Barker 1992; Flower 1993). Hence, although this is still pending of being experimentally tested, it is probable that, during longer

Fig. 6.5 Scanning Electron Microscope micrographs showing pristine (upper row) and dissolved (lower row) valves of diatom taxa representative of different genera. From left to right: *Surirella ovalis, Rhopalodia gibberula, Lemnicola hungarica, Cocconeis placentula* and *Cyclotella meneghiniana*

periods of time, exposure of valves to these salts and increased pHs could bias the composition of Pampean diatom assemblages towards robust taxa.

6.6 Implications for Paleoenvironmental Reconstructions

The strong relationship between diatom preservation and environmental variables observed in field-based and experimental studies enhanced the paleoenvironmental significance of preservation trends along the fossil record of diatoms in Pampean lakes. In the light of these new results, Hassan and De Francesco (2018) reexamined the previously published taphonomic history of *C. meneghiniana* in Holocene sediments of Nahuel Rucá and San Leoncio (Hassan et al. 2014), suggesting strong fluctuations on past salinity and hardness in these lakes. Increased dissolution values exhibited by *C. meneghiniana* indicate the occurrence of brackish environments in these lakes by ca. 1500 cal. years BP, being modern conditions developed during the last ca. 200 cal. years BP. (Hassan et al. 2014). Dissolution values for these levels were high, indicating the prevalence of highly destructive environments, suggesting salinity and hardness values lying on the upper reaches of the studied modern gradient. Accordingly, community composition corresponding to these moments of increased salinity should be interpreted with caution, as the inferred destructive conditions could have biased significantly the proportions of taxa towards resistant species. As suggested by Hassan and De Francesco (2018), dissolution data for those levels would be a more confident source of environmental data than community composition, supporting the importance of studying taphonomic histories in these lakes.

In a more detailed study, Hassan et al. (2018) compared the within-lake composition and preservation of modern and Holocene (last ca. 700 years BP) diatom assemblages from Nahuel Rucá lake through multivariate analyses. Comparison of taphonomic indices between contemporary and fossil assemblages allowed to identify modern analogs for past diatom preservation, as several fossil samples showed high similarities to within-lake modern assemblages when dissolution and fragmentation were considered. Among them, samples from the topmost zone of the core overlapped with modern planktonic and littoral samples, being the best preserved of the whole sequence. Samples from the middle of the sequence overlapped with open waters modern samples, as they showed intermediate dissolution and fragmentation values. A number of fossil levels, however, showed dissolution and fragmentation values higher than those recorded in modern assemblages of the lake, lacking analogs in the dataset. These poorly preserved fossil samples were comparable to diatom assemblages from three previously studied Pampean brackish lakes with average salinities between 4 and 13 ppt (Hassan and De Francesco, 2018), which proved to be good analogues of past taphonomic conditions for those levels with no analogues in the Nahuel Rucá within-lake dataset, suggesting the prevalence of brackish conditions towards ca. 700 years BP.

These brackish conditions, however, were only partially suggested when assemblages composition alone was considered, as the dominant taxa were mostly euryhaline species adapted to a wide range of salinity conditions (e.g., *C. meneghiniana, C. placentula, N. amphibia*) with the consequently reduced value for salinity bioindication (Hassan et al. 2011). In fact, only the presence of rare taxa (<1%), such as *Campylodiscus clypeus, Craticula halophila, Navicula salinicola, Nitzschia vitrea*, and *Synedra platensis*, suggested brackish conditions. Hence, as the high salinity conditions suggested by high dissolution values were coincident with the presence of halophilous taxa, the combination of both independent sources of evidence provided reliable data to propose higher salinities in the lake at ca. 700 cal. years BP (Hassan et al. 2018). On the whole, the joint analysis of compositional and taphonomic variations in the recent sedimentary record of Nahuel Rucá demonstrated the usefulness of taphonomic analyses to uncover subtle paleoenvironmental variations, which could be overlooked if only traditional compositional analyses are performed. Accordingly, both taphonomic analyses and a holistic consideration of ecologically meaningful taxa should be considered in order to improve paleoenvironmental reconstructions in environmentally heterogeneous shallow lakes.

6.7 Conclusions and Future Directions

The results reviewed in the present chapter demonstrate the significant role that taphonomic processes play in the structuration of sedimentary diatom assemblages. Overall, these factors can alter the structure and taphonomic signatures of assemblages, leading to biased abundances, lost of delicate taxa and low compositional fidelity. However, these studies also demonstrated the positive contribution that taphonomic signatures can provide to paleoenvironmental studies: as taphofacies were successfully related to local and regional environmental gradients, their potential as paleoenvironmental proxies arise as a useful independent tool of information. Moreover, the experimental demonstration of the role of water chemistry on diatom dissolution also supported the field observations, leading to an empirical demonstration of the processes structuring the patterns of preservation observed.

Accordingly, future directions in this project should tend to provide a more comprehensive picture of unknown aspects of diatom taphonomy. Particularly, long-term experiments on dissolution and fragmentation need to be conducted, in order to clarify the role of both water chemistry and physical factors in structuring the composition of assemblages in natural conditions. Previous studies have demonstrated that diatom preservation is linked to the particular characteristics of the frustule architecture (e.g. row pore frequency, width/length relation, size, the degree of silicification, etc.), being consequently strongly dependent on taxonomy (Haberyan 1985; Barker 1992; Ryves et al. 2009). These particular characteristics have led to the development of rankings of diatom taxa according to their preservation potential in other areas of the world, which proven to be useful in paleoenvironmental reconstructions (Ryves et al. 2009). As this data is still unavailable for Pampean diatom taxa, future research should focus on that direction, as this information would provide useful tools for the environmental interpretation of past diatom assemblages in Pampean shallow lakes. Moreover, experimental designs including fragmentation caused by sediment reworking and bioturbation are particularly necessary, as a way to elucidate the significance of the high fragmentation indices found in these shallow lakes. As the causes of frustule fragmentation have received little attention worldwide, this line of research constitutes an exciting future development that requires further attention. Finally, including new studies on diatom taphofacies that allow to expand the regional and local environmental gradients already studied would increase the applicability of these studies to paleoenvironmental research, allowing the reconstruction of a wider set of paleoenvironmental conditions.

Up to now, the usefulness of these actualistic taphonomic data become evident when applied to the reconstruction of the recent history of the freshwater Nahuel Rucá lake, as it allowed to identify environmental fluctuations not so clearly reflected by the composition of assemblages. Hereafter, if fossil studies are accompanied by local and regional analyses on actualistic taphonomy, the potential information provided by preservational profiles increases significantly. Hence, we conclude that increasing actualistic and experimental data should be a priority in future studies, as taphonomic information can be used to gain paleoecological information, and not simply to assess the loss of information from the fossil record.

Acknowledgements Financial support for this research was granted by the Agencia Nacional de Promoción Científica y Tecnológica (PICT 2727/2013), the Consejo Nacional de Investigaciones Científicas y Técnicas (PIP 0063/2010), and the Universidad Nacional de Mar del Plata (EXA 775/16). Special thanks to Pedro Urrutia (Estancia Nahuel Rucá), Santiago González Aguilar (Las Mostazas), Nicolás Bustamante (Camping Ruca Lauquen), and Héctor Sanabria (Los Carpinchos), for permission to access private land where the lakes are located. We also thank Gustava Bernava for performing the chemical analyses. D. Navarro, S. Stutz and M. Tonello helped at some stage of the research. Experimental studies are part of the Doctoral Thesis of M.C. Díaz at the University of Mar del Plata. G. Hassan and C.G. De Francesco are members of the Scientific Research Career of the Consejo Nacional de Investigaciones Científicas y Técnicas (CONICET).

References

Baars JWM (1981) Autecological investigations on marine diatoms. 2. Generation times of 50 species. Hydrobiol Bull 15:137–151

Barker P (1992) Differential diatom dissolution in Late Quaternary sediments from Lake Manyara, Tanzania: an experimental approach. J Paleolimnol 7:235–251

Barker P, Fontes JC, Gasse F, Druart J-C (1994) Experimental dissolution of diatom silica in concentrated salt solutions and implications for paleoenvironmental reconstruction. Limnol Oceanogr 39:99–110

Battarbee RW (1986) Diatom analysis. In: Berglund BE (ed) Handbook of Holocene Palaeoecology and Palaeohydrology. Wiley, NewYork, pp 527–570

Battarbee R, Jones V, Flower R, Cameron N, Bennion H, Carvalho L, Juggins S (2001) Diatoms. In: Smol J, Birks HJ, Last W, Bradley R, Alverson K (eds) Tracking environmental change using lake sediments. Springer, Netherlands, pp 155–202

Clarke KR, Warwick RM (1998) Quantifying structural redundancy in ecological communities. Oecologia 113:278–289

Cristini PA, Tietze E, De Francesco CG, Martínez DE (2017) Water geochemistry of shallow lakes from the southeastern Pampa plain, Argentina and their implications on mollusk shells preservation. Sci Total Environ 603:155–166

Diaz MC, Hassan GS (2016) Efecto de diferentes concentraciones de sales sobre la preservación de diatomeas: una aproximación experimental. 11° Congreso de la Asociación Paleontológica Argentina (General Roca), Resúmenes

Diaz MC, Hassan GS (2017) Evaluación experimental de la preservación de diatomeas en lagos someros pampeanos: efecto del pH e implicancias paleoambientales. Reunión de Comunicaciones de la Asociación Paleontológica Argentina, San Luis, 23–25 de noviembre de 2017, Resúmenes

Dodd JR, Stanton RJ (1990) Paleoecology. Concepts and applications., Second Edition edn. Wiley-Interscience Publication, New York, p 502

Fernández Cirelli A, Miretzky P (2004) Ionic relations: a tool for studying hydrogeochemical processes in Pampean shallow lakes (Buenos Aires, Argentina). Quat Intern 114:113–121

Flower R (1993) Diatom preservation: experiments and observations on dissolution and breakage in modern and fossil material. Hydrobiologia 269–270:473–484

Flower RJ, Nicholson AJ (1987) Relationships between bathymetry, water quality and diatoms in some Hebridean lochs. Fresh Biol 18:71–85

Flower RJ, Ryves DB (2009) Diatom preservation: differential preservation of sedimentary diatoms in two saline lakes. Acta Bot Croat 68:381–399

Fritz SC (2007) Salinity reconstructions from continental lakes. In: Elias SA (ed) Encyclopedia of quaternary science. Elsevier, pp 62–71

García-Rodríguez F, Piovano E, Puerto Ld, Inda H, Stutz S, Bracco R, Panario D, Córdoba F, Sylvestre F, Ariztegui D (2009) South American lake paleo-records across the Pampean region. PAGES News 17:115–118

Haberyan K (1985) The role of copepod fecal pellets in the deposition of diatoms in Lake Tanganyika. Limnol Oceanogr 30:1010–1023

Hassan GS (2013) Diatom-based reconstruction of Middle to Late Holocene paleoenvironments in Lake Lonkoy, southern Pampas, Argentina. Diat Res 28:473–486

Hassan GS (2015) On the benefits of being redundant: Low compositional fidelity of diatom death assemblages does not hamper the preservation of environmental gradients in shallow lakes. Paleobiology 41:154–173
Hassan GS (2018) Within vs. between-lake variability of sedimentary diatoms: the role of sampling effort in capturing assemblage composition in environmentally heterogeneous shallow lakes. J Paleolim. https://doi.org/10.1007/s10933-018-0038-8
Hassan GS, De Francesco CG (2018) preservation of *Cyclotella meneghiniana* Kützing (Bacillariophyceae) along a continental salinity gradient: implications for diatom-based paleoenvironmental reconstructions. Ameghiniana 55:263–276
Hassan GS, Espinosa MA, Isla FI (2008) Fidelity of dead diatom assemblages in estuarine sediments: how much environmental information is preserved? Palaios 23:112–120
Hassan G, Tietze E, De Francesco C (2009) Modern diatom assemblages in surface sediments from shallow lakes and streams in southern Pampas (Argentina). Aquatic Sci 71:487–499
Hassan GS, Tietze E, De Francesco CG, Cristini PA (2011) Problems and potentialities of using diatoms as paleoclimatic indicators in Central Argentina. In: Compton JC (ed) Diatoms: ecology and life cycle. Nova Science Publishers, New York, pp 185–216
Hassan GS, De Francesco CG, Peretti V (2012) Distribution of diatoms and mollusks in shallow lakes from the semiarid Pampa region, Argentina: their relative paleoenvironmental significance. J Arid Env 78:65–72
Hassan GS, Tietze E, Cristini PA, De Francesco CG (2014) Differential preservation of freshwater diatoms and mollusks in late Holocene sediments: paleoenvironmental implications. Palaios 29:612–623
Hassan GS, Rojas LA, De Francesco CG (2018) Incorporating taphonomy into community-based paleoenvironmental reconstructions: can diatom preservation discriminate between shallow lake sub-environments? Palaios 33:376–392
Julius ML, Theriot EC (2010) The diatoms: a primer. In: Smol JP, Stoermer EF (eds) The diatoms: applications for the environmental and earth sciences. Cambridge University Press, Cambridge, pp 8–22
Kidwell SM (2002) Time-averaged molluscan death assemblages: palimpsests of richness, snapshots of abundance. Geology 30:803–806
Kidwell SM (2013) Time-averaging and fidelity of modern death assemblages: building a taphonomic foundation for conservation palaeobiology. Palaeontology 56:487–522
Lewin JC (1961) The dissolution of silica from diatom walls. Geochim Cosm Acta 21:182–198
Mikkelsen N (1977) Silica dissolution and overgrowth of fossil diatoms. Micropaleontology 23:223–226
Parsons KM, Brett CE (1991) Taphonomic processes and biases in modern marine environments: an actualistic perspective on fossil assemblage preservation. In Donovan SK (ed) The Processes of Fossilization: Columbia University Press, New York, pp 22–65
Rojas LA, Hassan GS (2017) Distribution of epiphytic diatoms on five macrophytes from a Pampean shallow lake: host-specificity and implications for paleoenvironmental reconstructions. Diat Res 32:263–275
Round FE (1964) The diatom sequence in lake deposits, some problems of interpretation: Verhandlungen Internationale Vereinigung für Theoretische und angewandte Limnologie, vol 15, pp 1012–1020
Ryves DB (1994) Diatom dissolution in saline lake sediments: an experimental study in the Great Plains of North America. Ph.D. dissertation, Univ. College London, 307 p.
Ryves DB, Juggins S, Fritz SC, Battarbee RW (2001) Experimental diatom dissolution and the quantification of microfossil preservation in sediments. Palaeogeogr Palaeocl Palaeoecol 172:99–113
Ryves DB, Battarbee R, Juggins S, Fritz SC, Anderson NJ (2006) Physical and chemical predictors of diatom dissolution in freshwater and saline lake sediments in North America and West Greenland. Limnol Oceanogr 51:1355–1368

Ryves DB, Battarbee RW, Fritz SC (2009) The dilemma of disappearing diatoms: Incorporating diatom dissolution data into palaeoenvironmental modelling and reconstruction. Quat Sci Rev 28:120–136

Smol JP, Stoermer EF (2010) The diatoms: applications for the environmental and earth sciences. Cambridge University Press, Cambridge, p 667

Speyer SE, Brett CE (1986) Trilobite taphonomy and Middle Devonian Taphofacies: palaios, vol 1, pp 312–327, https://doi.org/10.2307/3514694

Staff GM, Powell EN (1990) Local variability of taphonomic attributes in a parautochthonous assemblage: can taphonomic signature distinguish a heterogeneous environment? J Paleontol 64:648–658

Stutz SM, Borel M, Fontana SL, Del Puerto L, Inda H, García-Rodríguez F, Tonello MS (2010) Late Holocene climate and environment of the SE Pampa grasslands, Argentina, inferred from biological indicators in shallow, freshwater Lake Nahuel Rucá. J Paleolim 44:761–775

ter Braak CJF (1994) Canonical community ordination. Part I: Basic theory and linear methods. Écoscience 1:127–140

Tietze E, De Francesco CG (2017) compositional fidelity and taphonomy of freshwater mollusks from three pampean shallow lakes of Argentina. Ameghiniana 54:208–223

Tomasových A, Kidwell SM (2009a) Fidelity of variation in species composition and diversity partitioning by death assemblages: time-averaging transfers diversity from beta to alpha levels. Paleobiology 35:94–118

Tomasových A, Kidwell SM (2009b) Preservation of spatial and environmental gradients by death assemblages. Paleobiology 35:119–145

Tomasových A, Kidwell SM (2011) Accounting for the effects of biological variability and temporal autocorrelation in assessing the preservation of species abundance. Paleobiology 37:332–354

Warnock JP, Scherer RP (2015) Diatom species abundance and morphologically-based dissolution proxies in coastal Southern Ocean assemblages. Cont Shelf Res 102:1–8

Wetzel R (2001) Limnology: lake and river ecosystems. Academic Press, San Diego, 1006 p

Chapter 7
Actualistic Taphonomy of Plant Remains in Tropical Forests of Southeastern Brazil

Fresia Ricardi-Branco, Francisco Santiago Rios and Sueli Yoshinaga Pereira

Abstract Studies in the accumulation of plants developed in tropical forests promote understanding the origin of elements that integrate the accumulation time of layers, and the physical and chemical processes that interfere in their formation. Thus, studies were conducted in meandering rivers located in the state of São Paulo. Accumulations of leaves found in point bars were collected in the Mogi-Guaçu, Capivari and Itanhaém (Branco, and Preto rivers) hydrographic basins and ^{14}C dated. The leaves were classified in morphotypes and taxonomically organized. In all assemblages, the leaves belong to arboreal vegetation near the deposit with well-preserved leaves, being in decimetric thickness layers deposited along the 1950s, and possibly trend associated with years of high precipitation. The preservation of the assemblages is related to changes in the level of the water table and the chemical composition of water.

Keywords Plant assemblages · Plant fossil associations · Riparian forest · Taphonomy · Quaternary

7.1 Introduction

Paleobotany seeks to reconstruct vegetation, interpret the environmental and climatic changes of different continental biomes over time, and try to understand the evolutionary and migratory patterns of vegetation (Burnham 1989; Wing and DiMichele 1995; Birks and Birks 2000) through the fossil record. However, the association of plant fossils did not preserve intact, due to the transport, burial, and diagenesis that occurs during the fossilization process (Greenwood 1991; Burnham 1989; Wing and DiMichele 1995; Birks and Birks 2000; Martín-Closas and Gomez 2004; Ricardi-Branco et al. 2009). Plant fossil assemblages occur in continental and marine environments, deposits containing plant assemblages associated with meandering river

F. Ricardi-Branco (✉) · F. S. Rios · S. Y. Pereira
Instituto de Geociências Universidade Estadual de Campinas-UNICAMP, Campinas, SP 13083-970, Brazil
e-mail: fresia@ige.unicamp

S. Martínez et al. (eds.), *Actualistic Taphonomy in South America*, Topics in Geobiology 48, https://doi.org/10.1007/978-3-030-20625-3_7

systems are frequent since those river systems become common from the establishment of embryophytes on the continents (Gibling et al. 2014).

The knowledge of how plant accumulation occurs is essential to improve the interpretations of plant fossil associations. Thus, actualistic taphonomy with regard to necrology and biostratinomy may be based on a wide variety of environments (Martín-Closas and Gomez 2004). Studying the meandering river systems is essential because they present great potential for preserving plant parts (Burnham 1994; Ricardi-Branco et al. 2009, 2011a, b; Ellis and Johnson 2013), mainly in the tropics where taphonomic phases present unique characteristics (Burnham 1989, 1994, 1997; Greenwood 1991, 1992; Steart et al. 2005; Ricardi-Branco et al. 2009; Ellis and Johnson 2013). Fossil associations not only allow the reconstruction of the vegetation but also their main physiognomic features. Paleobotanists have been using leaves of angiosperms to rebuild temperature and precipitation (Royer 2012; Ellis and Johnson 2013) of Cenozoic (Wilf et al. 1998; Jacobs and Herendeen 2004; Hinojosa et al. 2011).

The unique peculiarity tropical taphonomic offers to the continuous accumulation of high diversity of plant remains, allows to study the bias introduced by necrology and biostratinomy in accumulation of plant remains (Greenwood 1991, 1992; Burnham 1989, 1994, 1997; Steart et al. 2005; Ricardi-Branco et al. 2009, 2011b, 2015; Ellis and Johnson 2013). Unlike other plant macrofossils (such as seeds, branches or trunks), the associations of fossil leaves reflect the local vegetation with great accuracy, since they suffer short movement and have a great temporal resolution, usually on scale 10^{-1} to 101 years (Behrensmeyer et al. 2000; Burnham and Johnson 2004).

7.2 Geographical Setting

Ricardi-Branco et al. (2009, 2011a, b) carried out the first studies about leaf assemblages in Riparian Forest (also named as Restinga Forest) located in coastal meandering fluvial environments in the State of São Paulo, southeastern Brazil. Afterward, other studies were carried out on leaf assemblages of Riparian forests in ecotones between the Cerrado (Savanna) and Atlantic forests in continental meandering fluvial environments located in the State of São Paulo (Ricardi-Branco et al. 2015), southeastern Brazil, presenting similar results regarding the dominant species. The area was chosen due to the diversity of forests and their unique characteristics as corridors of plant biodiversity during the Quaternary (Oliveira-Filho et al. 2013). Packages containing plant remains composed of leaves/leaflets and branches were collected from point bar in active meanders.

7.2.1 *Characterization of Study Areas*

7.2.1.1 Country Side River Basins

The Capivari and Mogi-Guaçu river basins are located in the countryside of the State of São Paulo (Fig. 7.1). The substrate composed by permo-carboniferous sedimentary rocks (Itararé Group) and the geological structures (i.e. faults) provide the meandering river dynamics of the Capivari River, which has undergone significant changes over the years (Collares 2000). In the portion where the study was carried out, the rivers cross the southernmost Brazilian domain of the Cerrado Forest (Brazilian Savanna) explaining why it is influenced in its composition by the Semi-Deciduous Seasonal Forest, which belongs to the Atlantic Forest, and may be characterized by the presence of Semi-Deciduous Seasonal Forest, Savanna-Seasonal Forest, and Savanna (SMA/IF 2005). However, only small remnants of those forests are present in the area as a consequence of the expansion of agricultural activity (Collares 2000; SMA/IF 2005). On the other hand, the riparian forest that follows the courses of these rivers presents a mix of species of all these forests (Ribeiro and Walter 2008). The Riparian forest comprises about 33% of all the vegetal species known to the Cerrado Biome, despite the small area they occupy, approximately 5% in relation to other physiognomies, justifying its great importance in this biome plant diversity (Felfili et al. 2001).

The Capivari River Basin—Collection Site

The sampling site is on the left bank of Capivari River near its mouth on Tietê River, in Tietê Municipality, eastern region of the State of São Paulo (Fig. 7.1). Currently, small remnants of Riparian Forest are found on the banks of the Capivari River as a result of agricultural activity expansion. According to Köppen climatic classification, the collected area in Capivari basin presents Cwa climate, humid temperate climate.

The Capivari River basin presents dry winter and hot summer. The mean annual temperature (MAT) of approximately 22 °C, and mean annual precipitation (MAP) that varies between 1100 and 1200 mm (Collares 2000).

The Mogi Guaçu River Basin—Mogi-Guaçu Ecological Station (MGES)

The Mogi Guaçu Ecological Station (MGES) is located in the upper-middle portion of the Mogi Guaçu river basin, Mogi-Guaçu Municipality, northeast of the State of São Paulo (Fig. 7.1). The Riparian Forest that follows the river is characterized by a low and open canopy with dense woods (Passos 1998). It is relevant that the Riparian Forest is under the influence of the river through the hydric regime and temporary floods (Gibbs and Leitão-Filho 1978; Passos 1998; Ricardi Branco et al. 2015).

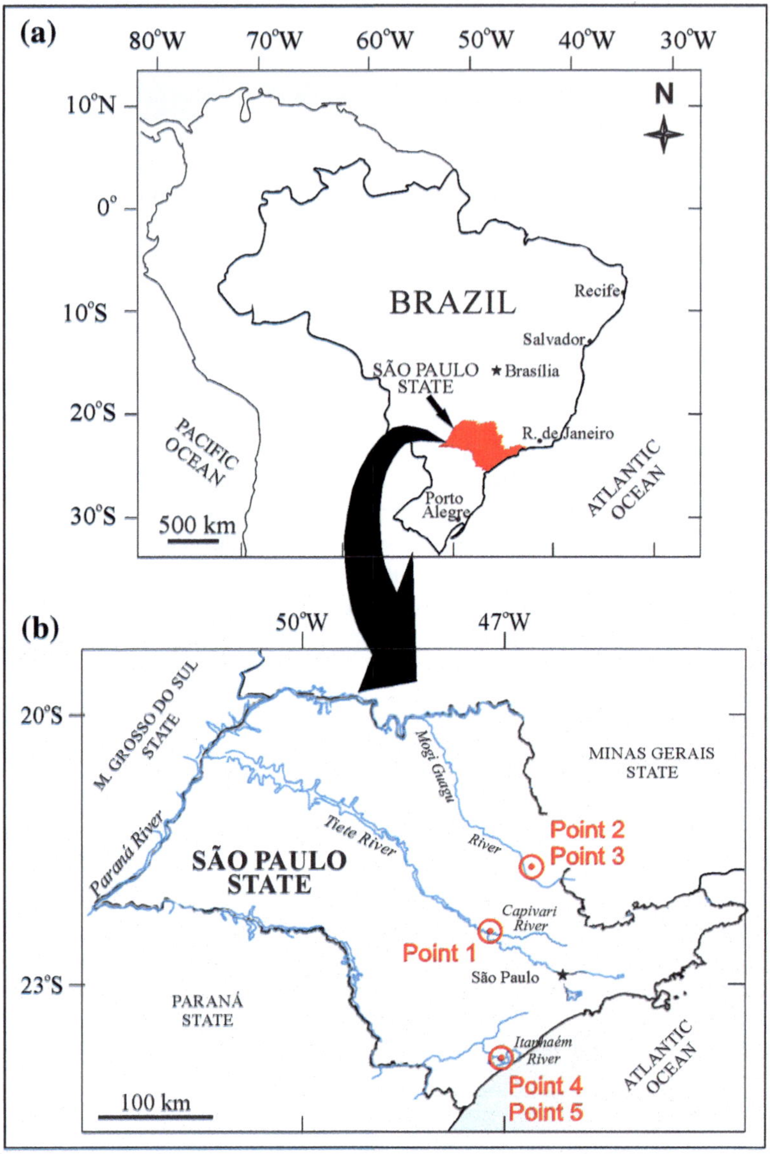

Fig. 7.1 Location maps; **a** Location of São Paulo State in Brazil; **b** Location of points with plant remain accumulations studied in the Capivari (Point 1), Mogi Guaçu (points 2 and 3), and Itanhaém (points 4 and 5) river basins

According to Köppen climatic classification, the MGES presents a Cwa climate, humid temperate with dry winter and hot summer (Sparovek et al. 2007).

According to the data collected by the meteorological station of the MGES from 1971 to 2010, the mean annual temperature (MAT) is 20.6 °C and the mean annual precipitation (MAP) is 1348 mm.

7.2.1.2 Coastal River Basin

The Itanhaém River Basin

The Itanhaém hydrological basin (Fig. 7.1) is located on the southern coast of the State of São Paulo, mainly in Itanhaém Municipality, and it is the second largest basin of the state coast with a surface area of 930 km^2 (Camargo et al. 2002). The basin is drained mainly by the Branco and Preto rivers and at its confluence originates the Itanhaém River (Amaral et al. 2006; Ricardi-Branco et al. 2009).

The Itanhaém basin lies on a quaternary plain that extends for approximately 50 km and has an average width of 15–16 km (Suguio and Martin 1978; Camargo et al. 2002; Suguio 2004; Amaral et al. 2006). Currently, the area is covered by fluvial sediments of tidal plain (Giannini and Santos 1996), and it preserves an important forest remnant of the Atlantic Forest (Garcia 2003; IF 2007). The Atlantic Forest is divided according to its topographic hierarchy in the Dense Montane Forest, Sub-montane Dense Ombrophylous Forest, and Dense Lowland Ombrophylous Forest (Veloso et al. 1991). In contrast, the tree/shrub-herbaceous vegetation is divided according to its environment in Mangrove and Restinga Forest (a mix of Lowland Rain Forest and Riparian Forest; IPT/PMI 2012). Therefore, there are at least six types of vegetation and the most important is the Restinga (Fig. 7.2b), where the study sites are located in the Branco and Preto rivers. According to Köppen climatic classification, the Itanhaém basin is influenced by a Cfa climate, hot-humid temperate climate with hot summer (Sparovek et al. 2007).

The Itanhaém river basin presents mean annual temperature (MAT) of 24.6 °C and mean annual precipitation (MAP) of 2230 mm (http://enos.cptec.inpe.br/).

The sampling sites located in the Branco and Preto rivers and the accumulations were studied by Ricardi-Branco et al. (2009, 2011a, b).

7.3 Materials and Methods

The methodology used for collecting leaf packages was proposed by Ricardi-Branco et al. (2009). A preliminary survey was conducted in the rivers for selection of the most promising point bars for the collection of blocks with plant remains.

Fig. 7.2 The Riparian forests. **a** and **b** Mogi Guaçu River at MGES and block collected at Point 2, red arrow shows leaves deposited at point bar; **c** Panoramic view of Riparian Forest (Restinga), Preto River, Itanhaém river basin

7.3.1 *Sample Collection*

7.3.1.1 Collection in the Capivari River

In the Point 1, near the outfall on Tiete River, accumulations of plant remains are abundant, and sampling was performed along August 2016. The assembly was collected (Point 1—22° 59′ 22.71″S\47° 45′ 18.87″W) on the convex side of the meandering curve and consisted of a block of 30 × 30 × 10 cm. Before packing the block, leaf samples were collected using gloves to avoid contamination at the top and bottom of it to perform ^{14}C dating.

7.3.1.2 Collection in the Mogi Guaçu River—MGES

In MGES the ciliary vegetation is well preserved, and two blocks of 30 × 30 × 15 cm were collected along April 2014. The first block was collected at Point 2 (22° 16′ 57.30″S\47° 11′ 57.80″W) on the concave side of the meandering curve (Fig. 7.1). The second block was collected at Point 3 (22° 16′ 30.30″S\47° 11′ 07.40″W) upstream, on the convex side of the meandering curve (Figs. 7.2a, b). Before packing the blocks for transport, leaf/leaflet samples were carefully collected at the top, middle and bottom of it to perform ^{14}C dating.

7.3.1.3 The Itanhaém River Basin

The leaf material from the Itanhaém river basin was previously collected and studied by Ricardi-Branco et al. (2009) in two accumulations of plant remains in Preto River (Point 4—24° 08′ 31.52″S/46° 53′ 55.03″W) and Branco River (Point 5—24° 05′ 03.97″S\46° 48′ 27.32″W). A total of 312 leaves/leaflets of angiosperms were collected from a block of 20 × 20 × 20 cm (Point 4). In the accumulation of remains from the surface of the Branco River (Point 5), 619 leaves/leaflets of angiosperms were collected in an area of 1 × 1 m. For the current research, all leaf material was re-studied with the purpose of improving the classification of leaves (Table 7.2).

7.3.2 *Treatment in Laboratory*

The number of species that are found in plant remains accumulations depends on the forest diversity that accompanies the river, the transport leaves/leaflets undergo before their sedimentation (Greenwood 1992; Ricardi-Branco et al. 2009; Ellis and Johnson 2013), the time required in their formation (Ricardi-Branco et al. 2009), the bias introduced by necrology, biostratinomy, sediment type, and the chemistry of the sub-aquatic environment. Yet, some losses occur while handling the samples (Greenwood 1992).

In the laboratory, the blocks collected in the Capivari river basin and in the Mogi Guaçu river basin were immersed in water, and manually we separated sediments from the plant fraction (leaves/leaflets, identifiable leaf/leaflet fragments, seeds, and branches). The leaf material was placed in water once again to remove clay and adhered colloidal material. All plant material was oven dried and pressed on sheets of paper, separated into packages according to their classification, origin, and date of collection. They are all stored in the Laboratory of Paleo-Hydrogeology at the Department of Geology and Natural Resources, Institute of Geosciences, University of Campinas.

7.3.3 *^{14}C Dating*

In order to date the accumulation time when the leaves were deposited, the eight samples collected for ^{14}C Center for Applied Isotope Studies of the University of Georgia, built by National Electrostatics Corporation, utilizes a Model 1.5SDH-1 Pelletron accelerator with a maximum terminal voltage of 0.5 MV (Cherkinsky et al. 2010). The samples were processed through the technique of accelerator mass spectrometry (AMS), and the ages obtained were calibrated with the radiocarbon calibration package CALIB 7.1 (Stuiver et al. 2017). All prior datings carried out by Ricardi-Branco et al. (2009) were incorporated into the study.

7.3.4 *Identification of Leaf Material*

The leaf material present in the blocks of plant remains collected in the Capivari river basin and in the Mogi Guaçu river basin were classified in morphotypes, and the material collected by Ricardi-Branco et al. (2009) in the Itanhaém river basin was reclassified into morphotypes. Classification/reclassification in morphotypes followed the Manual of Leaf Architecture by Ellis et al. (2009), and the Denver Museum of Nature Science (2011) classification scheme.

The leaf material of the Capivari river basin was identified with the aid of the reference collections performed in the Mogi Guaçu river basin and the Itanhaém river basin. The taxonomic identification of leaf material collected in the Mogi Guaçu river basin was made with the most representative species of the study area, as well as lists of vegetation (Gibbs and Leitão-Filho 1978; Gibbs et al. 1980; Mantovani et al. 1985; Passos 1998; Neto et al. 2012). The taxonomic identification of leaf material collected in the Itanhaém river basin was supported by the reference collection of the most representative species of the study area (Ricardi-Branco et al. 2009), as well as lists of vegetation (Lamberti 1969; Garcia 2003; IF 2007). During the identification process, virtual herbaria were used (http://fm1.fieldmuseum.org/vrrc/, http://www.herbariovirtualreflora.jbrj.gov.br/, http://www.tropicos.org/).

For each species identified, we counted the number of leaves/leaflets, and fragments of identifiable leaves/leaflets that presented more than 50% of their integrity preserved.

7.4 Results

7.4.1 The Capivari River Basin

Reworking plant remains is indicated by the inversion of ages between the basal portion and the top. The material from the base shows an age between 1957 and 1958 and the top material between 1260 and 1315 (Table 7.1). The top of the package did not allow other analysis, due to most plant remains were branches and fragmented leaves. We looked for other nearby locations to collect, however, no well-preserved packages were found. Thus, we decided to study only the base of the package collected in Point 1.

7.4.1.1 The Mogi Guaçu River Basin—MGES

The ages obtained by ^{14}C dating are presented in Table 7.1.

For Point 2, the dating shows reworking of plant remains indicated by the inversion of ages between the middle portion and the top. However, there is a temporal agreement between the base and the top of the block indicating that the package was deposited between the years 2002 and 2007 (Table 7.1). All inversions were considered for the results and discussion.

For Point 3, the dating shows a reworking between the middle and upper positions. However, there is an agreement between the base and the top indicating that the package was deposited between the years 1957 and 1958 (Table 7.1).

7.4.2 Characteristics of Leaf Material

7.4.2.1 Capivari River Basin

From Point 1, we recovered 91 leaves/leaflets that belong to 15 species. Five were not attributed to any known taxon (Table 7.2). Among the leaves/leaflets recovered, 64% belong to one species (*Inga vera* Willd. with 58 leaves), and the other 26% belong to 14 species. 70% of the species identified are arboreal, 20% climbing plants, and 10% shrub. Most species identified belong to Fabaceae and Sapindaceae families.

Table 7.1 ^{14}C dating of the plant remain accumulations of the Capivari (Point 1) and Mogi Guaçu (points 2 and 3) river basins. The results are corrected by natural isotopic fraction (−25‰) and presented at conventional age ^{14}C in percentage of Modern Carbon (pCM) in 1σ (68.3% probability), and in calibrated age (cal) in calendar years, 2σ

Capivari River Basin				Mogi Guaçu River Basin (MGES)							
Capivari River				Mogi Guaçu River							
Point 1				Point 2				Point 3			
Sample	Lab. number	pCM (%)	Calibrate age (yrs, 2σ)	Sample	Lab. number	PCM (%)	Calibrated age (yrs, 2σ)	Sample	Lab. number	pCM (%)	Calibrate age (yrs, 2σ)
Top	UGAMS#28360	–	1260–1315	Top	UGAMS#18353	104.70 ± 0.28	1957–1957	Top	UGAMS#18350	107.71 ± 0.29	2002–2006
–	–	–	–	Intermediate	UGAMS#18354	105.64 ± 0.29	2007–2011	Intermediate	UGAMS#18351	108.24 ± 0.29	2001–2004
Base	UGAMS#28364	105.32 ± 0.29	1957–1958	Base	UGAMS#18355	104.79 ± 0.29	1957–1958	Base	UGAMS#18352	107.30 ± 0.30	2003–2007

Table 7.2 Species identified and leaves/leaflet percentages in the plant remain accumulations collected in the Capivari, Mogi Cuaçu and Itanhaém (Preto and Branco rivers) river basins. The maximum values are in bold letters

Family	*Taxa*	Growth Habit	Leaves/leaflet counting amount (flora percentage)				
			Capivari River Basin	Mogi Guaçu River Basin		Itanhaém River Basin	
			Capivari River	Mogi Guaçu River		Preto River	Branco River
			Point 1 (1957–1958)	Point 2 (1957–1958)	Point 3 (2002–2007)	Point 4 (1957)	Point 5 (2004–2005)
Anacardiaceae	*Astronium graveolens* Jacq.	Tree		2 (0.15)			
	Tapirira sp.	Tree			3 (0.25)		
Annonaceae	*Duguetia lanceolata* A. St.-Hil.	Tree			2 (0.16)		
	Rollinia sp.	Tree			1 (0.08)		
Apocynaceae	*Forsteronia* aff. *pubescens* A. DC.	Vine		10 (0.76)	5 (0.41)		
	Hancornia speciosa Gomes	Tree		10 (0.76)			
Asteraceae	Asteraceae sp. 1	Shrub (?)	1 (1.10)				
	Asteraceae sp. 2	Shrub (?)		9 (0.69)			
Bignoniaceae	*Adenocalymma* aff. *bracteatum*	Vine		5 (0.38)			
	Adenocalymma paulistarum Bureau ex K. Schum.	Vine		6 (0.46)	3 (0.25)		
	Adenocalymma sp.	Vine		1 (0.08)			

(continued)

Table 7.2 (continued)

Family	*Taxa*	Growth Habit	Leaves/leaflet counting amount (flora percentage)				
			Capivari River Basin	Mogi Guaçu River Basin		Itanhaém River Basin	
			Capivari River	Mogi Guaçu River		Preto River	Branco River
			Point 1 (1957–1958)	Point 2 (1957–1958)	Point 3 (2002–2007)	Point 4 (1957)	Point 5 (2004–2005)
	Anemopaegma chamberlaynii (Sims) Bureau & K. Schum.	Vine		1 (0.08)	3 (0.25)		
	Arrabidaea sp. 1	Vine		4 (0.30)			
	Arrabidaea sp. 2	Vine		2 (0.15)			
	Bignonia aff. *exoleta* Vell.	Vine		1 (0.08)			
	Bignoniaceae 1	Vine (?)			1 (0.08)		
Boraginaceae	*Cordia sellowiana* Cham.	Tree			4 (0.33)		
Calophyllaceae	*Calophyllum brasiliense* Cambess.	Tree				19 (6.10)	
Chrysobalanaceae	*Parinari* sp.	Tree				4 (1.28)	
Euphorbiaceae	*Actinostemon concepcionis* (Chodat & Hassl.) Hochr.	Tree			2 (0.16)		
	Alchornea glandulosa Poepp.	Tree		5 (0.38)	1 (0.08)		

(continued)

Table 7.2 (continued)

Family	*Taxa*	Growth Habit	Leaves/leaflet counting amount (flora percentage)				
			Capivari River Basin	Mogi Guaçu River Basin		Itanhaém River Basin	
			Capivari River	Mogi Guaçu River		Preto River	Branco River
			Point 1 (1957–1958)	Point 2 (1957–1958)	Point 3 (2002–2007)	Point 4 (1957)	Point 5 (2004–2005)
	Alchornea triplinervia (Spreng.) Müll. Arg	Tree				5 (1.60)	
	Croton floribundus Spreng.	Tree		4 (0.30)			
	Sebastiania commersoniana (Baillon) Smith & Downs	Tree		4 (0.30)			
Fabaceae	*Andira fraxinifolia* Benth.	Tree		3 (0.23)			
	Bowdichia virgilioides Kunth	Tree		39 (2.97)	1 (0.08)		
	Centrolobium aff. *tomentosum* Guillemin ex Benth.	Tree		2 (0.15)			
	Copaifera langsdorffii Desf.	Tree	2 (2.20)	**137 (10.44)**	39 (3.20)		

(continued)

Table 7.2 (continued)

Family	*Taxa*	Growth Habit	Leaves/leaflet counting amount (flora percentage)				
			Capivari River Basin	Mogi Guaçu River Basin		Itanhaém River Basin	
			Capivari River	Mogi Guaçu River		Preto River	Branco River
			Point 1 (1957–1958)	Point 2 (1957–1958)	Point 3 (2002–2007)	Point 4 (1957)	Point 5 (2004–2005)
	Cyclolobium brasiliense Benth	Tree			6 (0.49)		
	Enterolobium aff. contortisiliquum (Vell.) Morong	Tree	**8 (8.78)**	6 (0.46)			
	Hymenaea courbaril L.	Tree		23 (1.75)	34 (2.79)		
	Inga vera Willd.	Tree	**58 (63.73)**	**679 (51.75)**	**902 (73.93)**	**129 (41.35)**	**222 (35.86)**
	Lonchocarpus guillemineanus (Tul.) Malme	Tree		2 (0.15)			
	Lonchocarpus sp.	Tree		2 (0.15)			
	Ormosia arborea (Vell.) Harms	Tree		2 (0.15)	1 (0.08)		
	Platymiscium sp.	Tree				8 (2.58)	
	Platypodium elegans Vogel	Tree		14 (1.07)	1 (0.08)		
	Pterogyne nitens Tul.	Tree		1 (0.08)			

(continued)

Table 7.2 (continued)

Family	*Taxa*	Growth Habit	Leaves/leaflet counting amount (flora percentage)				
			Capivari River Basin	Mogi Guaçu River Basin		Itanhaém River Basin	
			Capivari River	Mogi Guaçu River		Preto River	Branco River
			Point 1 (1957–1958)	Point 2 (1957–1958)	Point 3 (2002–2007)	Point 4 (1957)	Point 5 (2004–2005)
	Senna sp.	Tree					**140 (22.63)**
	Senna aff. multijuga (Rich.) H.S. Irwin & Barneby	Tree		23 (1.75)			
	Fabaceae sp. 1	Tree (?)	4 (4.38)				
	Fabaceae sp. 2	Tree		1 (0.08)			
	Fabaceae sp. 3	Tree			1 (0.08)		
Lauraceae	*Nectandra megapotamica* (Spreng.) Mez	Tree		**151 (11.51)**	**120 (9.84)**		
	Nectandra sp. 1	Tree	3 (3.30)				
	Nectandra sp. 2	Tree					**121 (19.55)**
	Ocotea sp.	Tree				4 (1.28)	
Lecythidaceae	*Cariniana estrellensis* (Raddi) Kuntze	Tree		3 (0.23)	1 (0.08)		
Malvaceae	*Luehea divaricata* Mart.	Tree		7 (0.53)			

(continued)

Table 7.2 (continued)

Family	*Taxa*	Growth Habit	Leaves/leaflet counting amount (flora percentage)				
			Capivari River Basin	Mogi Guaçu River Basin		Itanhaém River Basin	
			Capivari River	Mogi Guaçu River		Preto River	Branco River
			Point 1 (1957–1958)	Point 2 (1957–1958)	Point 3 (2002–2007)	Point 4 (1957)	Point 5 (2004–2005)
Malpighiaceae	*Heteropteris* sp.	Vine		6 (0.46)	4 (0.33)		
Melastomataceae	*Miconia cinnamomifolia* (DC.) Naudin	Tree					25 (4.04)
	Tibouchina sellowiana (Cham.) Cogn.	Tree					72 (11.63)
Meliaceae	*Guarea guidonia* (L.) Sleumer	Tree		16 (1.22)	11 (0.90)		
	Trichilia catigua A. Juss.	Tree		3 (0.23)			
	Trichilia pallida Sw.	Tree			4 (0.33)		
	Trichilia sp.	Tree		2 (0.15)			
Myrtaceae	*Calyptranthes lucida* Mart. ex DC.	Tree		4 (0.30)			
	Eugenia moraviana O. Berg	Tree		16 (1.22)			

(continued)

Table 7.2 (continued)

Family	*Taxa*	Growth Habit	Leaves/leaflet counting amount (flora percentage)				
			Capivari River Basin	Mogi Guaçu River Basin		Itanhaém River Basin	
			Capivari River	Mogi Guaçu River		Preto River	Branco River
			Point 1 (1957–1958)	Point 2 (1957–1958)	Point 3 (2002–2007)	Point 4 (1957)	Point 5 (2004–2005)
	Eugenia sp.	Tree		14 (1.07)	14 (1.15)		
	Myrcia sp. 1	Tree		4 (0.30)	2 (0.16)		
	Myrcia sp. 2	Tree				12 (3.85)	
	Syzygium cumini (L.) Skeels	Tree			2 (0.16)		
	Myrtaceae sp. 1	Tree			1 (0.08)		
	Myrtaceae sp. 2	Tree					7 (1.13)
Phytolaccaceae	*Gallesia integrifolia* (Spreng.) Harms	Tree		14 (1.07)			
Poaceae	Bambusoidae sp.	Herb				3 (0.96)	
	Poaceae sp.	Herb					5 (0.81)
Rubiaceae	*Faramea* sp.	Tree				25 (8.01)	
Salicaceae	*Casearia sylvestris* Sw.	Tree		4 (0.30)			
	Xylosma glaberrima Sleumer	Tree		1 (0.08)			
	Salicaceae sp.	Tree (?)	1 (1.10)				

(continued)

Table 7.2 (continued)

Family	*Taxa*	Growth Habit	Leaves/leaflet counting amount (flora percentage)				
			Capivari River Basin	Mogi Guaçu River Basin		Itanhaém River Basin	
			Capivari River	Mogi Guaçu River		Preto River	Branco River
			Point 1 (1957–1958)	Point 2 (1957–1958)	Point 3 (2002–2007)	Point 4 (1957)	Point 5 (2004–2005)
Sapindaceae	*Cupania vernalis* Cambess.	Tree	1 (1.1)	2 (0.15)	2 (0.16)		
	Matayba elaeagnoides Radlk.	Tree		4 (0.30)	12 (0.98)		
	Paullinia elegans Cambess.	Vine		19 (1.45)	9 (0.74)		
	Paullinia sp.	Vine				**73 (23.40)**	
	Serjania lethalis A. St.-Hil.	Vine		13 (0.99)	12 (0.98)		
	Serjania sp. 1	Vine	3 (3.30)				
	Serjania sp. 2	Vine	1 (1.10)				
	Serjania sp. 3	Vine		2 (0.15)			
	Serjania sp. 4	Vine			4 (0.33)		
	Serjania sp. 5	Vine			7 (0.57)		
	Urvillea ulmacea Kunth	Vine		13 (0.99)			

(continued)

Table 7.2 (continued)

Family	*Taxa*	Growth Habit	Leaves/leaflet counting amount (flora percentage)				
			Capivari River Basin	Mogi Guaçu River Basin		Itanhaém River Basin	
			Capivari River	Mogi Guaçu River		Preto River	Branco River
			Point 1 (1957–1958)	Point 2 (1957–1958)	Point 3 (2002–2007)	Point 4 (1957)	Point 5 (2004–2005)
Smilacaceae	*Smilax campestris* Griseb.	Vine		1 (0.08)			
	Smilax sp.	Vine			1 (0.08)		
Violaceae	*Hybanthus atropurpureus* (A. St.-Hil.) Taub.	Vine		6 (0.46)			
	Undetermined 1	(?)	3 (3.30)				
	Undetermined 2	(?)	3 (3.30)				
	Undetermined 3	(?)	1 (1.10)				
	Undetermined 4	(?)	1 (1.10)				
	Undetermined 5	(?)	1 (1.10)				
	Undetermined 6	(?)		1 (0.08)			
	Undetermined 7	(?)		1 (0.08)			
	Undetermined 8	(?)		3 (0.23)			
	Undetermined 9	(?)		4 (0.30)			

(continued)

Table 7.2 (continued)

Family	*Taxa*	Growth Habit	Leaves/leaflet counting amount (flora percentage)				
			Capivari River Basin	Mogi Guaçu River Basin		Itanhaém River Basin	
			Capivari River	Mogi Guaçu River		Preto River	Branco River
			Point 1 (1957–1958)	Point 2 (1957–1958)	Point 3 (2002–2007)	Point 4 (1957)	Point 5 (2004–2005)
	Undetermined 10	(?)			1 (0.08)		
	Undetermined 11	(?)			1 (0.08)		
	Undetermined 12	(?)			1 (0.08)		
	Undetermined 13	(?)			1 (0.08)		
	Undetermined 14	(?)				1 (0.32)	
	Undetermined 15	(?)				1 (0.32)	
	Undetermined 16	(?)				1 (0.32)	
	Undetermined 17	(?)				2 (0.64)	
	Undetermined 18	(?)				2 (0.64)	
	Undetermined 19	(?)				3 (0.96)	
	Undetermined 20	(?)				3 (0.96)	
	Undetermined 21	(?)				3 (0.96)	
	Undetermined 22	(?)				4 (1.28)	

(continued)

Table 7.2 (continued)

Family	*Taxa*	Growth Habit	Leaves/leaflet counting amount (flora percentage)				
			Capivari River Basin	Mogi Guaçu River Basin		Itanhaém River Basin	
			Capivari River	Mogi Guaçu River		Preto River	Branco River
			Point 1 (1957–1958)	Point 2 (1957–1958)	Point 3 (2002–2007)	Point 4 (1957)	Point 5 (2004–2005)
	Undetermined 23	(?)				4 (1.28)	
	Undetermined 24	(?)				6 (1.92)	
	Undetermined 25	(?)					1 (0.16)
	Undetermined 26	(?)					1 (0.16)
	Undetermined 27	(?)					1 (0.16)
	Undetermined 28	(?)					2 (0.32)
	Undetermined 29	(?)					2 (0.32)
	Undetermined 30	(?)					3 (0.48)
	Undetermined 31	(?)					3 (0.48)
	Undetermined 32	(?)					4 (0.65)
	Undetermined 33	(?)					5 (0.81)
	Undetermined 34	(?)					5 (0.81)
Total number of Leaves/leaflet			91	1312	1220	312	619
Total number of *taxa*			15	54	38	21	17
Total number of indentified *taxa*			10	50	34	10	7
Percentages trees/percentage Vines			70/20	68/30	71/29	80/10	86/0

7.4.2.2 Mogi Guaçu River Basin—MGES

From Point 2, we recovered 1312 leaves/leaflets that belong to 54 species. Four were not attributed to any known taxon (Table 7.2). 74% of the leaves/leaflets belong to three species [*Inga vera* with 679 leaves, *Nectandra megapotamica* (Spreng.) Mez with 151 leaves, and *Copaifera langsdorffii* Desf. with 137 leaves], and the other 26% belong to 50 species (Table 7.2). Sixty-eight percent of the species are arboreal, 30% climbing plants, and 2% shrub. Most species identified belong to Bignoniaceae, Fabaceae and Sapindaceae families.

From Point 3, we recovered 1220 leaves/leaflets that belong to 38 species. Five were not attributed to any known taxon (Table 7.2). 84% of the leaves/leaflets belong to two species (*Inga vera* with 902 leaves, and *Nectandra megapotamica* with 120 leaves), and the other16% belong to 35 species. Seventy-one percent of the identified species are arboreal, and 29% climbing plants. Most species identified belong to Fabaceae and Sapindaceae families.

7.4.2.3 Itanhaém River Basin

From Point 4 we collected 585 leaves/leaflets at point bar from the meander bend of Preto River (Ricardi-Branco et al. 2009, 2015). From these, 312 leaves/leaflets were recovered and classified, belonging to 21 species. Eleven leaf morphotypes were not attributed to any known taxon. Sixty-five percent of the leaves/leaflets belong to two species (*Inga vera* with 129 leaves, and *Paullinia* sp. with 73 leaves), and the other 35% belong to 19 species. 71% of the identified species are arboreal, and 29% are climbing plants. Most species identified belong to Fabaceae and Sapindaceae families.

For Branco River (Point 5), we recovered 619 leaves/leaflets that belong to 17 species. Ten leaf morphotypes were not attributed to any known taxon. Ninety percent of the leaves/leaflets belong to four species [*Inga vera* with 222 leaves, *Senna* sp. with 140 leaves, *Nectandra* sp. 2 with 121 leaves, and *Tibouchina sellowiana* (Cham.) Cogn. with 72 leaves], and the remaining 10% belong to 13 species. Eighty-six percent of the identified species are arboreal, and 14% are herbs. Most identified species belong to Fabaceae.

7.5 Discussion

7.5.1 Characteristics of the Plant Remain Accumulations

The accumulation of plant remains are composed of leaf material of parautochthonous origin, considering the distance traveled from the tree to the soil in the MGES fluvial plain in the Mogi Guaçu river basin, in the Capivari river basin, and the Itanhaém river basin (Greenwood 1991; Behrensmeyer et al. 2000; Martín-Closas and Gomez 2004). The leaf material is associated with the current vegetation of the Riparian forest, as the oldest accumulations belong from 59 to 60 years ago (i.e. Point 2), indicating that these plant accumulations were limited to their geographical origin area. These characteristics seem to be typical to plant macro-remain accumulations that are found in the meandering rivers studied here.

In the accumulations of sampled plant remains, approximately 63–84% of the leaf material belongs to 1–3 species (Table 7.2). These species dominate some areas along the Riparian Forest with many individuals (Mantovani and Martins 1988; Passos, 1998). In general, these dominant species correspond to the emerging plant, and most important component of the canopy (Gibbs and Leitão-Filho 1978; Gibbs et al. 1980; Mantovani et al. 1985; Passos 1998). In tropical forests, the most important canopy species present small leaves/leaflets (Greenwood 1992; Martín-Closas and Gomez 2004; Steart et al. 2005), and provide the largest amount of leaf material in the forest (Delitti 1984; Burnham 1989; Greenwood 1991, 1992), because they are directly exposed to wind action (Martín-Closas and Gomez 2004). Therefore, in accumulations of plant remains, we may see the most important canopy species (Greenwood 1991, 1992).

On the other hand, in the accumulations of sampled remains studied here, a small proportion of leaf material belongs to a large set of species (Table 7.2). Those that do not dominate the leaf accumulations may be considered rare because they are masked by a large proportion of leaves/leaflets from the most important species of the canopy (Steart et al. 2005). However, these taxa occur with low density in the area, and their leaves/leaflets are less abundant (Gibbs and Leitão-Filho 1978; Mantovani et al. 1985; Neto et al. 2012).

Most species identified in plant accumulations are associated with the families Fabaceae, Sapindaceae and Bignonaceae (Table 7.2). Among these families Fabaceae present of 20–40% of identified species (Table 7.2). The species *Inga vera* Willd (Fabaceae) form a large amount of leaf material in the Riparian Forest of the studied rivers, increasing their probability of representation in accumulations (Burnham 1989; Mantovani et al. 1985). In addition, the Fabaceae present the greatest wealth of arboreal species in extant Riparian forests associated with the studied accumulations (Mantovani et al. 1985; Passos 1998).

The Bignoniaceae and Sapindaceae families may stand out in accumulations of plant remains because they concentrate the greatest richness of climbing species in extant Riparian forests (Neto et al. 2012). Thus, 68–85% of the identified species are arboreal and 10–30% are climbing (Table 7.2), similarly to leaf litter accumulated in tropical forests (Burnham 1994, 1997). Climbing plants easily access the accumulations because they grow on the highest parts of the canopy (Neto et al. 2012) and exposed to high light due to river banks (Mantovani et al. 1985).

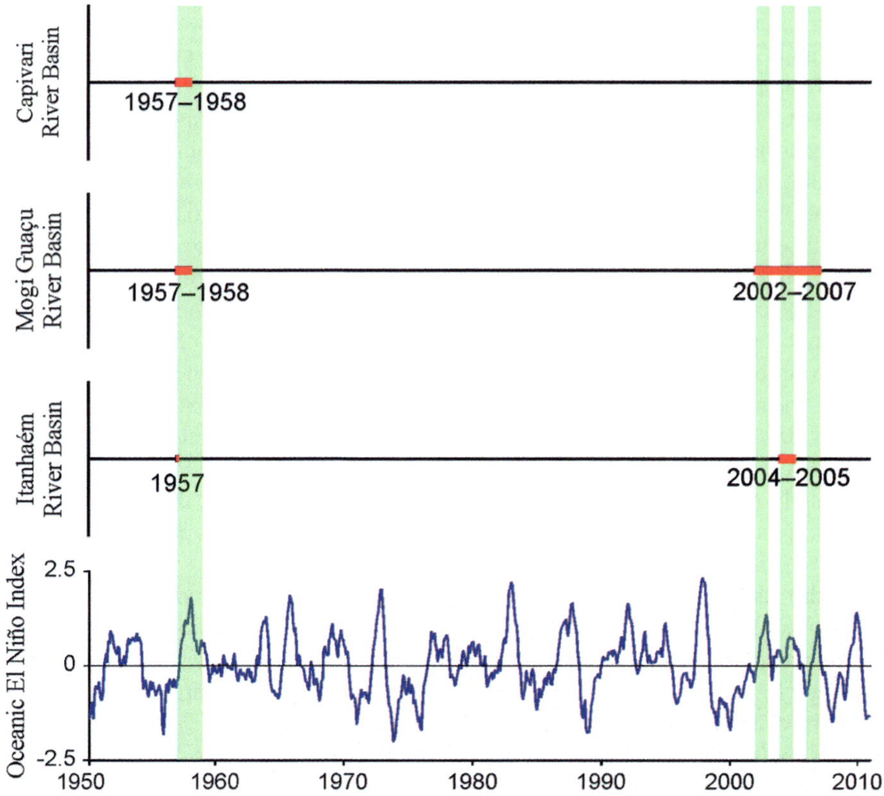

Fig. 7.3 Likely trend suggested relationship among studied plant remain accumulations (ages in red lines) with El Niño. The positive values correspond to the hot phase of EL Niño, and negative values to the cold phase, la Niña (www.esrl.noaa.go)

7.5.2 *Accumulation of Plant Remains and the Likely Influence to Global Climatic El Niño-Southern Oscillation Phenomenon in Its Deposition*

The datings carried out in the studied leaf packages matched with years of different intensity episodes of the El Niño-Southern Oscillation phenomenon (Fig. 7.3) as follows:

– the accumulations collected in Points 1, 2 and 4 were deposited during the intense El Niño episode. Its deposition is associated with the hot phase of the El Niño-Southern Oscillation phenomenon that happened between 1957 and 1959 (http://enos.cptec.inpe.br/);

- the accumulation of Point 5 (Ricardi-Branco et al. 2009) was deposited during the weak episode of El Niño that occurred in the period 2004–2005 (http://enos.cptec.inpe.br/).
- the accumulation of Point 3 was deposited during three successive El Niño episodes. A moderate episode that happened in the period 2002–2003, and two weak episodes that happened in the periods 2004–2005 and 2006–2007 (http://enos.cptec.inpe.br/).

As presented above in tropical forests, leaf production has high seasonality throughout the year, and they respond immediately to climatic variations (Delitti 1984; Matos and Costa 2012; Hofhansl et al. 2014), such as El Niño (Hofhansl et al. 2014). When episodes of El Niño occur, temperatures are higher than usual, and the wind blows stronger than in other periods, causing an intense fall of leaves, and therefore their accumulation in leaf litter (Wood et al. 2009; Costa et al. 2014). In addition, the occurrence of extreme climatic events, such as occasional storms, may cause an increase in leaf litter deposition (Schumacher et al. 2008).

The coincidence between the age of leaves and the deposition of accumulations studied may be further tested based on detailed sedimentological studies. The leaves in tropical forests obey a seasonal pattern of production, and the regular years of El Niño phenomenon suggest a possible trend of correlation between the deposition of the assemblies.

7.6 Conclusions

- In all plant accumulations studied here, the short transport of plants that compose them is evident.
- The leaf accumulations studied here reflect the dominant arboreal (i.e. *Inga vera* Willd) and climbing (i.e. *Serjania lethalis*, *Paullinia* sp.) species of the vegetation present in the Riparian forest. The taxa present a canopy free of obstacles so that the leaves are directly deposited in the river course or its banks;
- Assemblages are mostly composed of leaves and branches, hardly any seeds;
- The cause of deposition of the accumulations studied here may have some trend of correspondence with rainier years, such as periods of El Niño activity. However, more detailed sedimentological studies may confirm or decline this relation.

Acknowledgements The authors of the current chapter thank the São Paulo Research Foundation—FAPESP for the funding support provided to carry out the researches presented through FAPESP Processes 01/09881-2. "Environmental Study in the Itanhaém River estuary, South Coast of the State of São Paulo"; FAPESP 2007/07190-9. "Characterization of accumulations of plant remains in Rio Preto, Itanhaém, SP: elaboration of a distribution model using geotechnology tools"; FAPESP 2010/20379-6. "Quaternary paleoenvironmental analysis of the Cerrado/Atlantic Forest in the Municipality of Mogi-Guaçú, SP"; FAPESP 2013/22729-2. "Evolution of the quaternary plain and groundwater dynamics of the Mogi-Guaçu river, SP", and FAPESP 2016/20927-0 "Studies of modern accumulations and bioclast fossils in continental and coastal environments". We are also grateful for the collaboration of the Forestry Institute for the research at the Mogi-Guaçu

Experimental Station (MGES), the National Council for Scientific and Technological Development (CNPq) for granting research fellowships (303527/2017-0), and the Coordination of Improvement of Higher Education Personnel (CAPES) for the doctoral fellowship awarded. Finally we would like to thank the referees for their suggestions that helped to improve this chapter.

References

Amaral PGC, Ledru MP, Ricardi-Branco F et al (2006) Late Holocene development of a mangrove ecosystem in southeastern Brazil (Itanhaém, state of São Paulo). Paleogeogr Palaeoclimatol Palaeoecol 241:608–820. https://doi.org/10.1016/j.paleo.2006.04.010

Behrensmeyer AK, Kidwell SM, Gastaldo RA (2000) Taphonomy and paleobiology. Paleobiology 26:103–147. https://doi.org/10.1666/0094-8373(2000)26%5b103:TAP%5d2.0.CO;2

Birks HH, Birks HJB (2000) Future uses of pollen analysis must include plant macrofossils. J Biogeogr 27:31–35. https://doi.org/10.1046/j.1365-2699.2000.00375.x

Burnham RJ (1989) Relationships between standing vegetation and leaf litter in a paratropical forest: implications for paleobotany. Rev Palaeobot Palyn 58:5–32. https://doi.org/10.1016/0034-6667(89)90054-7

Burnham RJ (1994) Patterns in tropical leaf litter and implications for angiosperm. Paleobot Rev Palaeobot Palyn 81:99–113. https://doi.org/10.1016/0034-6667(94)90129-5

Burnham RJ (1997) Stand characteristics and leaf litter composition of a dry forest hectare in Santa Rosa National Park, Costa Rica. Biotropica 29:384–395. https://doi.org/10.1111/j.1744-7429.1997.tb00034.x

Burnham RJ, Johnson KR (2004) South American paleobotany and the origins of neotropical rainforests. Phil Trans Royal Soc London B 359:1595–1610

Camargo AFM, Pereira LA, Pereira AMM (2002) Ecologia da bacia hidrográfica do rio Itanhaém. In: Schiaretti A, Camargo AFM (eds) Conceitos de Bacias Hidrográficas: Teorias e Aplicações, 1st edn. Editora da Universidade Estadual de Santa Cruz, Ilhéus, pp 239–256

Cherkinsky A, Culp RA, Dvoracek DK, Noakes JE (2010) Status of the AMS facility at the University of Georgia. Nucl Instrum Methods Phys Res, Sect B 268:867–870. https://doi.org/10.1016/j.nimb.2012.09.030

Collares EG (2000) Avaliação de alterações em redes de drenagem de microbacias como subsídio ao zoneamento geoambiental de bacias hidrográficas: aplicação na bacia hidrográfica do Rio Capivari-SP. Ph.D. thesis, Universidade de São Paulo, São Carlos

Costa MC, Costa ACL, Coelho LTS, Silva TML, Azevedo AF (2014) Correlação entre precipitação pluviométrica e umidade do solo na produção de serapilheira em Caxiuanã (PA). Rev Ibero-Amer Ciênc Amb 5:170–179. https://doi.org/10.6008/SPC2179-6858.2014.001.0012

Delitti WBC (1984) Aspectos comparativos da ciclagem de nutrientes minerais na Mata Ciliar, no Campo Cerrado e na floresta implantada de *Pinus elliotti* Engelm. var. *elliotti* (Mogi-Guaçu, SP). PhD thesis, Universidade de São Paulo, São Paulo

Denver Museum of Nature Science (DMNS): Guide to morphotyping fossil floras. http://www.paleobotanyproject.org/morphotyping.aspx. Accessed 15 Jan 2011

Ellis B, Daly DC, Hickey LJ et al (2009) Manual of leaf architecture. Cornell University Press, Ithaca, NY

Ellis B, Johnson KR (2013) Comparison of leaf samples from mapped tropical and temperate forests: implications for interpretations of the diversity of fossil assemblages. Palaios 28:163–177. https://doi.org/10.2110/palo.2012.p12-073r

Felfili JM, de Mendoça RC, Walter BMT et al (2001) Flora fanerogâmica das Matas de Galeria e Ciliares do Brasil Central. In: Ribeiro JF, da Fonseca CEL, Sousa-Silva JC (eds) Cerrado: Caracterização e recuperação de Matas de Galeria. Embrapa Cerrados, Planaltina, DF, pp 195–263

Garcia RJF (2003) Estudo florístico dos campos alto-montanos e matas nebulares do Parque Estadual da Serra do Mar, Núcleo Curucutu, São Paulo, SP, Brasil. PhD thesis, Universidade de São Paulo, São Paulo

Giannini PCF, Santos ER (1996) Caracterização sedimentológica da lama negra de Peruíbe (SP). Anais of III congresso brasileiro de termalismo. Universidade Federal de Santa Catarina, Santa Catarina, 1996

Gibbs PE, Leitão-Filho HF (1978) Floristic composition of an area of gallery forest near Mogi Guaçu, state of São Paulo, S.E. Brazil Rev Bras Bot 1:151–156

Gibbs PE, Leitão-Filho HF, Abbott RJ (1980) Application of the point-centred quarter method in a floristic survey of an area of gallery forest at Mogi-Guaçu, SP, Brazil. Rev Bras Bot 3:17–22

Gibling MR, Davies NS, Falcon-Lang HJ et al (2014) Palaeozoic co-evolution of rivers and vegetation: a synthesis of current knowledge. Proc Geol Assoc 125:524–533. https://doi.org/10.1016/j.pgeola.2013.12.003

Greenwood DR (1991) The taphonomy of plant macrofossils. In: Donovan SK (ed) The processes of fossilization. Belhaven Press, London, pp 145–169

Greenwood DR (1992) Taphonomic constraints on foliar physiognomic interpretations of Late Cretaceous and Tertiary Palaeoclimates. Rev Palaeobot Palyn 71:149–190. https://doi.org/10.1016/0034-6667(92)90161-9

Hinojosa LF, Pérez F, Gaxiola A et al (2011) Historical and phylogenetic constraints on the incidence of entire leaf margins: insights from a new South American model. Global Ecol Biog 20:380–390. https://doi.org/10.1111/j.1466-8238.2010.00595.x

Hofhansl F, Kobler J, Ofner J et al (2014) Sensitivity of tropical forest above ground productivity to climate anomalies. SW Costa Rica. Global Biogeoch Cycles 28:1437–1454. https://doi.org/10.1002/2014GB004934

IF—Instituto Florestal (2007) Inventário florestal da vegetação natural do Estado de São Paulo: Regiões Administrativas de São José dos Campos (Litoral), Baixada Santista e Registro. SMA/Imprensa Oficial, São Paulo

IPT/PMI—Instituto de Pesquisas Tecnológicas do Estado de São Paulo/Prefeitura Municipal de Itanhaém (2012) Atlas ambiental do município de Itanhaém: Imprensa Oficial, São Paulo

Instituto Nacional de Pesquisas Espaciais. Centro de Previsão de Tempo e Estudos Climáticos. http://enos.cptec.inpe.br/. Accessed May 2018

Jacobs BF, Herendeen PS (2004) Eocene dry climated and woodland vegetation in tropical Africa reconstructed from fossil leaves from northern Tanzania. Palaeogeogr Palaeoclimatol Palaeoecol 213:115–123. https://doi.org/10.1016/j.palaeo.2004.07.007

Lamberti A (1969) Contribuição ao conhecimento da ecologia das plantas do Manguezal de Itanhaém. Boletim da Faculdade de Filosofia, Ciênc Let Univ São Paulo 23:1–217

Mantovani W, Martins FR (1988) Variações fenológicas das espécies do cerrado da Reserva Biológica de Moji Guaçu, Estado de São Paulo. Rev Bras Bot 11:101–112

Mantovani W, Leitão Filho HF, Martins FR (1985) Chave baseada em caracteres vegetativos para identificação de espécies lenhosas do cerrado da Reserva Biológica de Moji-Guaçu, Estado de São Paulo. Hoehnea 12:35–56

Martín-Closas C, Gomez B (2004) Taphonomiedes plantes et interprétations paléoécologiques. Une synthèse. Geobios 37:65–88. https://doi.org/10.1016/j.geobios.2003.01.006

Matos BRM, Costa ACL (2012) Efeito da deficiência hídrica na produção dos componentes da liteira vegetal em floresta tropical nativa na Flona Caxiuanã-Pará. Rev Biol Neotrop 9:24–36. https://doi.org/10.5216/rbn.v9i2.22520

Neto SR, Godoil JV, Villagra BLP et al (2012) Caracterização florística, fitossociológica e fenológica de trepadeiras de Mata Ciliar da Fazenda Campininha, Mogi Guaçu, SP, Brasil. Hoehnea 39:145–155. https://doi.org/10.1590/S2236-89062012000100009

Oliveira-Filho AT, Budke JC, Jarenkow JÁ et al (2013) Delvind into variations in tree species composition and richness across South American subtropical Atlantic and Pampean forest. J Plant Ecol 8:242–260. https://doi.org/10.1093/jpe/ett058

Passos MJ (1998) Estrutura da vegetação arbórea e regeneração natural em remanescentes de Mata Ciliar do Rio Mogi Guaçu-SP. Dissertation, Universidade de São Paulo, Piracicaba
Ribeiro JF, Walter BMT (2008) As principais fitofisionomias do Bioma Cerrado. In Sano SM, Almeida SP, Ribeiro JF (eds.) Cerrado: Ecologia e Flora: Embrapa Cerrados, Brasília, pp 152–212
Ricardi-Branco F, Branco FC, Garcia RF et al (2009) Plant accumulations along the Itanhaém River Basin, Southern coast of São Paulo state, Brazil. Palaios 24:416–424. https://doi.org/10.2110/palo.2008.p08-079r
Ricardi-Branco F, Ianniruberto M, Silva AM et al (2011a) Plant debris accumulations in the Preto river subbasin, Itanhaém, São Paulo, Brazil: insights from geotechnology. Palaios 26:264–274. https://doi.org/10.2110/palo.2010.p10-125r
Ricardi-Branco F, Pereira SY, Branco F et al (2011b) Accumulation of bio debris and its relation with the underwater environment in the estuary of Itanhaém River, São Paulo State, Brazil. In: Dar IA, Dar MA (eds) Earth and Environmental Sciences/Book 2: 565–590. InTech Publisher. http://www.intechweb.org/
Ricardi-Branco F, Pereira SY, Souza MM et al (2015) Relationships among subaquatic environment and Leaf/Palynomorph assemblages of the quaternary Mogi-Guaçú River alluvial plain, SP, Brazil. In: Ramkumar M, Kumaraswamy K, Mohanraj R (eds) Environmental management of river Basin ecosystems. Springer International Publishing, Switzerland, pp 667–705. https://doi.org/10.1007/978-3-319-13425-3_30
Royer DL (2012) Climate reconstruction from leaf size and shape: new developments and challenges. In: Ivany LC, Huber BT (eds) Reconstructing Earth's deep-time climate–the state of the art in 2012. Pal Soc Papers 18:195–212
Schumacher MV, Viera M, Witschorec, R (2008) Produção de serapilheira e transferência de nutrientes em área de segunda rotação com floresta de *Pinus taeda* L. no município de Cambará do Sul, RS. Ciênc Florestal 18:471–480
SMA/IF—Secretaria do Meio Ambiente/Instituto Florestal (2005) Inventário florestal da vegetação natural do Estado de São Paulo: Imprensa Oficial, São Paulo
Spavorek G, Van Lier QDJ, Neto DD (2007) Computer assisted Koeppen climate classification: a case study for Brazil. Int J Climatol 27:257–266. https://doi.org/10.1002/joc.1384
Steart DC, Greenwood DR, Boon PI (2005) Paleoecological implications of differential biomass and litter production in canopy trees in Australian *Nothofagus* and *Eucalyptus* Forests. Palaios 20:452–462. https://doi.org/10.2110/palo.2004.P04-57
Stuiver M, Reimer PJ, Reimer RW (2017) CALIB 7.1. http://www.calib.org. Accessed 15 April 2017
Suguio K (2004) O papel das variações do nível relativo do mar durante o Quaternário tardio na origem da baixada litorânea de Juréia, SP. In: Marques OAV, Duleba W (eds.) Estação Ecológica Juréia-Itatins: Ambiente Físico, Flora e Fauna: Holos Editora, Ribeiraõ Preto, pp 34–41
Suguio K, Martin L (1978) Formações quaternárias marinhas do litoral paulista e sul fluminense. Boletim, Instituto de Geociências, Universidade São Paulo, Sociedade Brasileira de Geologia (IG/USP/SBG) Edição especial, pp 1–55
Veloso HP, Rangel-Filho ALR, Lima JCA (1991) Classificação da vegetação brasileira adaptada a um sistema universal. Instituto Brasileiro de Geografia e Estatística, Departamento de Recursos Naturais e Estudos Ambientais, Rio de Janeiro, p 124p
Wilf P, Wing SL, Greenwood DR, Greenwood CL (1998) Using fossil leaves as paleoprecipitation indicators: an Eocene example. Geology 26:203–206. https://doi.org/10.1130/0091-7613(1998)026%3c0203:UFLAPI%3e2.3.CO;2
Wing SL, DiMichele W (1995) Conflict between local and global changes in plant diversity through geological time. Palaios 10:551–564. https://doi.org/10.2307/3515094
Wood TE, Lawrence D, Clark DA et al (2009) Rain forest nutrient cycling and productivity in response to large-scale litter manipulation. Ecology 90:109–121

Chapter 8
Recent Root Damages of Fossilized Vertebrate Remains from New Mexico, USA

Heitor Francischini, Spencer G. Lucas, Paula Dentzien-Dias and Cesar L. Schultz

Abstract Damages produced by plant roots are one of the biological processes that may modify or destroy vertebrate remains (such as bones, scutes, scales, and teeth) during the biostratinomic stage of fossilization. These damages can be produced in fresh remains (pre-fossilization or pre-diagenesis phase) and/or in fossil remains (post-fossilization or post-diagenesis phase) and differentiate the traces (biological modification of the remains) produced in these different stages is not an easy task. In this contribution, we describe four morphotypes of traces on vertebrate remains collected in the Eocene San Juan Formation (northeastern New Mexico, USA), some of which had recent roots adhered in their surface. The morphological differences between traces of each morphotype are related to the tissue organization of the substrates. Since we could relate the morphology to a specific producer, the traces described here can be used as proxies to recognize recent root damages of vertebrate remains found in other localities and time. Recent root traces made in fossils indicate that these remains suffered at least one cycle of exhumation-modification-burial, being important signatures that allow to better understanding the taphonomic history of a fossil assemblage.

Keywords Vertebrate taphonomy · Ichnology · Bioerosion · Corrosichnia · Root corrosion

---

H. Francischini (✉) · C. L. Schultz
Laboratório de Paleontologia de Vertebrados, Programa de Pós-Graduação em Geociências, Instituto de Geociências, Universidade Federal do Rio Grande do Sul, Porto Alegre, RS, Brazil
e-mail: heitorfrancischini@hotmail.com

C. L. Schultz
e-mail: cesar.schultz@ufrgs.br

S. G. Lucas
New Mexico Museum of Natural History and Science, Albuquerque, NM, USA
e-mail: spencer.lucas@state.nm.us

P. Dentzien-Dias
Laboratório de Geologia e Paleontologia, Instituto de Oceanografia, Universidade Federal do Rio Grande, Rio Grande, RS, Brazil
e-mail: pauladentzien@gmail.com

S. Martínez et al. (eds.), *Actualistic Taphonomy in South America*,
Topics in Geobiology 48, https://doi.org/10.1007/978-3-030-20625-3_8

8.1 Introduction

In continental settings, plant roots are one of the main elements that can produce biological damage in lithic, mineralized or consolidated substrates, which include igneous, metamorphic and lithified sedimentary rocks, invertebrate and vertebrate skeletons, and wood, among others (e.g., Neumann 1966; Sarjeant 1975; Behrensmeyer 1978; Klappa 1980; Mikuláš 1999, 2001; Mikuláš and Žítt 1999; Fernández-Jalvo et al. 2002; Lucas 2016; Tjelldén et al. 2016; Paes-Neto et al. 2018). Since these kinds of substrates have a high potential of preservation, they can be found often in both archeological and paleontological contexts (e.g., Mikuláš 2001; Fernández-Jalvo and Andrews 2016). Roots attach to hard substrates mainly to provide support or for the acquisition of essential nutrients for the development of the plant (Sarjeant 1975; Raven 2009). The growth of plant roots on or around bones occurs when these vertebrate remains lie on the ground, exposed for a reasonable time in the soil, before or just after their burial, where the bones also suffer other destructive biostratinomic processes (e.g., fracture, abrasion, dissolution, spatial mixing) caused by other biological (e.g., predators, scavengers, osteophagic animals, etc.) and physicochemical agents (e.g., variation in temperature, precipitation, oscillations in the soil pH, etc.) (Behrensmeyer 1978; Fernández-Jalvo et al. 2002; Noto 2011; Fernández-Jalvo and Andrews 2016; Paes-Neto et al. 2018). In the case of root-etched bone that is buried, it will pass through the diagenetic process and will be fossilized, together with the traces previously produced on its surface. On the other hand, erosive processes can act after fossilization, exposing the remains again to the action of the same destructive, biostratinomic agents, which may act on a lithic substrate (the fossilized bone) instead of an organic one (the fresh bone). In this case, even if osteophagic vertebrates and invertebrates do not act as biostratinomic agents, the fossil bones may be damaged by the direct action of other abiogenic agents and organisms that may attach to its surface, such as plants, algae, fungi and lichens (Sarjeant 1975; Mikuláš 2001).

Plant-made traces were classified by Mikuláš (1999) according to their ethological categories. Among them, traces produced by roots during the biochemical dissolution of the surface of hard substrates are classified within the category Corrosichnia (Mikuláš 1999). Plants absorb ions dissolved in the liquid phase of the soil, and the availability and dissolution of these ions are dependent on several factors, mainly rainfall and organic matter decomposition rates, which may affect the pH of the soil (Taiz and Zeiger 2010). Low pH acts on the weathering of rocks and other lithified substrates, making several ions available to absorption by roots, such as carbonates, silicates, and phosphates (Taiz and Zeiger 2010). Among the essential nutrients, only Carbon, Oxygen and Hydrogen can be acquired from the atmosphere and water, and all other elements necessary for plant development absorbed are by the roots from the soil (e.g., Raven 2009). Even if several factors are known to act on the weathering of lithic substrates and, consequently, the availability of nutrients for plants, these organisms can accelerate the weathering process by producing exudates that can change the substrate chemistry (Keller and Frederickson 1952; Bais et al. 2006;

Noto 2011). This biochemical dissolution may modify the chemical composition of the surrounding clasts, disincorporating and/or etching the surfaces of them.

Differentiating traces produced by roots in non-fossilized bones from traces produced on fossil bones is not an easy task, mainly because there are not many cases of traces produced in fossilized bones by recent roots, and specific criteria for distinguishing them are still in an incipient state of interpretation (e.g., Mikuláš and Žítt 1999; Montalvo 2002; Bader et al. 2009). Therefore, in order to contribute to this issue, we describe the presence of damage in vertebrate fossilized remains (ganoid scales, turtle shells, crocodilian scutes, fragmented bones, and teeth) from the lower Eocene San Jose Formation (New Mexico, USA) produced by recent plant roots. The main goals of this contribution are to document the occurrence of differences between traces produced by the same trace-maker (i.e., plant roots) on different substrates (i.e., different vertebrate tissues) and discuss the importance of recognizing this type of taphonomic signature in the fossil record.

8.2 Materials and Methods

The materials described here come from the NMMNH (New Mexico Museum of Natural History and Science) locality 11,887, near the city of Cuba (Sandoval County), northwestern New Mexico, USA, where the San Jose Formation (San Juan Basin) crops out (Fig. 8.1). The San Jose Formation is a fluvially-deposited unit of early Eocene (Wasatchian) age that yields an extensive record of vertebrate fossils, primarily of mammals (e.g., Lucas 2015).

Part of the studied material was collected *in situ*, directly from a conglomerate level, which was exposed and colonized by a diverse array of grasses and bushes, which compose the Mixed Grassland Association of the Grassland Biome (*sensu* Castetter 1956). Because some of the vertebrate remains still had roots attached to them (Fig. 8.2), part of the bone-bearing matrix was also collected and analyzed in the laboratory. Materials *ex situ* were also collected, and their exact origin could be traced to the same point of the *in situ* remains. All the collected material was sieved (425 μm sieve) in the laboratory, and only fragments larger than 1 mm were considered in this study. A total of 649 specimens were collected, and all of them were housed in the New Mexico Museum of Natural History and Science (Albuquerque, NM, USA) under the catalog number NMMNH-P-75177.

The traces were analyzed under a stereomicroscope, measured using the software ImageJ®, and their colors were defined using the Munsell color system. The description followed the nomenclature and ichnotaxobases proposed by Britt et al. (2008) and Pirrone et al. (2014). Five elements (one of each category: ganoid scale, crocodilian scute, turtle shell fragment, indeterminate bone fragment, and tooth) were mounted on aluminum stubs, coated with gold and analyzed under a JEOL JSN-6610LV Scanning Electron Microscope (SEM) in the Centro de Microscopia Eletrônica da Zona Sul (CEME-SUL) of the Universidade Federal do Rio Grande

Fig. 8.1 Location of the San Juan Basin at northwestern New Mexico (USA), where the San Jose Formation crops out. Modified from Kondrashov and Lucas (2015)

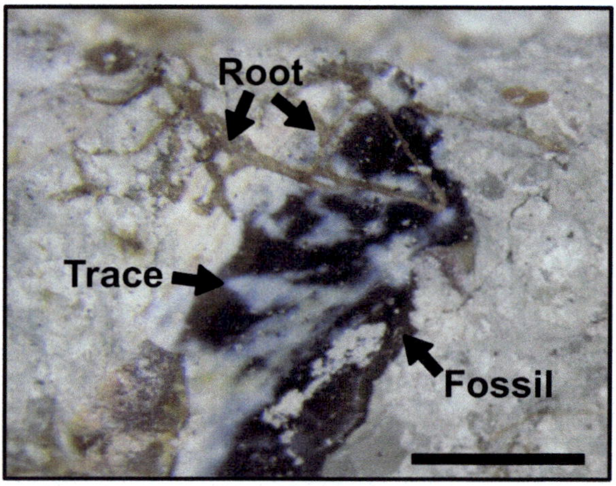

Fig. 8.2 Recent root attached to the surface of an indeterminate fossil bone from the San Jose Formation (Eocene of New Mexico, USA) and the derived trace (the white mottle). Scale: 2 mm

(Rio Grande, RS, Brazil). They were also subjected to an Energy-dispersive X-ray spectroscopy (EDS) analysis using the same SEM.

Among the bone remains (Table 8.1), we could identify ganoid scales of lepisosteiform fishes (N = 61, 9.40%), crocodilian osteoderms (N = 30, 4.62%), turtle shell fragments (N = 110, 16.95%), and mammalian and crocodilian teeth (N = 36, 5.54%). Indeterminate bone fragments total 412 pieces (63.48%). None of the specimens were articulated and most of them have indications of pre-fossilization

Table 8.1 Number of sampled specimens (with or without root traces) from the San Jose Formation (Eocene of New Mexico, USA)

Material	Without traces	With traces	Total
Scales	53	8	61
Teeth	30	6	36
Turtle shell fragments	67	43	110
Scutes	28	2	30
Indeterminate bones	340	72	412
Total	518	131	649

transport and reworking (such as fragmentation, weathering fractures and *in situ* random orientation). Because of their fragmented preservation, only some of the entire mammal teeth allowed a taxonomic identification and were assigned to the pantodont *Coryphodon* and the perissodactyl *Hyracotherium*.

8.3 Results

A total of 649 fossil fragments was analyzed, 518 (79.81%) of them devoid of traces and 131 (20.19%) of them damaged by the action of recent roots (Table 8.1; Figs. 8.2 and 8.3). The fossil bones vary in color from brown (Hue 7.5 YR 4/6) to light yellow (Hue 10 YR 8/3) and the scales and teeth are mainly dark brown (Hue 7.5 YR 3/4). The root traces are very light in color, and their color varies from light gray (Hue 2.5 Y 8/1) to yellow (Hue 2.5 Y 8/6). Only the traces produced in the ganoid scale present a light bluish gray (Hue 10 BG 7/1) color. Four morphotypes of traces could be recognized, being primarily related to the type of substrate (i.e., the tissue organization of each type of element), and are described as follows:

Morphotype 1 (Damage to indeterminate bones and turtle shells) Traces in bones and turtle shells are very shallow furrows with a very well defined shape (Fig. 8.3a–c). In plan view, these furrows are straight, gently curved or meandering, with a very well defined border. Haloes of chemical activity can occur, but they are rarer than in other morphotypes. Because the trace is very shallow, it does not form walls. Under the stereomicroscope, the bottom of the trace seems to be smooth, but the SEM analysis showed that it is rougher than the bone surface (Fig. 8.4a). These furrows are composed of the main axis from which secondary branches split at angles less than 90° (range: 34°–86°). Tertiary branches can split from the secondary ones (Fig. 8.3a). The Morphotype 1 traces do not penetrate to the spongy bone, but they are randomly spread on the bone surface, covering almost all the sides of the cortical bone (but they can also occur in the spongy bone of the fractured side). In some bone fragments, the traces overlap each other, forming a net-like organization of traces (Fig. 8.3b).

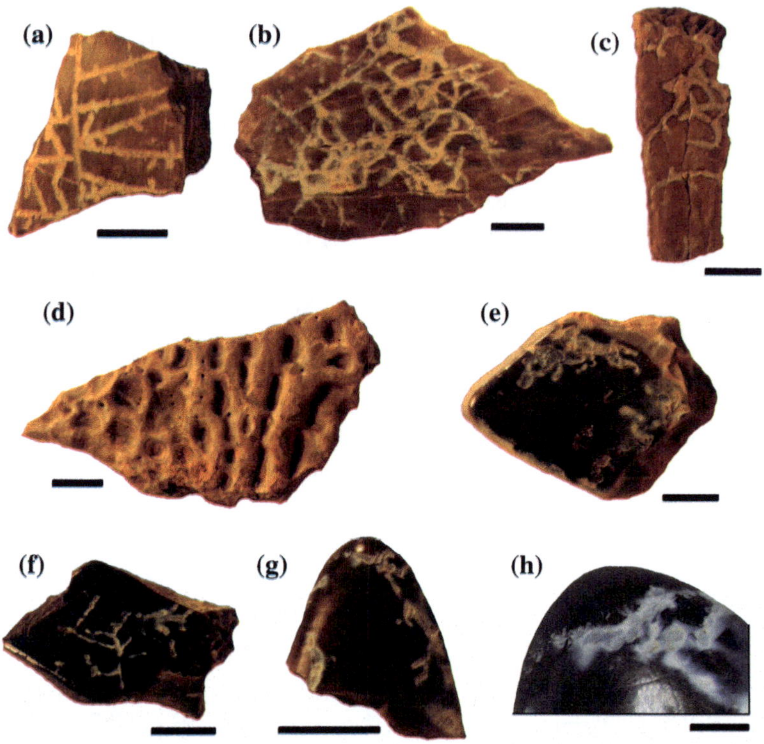

Fig. 8.3 Recent root traces in vertebrate remains from the San Jose Formation (Eocene of New Mexico, USA). **a** Turtle shell fragment. **b, c**: Indeterminate bone fragments. **d** Crocodilian scute. **e**, **f** Lepisosteiform fish scales. **g**, **h** Crocodilian teeth. Scales: 5 mm (except h: 1 mm)

Small irregular pits of etching can occur within or close to the trace (Fig. 8.4a). The width of the traces ranges from 0.275 to 0.525 mm.

Morphotype 2 (Damage of crocodilian osteoderms) As in Morphotype 1, damage of crocodilian osteoderms has a very well defined shape, with the rare occurrence of haloes (Fig. 8.3d). However, these furrows are more curved (bypassing the natural ornamentation of the scute) and predominate in the higher relief parts of the scute (the occurrence of damage inside the ornamentation pits is rare). Although several small, anatomical foramina are present in the external surface of the scute, the roots do not enter into these natural spaces. No overlapping of traces was observed. This morphotype of traces also does not form walls, and the floor of the trace is rougher than the bone surface. Similar to the Morphotype 1 traces, branching occurs, as well as associated erosive pits. The width range is 0.231–0.611 mm.

Morphotype 3 (Damage to ganoid scales) Traces made in the dorsal (external) face of the ganoid scales are very irregular. In plan view, some of the traces can be similar to the Morphotype 1 by being composed of a main axis that may split into

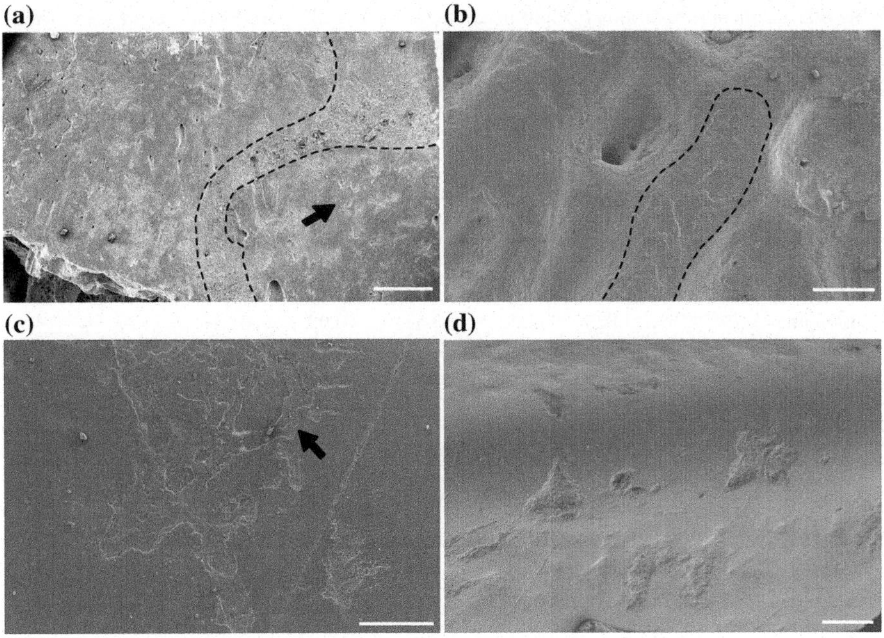

Fig. 8.4 Morphotypes of root traces under the SEM. **a** Morphotype 1 on a turtle shell fragment. The dashed lines show the outline of the root trace. The black arrow points to erosive pits. **b** Morphotype 2 on a crocodilian scute fragment. The dashed line shows the outline of the root trace. **c** Morphotype 3 on a ganoid scale. The black arrow points to the ganoin (enamel) desquamation. **d** Morphotype 4 on a crocodilian tooth. The smooth area in the center of the image is the tooth carina. Scales: a–b: 1 mm; c–d: 500 μm

branches (Fig. 8.3f). These splits occur frequently at angles smaller than 90°, but higher angles can also be found. Traces from other materials are more irregular, and branches could not be recognized (Fig. 8.3e). In both types of preservation, it is possible to identify etching holes (larger than the erosive pits of morphotypes 1 and 2), which are frequently associated with discoloration of the ganoine (enamel). The discoloration of the ganoine is not subtle but forms whitish, yellowish or bluish halos. In addition, these holes deepen by the successive desquamation of the ganoine layers, giving a "stair step" shape to the trace when viewed in cross-section (Fig. 8.4c). The walls of the traces are irregular and scalloped, but the bottom of each ganoine layer is almost smooth and straight. The width of the Morphotype 3 traces ranges from 0.164 to 0.450 mm. Traces made in the ventral (internal) face of the scales are similar to Morphotype 1.

Morphotype 4 (Damage to teeth) These traces are composed of irregular-shaped mottles of discoloration associated with erosive holes (Fig. 8.3g). These holes are deeper than those pits produced in bones and scales and have both the walls and floors scalloped. The shape of these holes is completely irregular, encompassing

Table 8.2 Element proportions (weight percentage) recovered by EDS analyzes in both original surfaces and root-etched surfaces of fossils with root traces of each morphotype

Element	Morphotype 1		Morphotype 2		Morphotype 3		Morphotype 4	
	Etched surface	Original surface	Etched surface	Original surface	Etched surface	Original surface	Etched surface	Original surface
Oxygen	49.36	45.21	43.16	50.76	40.62	41.44	35.84	38.88
Calcium	20.90	23.32	32.91	16.62	32.99	36.57	41.85	41.37
Silicon	9.62	8.46	3.79	13.52	2.46	0.89	3.84	0.97
Phosphorus	7.32	7.45	11.02	5.35	13.31	14.92	8.94	15.35
Aluminum	4.14	3.88	2.00	5.52	0.98	0.21	2.77	0.16
Iron	3.22	6.66	–	3.19	–	–	–	–
Fluorine	2.18	1.57	3.41	0.86	0.15	1.41	–	0.32
Carbon	1.36	1.49	2.32	1.38	6.54	3.13	3.78	2.06
Sodium	0.66	0.65	0.75	0.76	0.18	1.15	0.92	0.59
Magnesium	0.65	0.77	0.39	0.92	–	–	0.57	–
Potassium	0.59	0.56	0.24	1.11	0.73	–	0.90	–
Chlorine	–	–	–	–	1.03	0.28	0.60	0.30

cross-sections of nearly square shapes to V shapes (Fig. 8.3h). Unlike Morphotype 3, the erosional traces do not form layers. The color of these traces is very diffuse and less opaque than the other morphotypes (Figs. 8.3g, h and 8.4d). The borders are irregular and poorly defined, mainly because of the presence of an extensive erosive halo.

Regarding the EDS analyses, Oxygen, Calcium, Silicon and Phosphorus are the main elements recovered from all the examined fossilized vertebrate remains (Table 8.2). Other elements (Al, C, Cl, F, Fe, K, Mg, and Na) were also found, but in minor proportions. Analyses made in both the original surface and the root etched surface indicate that the chemical activity of the roots does not deplete any specific element, at least not influencing the proportions of elements. Therefore, we did not found any evidence that correlates the discoloration of the fossil remains to a decreased proportion of a specific chemical element.

8.4 Discussion and Conclusions

Based on the above-described data and on the references in the bibliography, it is possible to identify two main criteria for the recognition of recent root damage on fossils. Because the traces described here were found in association with recent roots (Fig. 8.2), they may serve as proxies for cases in which the traces were found dissociated from their context (trace makers).

The first criterion that should be used is the extensiveness of the trace occurrence. Both the biochemical activity of plant roots and the oscillation of the soil pH modify (fossil and non-fossilized) buried vertebrate remains. Notwithstanding, it is expected that root activity may be less extensive, acting only in the etching or dissolution of clasts located on or near the rhizosphere. Therefore, traces produced by roots should occur more strictly, only on part of a fossil assemblage, in opposition to the expected more widely dispersed marks derived from soil pH oscillations. In the sampled vertebrate remains from the San Jose Formation, only 20.19% have evidence of recent root activities. However, this abundance of root traces may vary according to the quantity of vegetation growing on the fossil bearing-unit, which is related to several other climatic, geographic, biogeographic and ecological factors.

According to Bader et al. (2009), modern root traces produce "haloes of chemical activity" in fossil remains, leaving discolored mottles or traces, mainly whitish, yellowish or grayish in color. These features were described by Retallack (1990) and Mikuláš (2001) as "drab-haloed traces." All the four morphotypes of traces described here present this type of preservation, despite the differences between them. Other authors also recognize the presence of mottles and discoloration derived from the activity of root traces on fossil remains (e.g., Mikuláš and Žítt 1999; Montalvo Montalvo 2002; Fernández-Jalvo and Andrews 2016). On the other hand, root traces produced on fresh, non-fossilized bones are often described as grooves with a U-shaped cross-section (Fisher 1995; Montalvo 2002; Bader et al. 2009; Fernández-Jalvo and Andrews 2016; Paes-Neto et al. 2018). These differences can be related to the fact that fossilized bones are in fact lithoclasts because several processes can modify the original bone and tooth bioapatite during diagenesis (e.g., Noto 2011), and constitute the second criterion for recognizing traces derived from the activity of recent roots on fossil remains.

Although it is a common sense observation that mottles, haloes, discoloration, and changes in the original colors of fossils are produced by geochemical variations during diagenesis, pedogenesis and/or biochemical activity, the data obtained by the EDS analyses from the San Jose traces do not allow us to speculate more about their formation, because the changes of the elemental proportions do not form a coherent pattern.

However, it is not obvious how to predict the real extent of root damage to fossils, because it will depend on: (i) the type of fossilization; (ii) the amount of time in which the fossils remained subjected to root activity; and (iii) various conditions that may accelerate or delay the root-damaging process, such as climate, precipitation, and oscillations of the water table, among others.

The root traces described here occur in all the types of vertebrate hard tissue (i.e., bones, scales, and teeth), independently of the taxa or region of the body to which pertains. However, their morphology changes according to the tissue organization of each substrate. Root traces in bones are very shallow, discolored grooves, while damages to scales and teeth are formed by deeper holes with mottles and more diffuse discoloration. Although we do not have enough data to completely understand this variation, it also occurs when we compare our traces with those described in the literature.

For example, Bader et al. (2009) recognized modern root etchings in bones of the hadrosaurian dinosaur *Maiasaura* from the Late Cretaceous of Montana. According to them, the chemical etching produced by roots has a smooth texture. In contrast, the traces produced on the bones from the San Jose Formation are rough and discolored, often associated with small erosive pits. Fernández-Jalvo and Andrews (2016) also stated that discoloration of fossil and fresh bones should occur accompanied by other types of modification. The differences in preservation between the traces mentioned are still understudied. Future controlled experiments should shed light on the ichnogeny of these traces (which means, how the traces evolved, from the early attachment of the roots through the complete destruction of the remains) or even how different plant exudates can affect the fossils.

Recent root traces on fossil bones and other remains have a particular importance for understanding their taphonomic history. Because these traces are produced on fossil bones rather than on fresh, "green" bones, they indicate that these fossils have been passed by diagenetic processes before being exhumed and modified. Theoretically, cycles of exhumation-modification by organisms-burial can occur many times before the collection of the fossils by paleontologists, and bioerosion traces are one of the main taphonomic signatures to recognize them.

For example, in marine-deltaic settings, Tapanila et al. (2004, 2008) and Dentzien-Dias et al. (2018) described the presence of *Gastrochaenolites* on the surface of vertebrate coprolites that were produced after the fossilization of the dung. Because coprophagous animals do not leave traces with similar clavate morphology, and boring bivalves are often found in association with the traces, *Gastrochaenolites* borings are typically related to these mollusks. Therefore, this ichnogenus is an important taphonomic signature for periods of exhumation-modification-burial, when the fossils are used as substrates by the boring fauna just like any other lithic clast. Root traces act analogously in terrestrial environments, attaching and etching the fossils that were exhumed independent of their age, taxon or skeletal topology. In arid and semi-arid environments, where physicochemical processes take place at a slower rate, the biochemical activity of roots can be the only evidence of post-fossilization weathering of the fossils. The traces described here (from northwestern New Mexico) are a very good example of that.

Since several cycles of exhumation-modification-burial can occur during the taphonomic history of certain fossils, dating this type of trace is difficult. The presence of roots attached to the modified surface is the only reliable evidence that points to a Recent age. In cases in which this does not occur, only an age younger than the fossil age can be properly considered for the traces (unless there is other evidence, such as the presence of fossil taxa that indicate a huge temporal mixing). For example, root traces on Turonian (Late Cretaceous) vertebrate remains described by Mikuláš and Žítt (1999) are considered post-Turonian in age.

Lastly, the description of recent root traces in fossil vertebrate remains from the San Jose Formation led us to extend this phenomenon to other fossils that bear similar traces. Among them, we can mention an isolated abelisaur theropod tooth (Specimen: UFRGS-PV-0038-K) from the Late Cretaceous of central Brazil (Sales et al. 2018, Fig. 4) and a left ramus of a *Tyrannosaurus rex* (Specimen: NMMNHS-P-3698) from the McRae Formation of New Mexico (Sullivan and Lucas 2015, Fig. 7b). Blumenschine et al. (2007, Fig. 1c) described very similar traces produced on a fossilized metapodial of a bovid found in the Olduvai Gorge but attributed them to microorganisms. However, the morphology and size are very similar to the San Jose traces, and the authors did not find evidence of microorganisms attached to the fossils, suggesting that the traces also could be produced by recent roots. Hospitaleche et al. (2011, Fig. 3) described traces of recent lichens on bones and teeth from the Miocene and Eocene of Argentina and Antarctica. Although they were produced by different organisms, their genesis and taphonomic implications are the same.

Acknowledgements We would like to thank Asher J. Lichtig and the "Friends of Paleontology" group for the fieldwork assistance. This study was possible due to the grants aided by CNPq (Process: 150623/2018-6) and CAPES (Process: PDSE-88881.133764/2016-01; Finance Code 001) to HF.

References

Bader KS, Hasiotis ST, Martin LD (2009) Application of forensic science techniques to trace fossils on dinosaur bones from a quarry in the Upper Jurassic Morrison Formation, Northeastern Wyoming. Palaios 24(3):140–158

Bais HP, Weir TL, Perry LG, Gilroy S, Vivanco JM (2006) The role of root exudates in rhizosphere interactions with plants and other organisms. Annu Rev Plant Biol 57:233–266

Behrensmeyer AK (1978) Taphonomic and ecologic information from bone weathering. Paleobiology 4(2):150–162

Blumenschine RJ, Prassack KA, Kreger CD, Pante MC (2007) Carnivore tooth-marks, microbial bioerosions, and the invalidation of Domínguez-Rodrigo and Barba's (2006) test of Oldowan hominin scavenging behavior. J Hum Evol 53:420–426

Britt BB, Scheetz RD, Dangerfield A (2008) A suite of dermestid beetle traces on dinosaur bone from the Upper Jurassic Morrison Formation, Wyoming, USA. Ichnos 15(2):59–71

Castetter EF (1956) The vegetation of New Mexico. New Mexico Quart 26(3):257–288

Dentzien-Dias P, Carrillo-Briceño JD, Francischini H, Sánchez R (2018) Paleoecological and taphonomical aspects of the Late Miocene vertebrate coprolites (Urumaco Formation) of Venezuela. Palaeog Palaeocl Palaeoecol 490:590–603

Fernández-Jalvo Y, Andrews P (2016) Atlas of taphonomic identifications. Springer, London, p 359

Fernández-Jalvo Y, Sánchez-Chillón B, Andrews P, Fernández-López S, Alcalá-Martínez L (2002) Morphological taphonomic transformations of fossil bones in continental environments, and repercussions on their chemical composition. Archaeometry 44(3):353–361

Fisher JW Jr (1995) Bone surface modifications in Zooarchaeology. J Archaeol Method Theor 2(1):7–68

Hospitaleche CA, Márquez G, Pérez LM, Rosato V, Cione AL (2011) Lichen bioerosion on fossil vertebrates from the Cenozoic of Patagonia and Antarctica. Ichnos 18(1):1–8

Keller WD, Frederickson AF (1952) Role of plants and colloidal acids in the mechanism of weathering. Am J Sci 250:594–608

Klappa CF (1980) Rhizoliths in terrestrial carbonates: classification, recognition, genesis and significance. Sedimentology 27:613–629.

Kondrashov PE, Lucas SG (2015) Paleocene vertebrate faunas of the San Juan Basin, New Mexico. New Mexico Mus Nat Hist Sci Bull 68:131–148

Lucas SG (2015) Eocene fossil vertebrates of New Mexico. New Mexico Mus Nat Hist Sci Bull 68:149–157

Lucas SG (2016) Two new, substrate controlled nonmarine ichnofacies. Ichnos 23(3–4):248–261

Mikuláš R (1999) Notes on the concept of plant trace fossils related to plant-generate sedimentary structures. Věstník Českého Geologického Ústavu 74(1):39–42
Mikuláš R (2001) Modern and fossil traces in terrestrial lithic substrates. Ichnos 8(3–4):177–184
Mikuláš R, Žítt J (1999) Fossil corrosive root traces on rock surfaces and bioclasts (Bohemian Cretaceous Basin, Czech Republic). Věstník Českého Geologického Ústavu 74(3):289–292
Montalvo CI (2002) Root traces in fossil bones from the Huayquerian (Late Miocene) faunal assemblage of Telén, La Pampa, Argentina. Acta Geol Hisp 37(1):37–42
Neumann AC (1966) Observations on coastal erosion in Bermuda and measurements of the boring rate of the sponge, *Cliona lampa*. Limnol Oceanogr 11(1):92–108
Noto CR (2011) Hierarchical control of terrestrial vertebrate taphonomy over space and time: discussion of mechanisms and implications for vertebrate paleobiology. In: Allison PA and Bottjer DJ (eds) Taphonomy. Process and bias through time. 2nd Edition. Springer, New York, pp. 287–336
Paes-Neto VD, Francischini H, Martinelli AG, Marinho TS, Ribeiro LCB, Soares MB, Schultz CL (2018) Bioerosion traces on titanosaurian sauropod bones from the Upper Cretaceous Marília Formation of Brazil. Alcheringa 42(8):415–426
Pirrone CA, Buatois LA, Bromley RG (2014) Ichnotaxobases for bioerosions trace fossils in bones. J Paleontol 88(1):195–203
Raven PH (2009) Biology of plants. 6th Edition. W. H. Freeman & Company, p 944
Retallack GJ (1990) Soils of the past: an introduction to paleopedology. Allen & Unwin, London, p 520
Sales MAF, Martinelli AG, Francischini H, Rubert RR, Marconato LP, Soares MB, Schultz CL (2018) New dinosaur remains and the tetrapod fauna from the Upper Cretaceous of Mato Grosso State, central Brazil. Hist Biol 30(5):661–676
Sarjeant WAS (1975) Plant trace fossils. In: Frey RW (ed) The study of trace fossils: a synthesis of principles, problems and procedures in ichnology. Springer-Verlag, New York, pp. 163–179
Sullivan RM, Lucas SG (2015) Cretaceous vertebrates of New Mexico. New Mexico Mus Nat Hist Sci Bull 68:105 129
Taiz L, Zeiger E (2010) Plant physiology, 5th edn. Sinauer Associates, Sunderland, p 782
Tapanila L, Roberts EM, Bouaré ML, Sissoko F, O'Leary MA (2004) Bivalve borings in phosphatic coprolites and bone, Cretaceous-Paleogene, Northeastern Mali. Palaios 19(6):565–573
Tapanila L, Roberts EM, Bouaré ML, Sissoko F, O'Leary MA (2008) Phosphate taphonomy of bone and coprolite conglomerates: a case study from the Eocene of Mali, NW Africa. Palaios 23(3):139–152
Tjelldén AKE, Kristiansen SM, Matthiesen H, Pedersen O (2016) Impact of roots and rhizomes on Wetland Archaeology: a review. Cons Manag Archaeol Sites 17(4):370–391

Chapter 9
Natural Shell Deposits from a Río de la Plata Estuarine Beach, Uruguay: Formation Processes and Archaeological Implications

Laura Beovide and Sergio Martínez

Abstract The study of the current deposits of mollusk shells represents a very important source of analogy to distinguish between the natural and anthropic formation of shell archaeological deposits. This work focuses on the formation dynamics of modern mollusk shell deposits on a Río de la Plata beach, and the main processes involved, both natural and anthropic. The deposits of Penino Beach are located on the left bank of the Rio de la Plata, near the mouth of the Santa Lucia River, in Uruguay. Topographic and stratigraphic aspects were taken into account for the study of the deposits, as well as the composition of the samples and the shell deposition dynamics. Two estuarine mollusks were clearly predominant (*Erodona mactroides* and *Heleobia* sp.). Differential composition could be distinguished in the attributes of the samples, related to different shell concentration formation events (storms, drag-by-sea movements, anthropic use of space, and burial of valves by natural events). As a consequence, we obtained reference elements that allow for the discrimination of shell accumulations that were formed by the same species, but in different ways, in the past.

Keywords Shell middens · Formation processes · Shells deposits · Zooarcheology

9.1 Introduction

All the events, activities, and processes that affect objects after their initial use in a particular type of activity are considered the formation processes of the archaeological record (Schiffer 1972, 1986; Binford 1980, 2001). Work on this topic has been developed in Uruguay since the 1990s, mainly in relation to the study of pre-hispanic

---

L. Beovide (✉)
Centro de Investigación Regional Arqueológica y Territorial (CIRAT), Dirección Para el Desarrollo de la Ciencia y el Conocimiento, MEC, San José, Uruguay
e-mail: beovide@mec.gub.uy

S. Martínez (✉)
Departamento de Paleontología, Facultad de Ciencias, Universidad de la República, Montevideo, Uruguay
e-mail: smart@fcien.edu.uy

S. Martínez et al. (eds.), *Actualistic Taphonomy in South America*,
Topics in Geobiology 48, https://doi.org/10.1007/978-3-030-20625-3_9

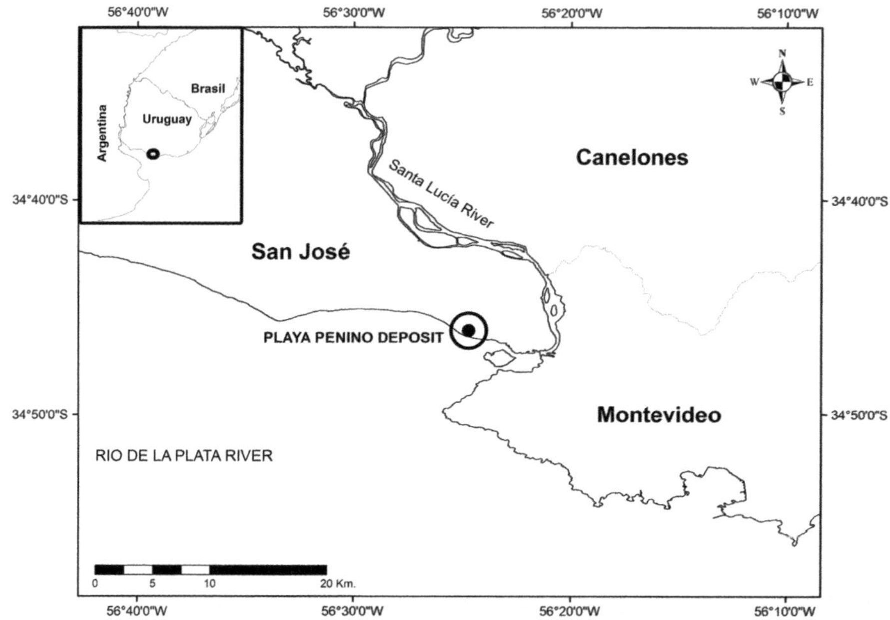

Fig. 9.1 Location of the shell deposits of Penino Beach

mound-building societies (Bracco and Ures 1999; Bracco et al. 2000; Castiñeira et al. 2001; López 2001; Iriarte 2006; Gianotti et al. 2009; López et al. 2009; Gianotti and Bonomo 2013; Suárez and Gianotti 2013; Inda et al. 2017, among others) or related to historical sites (Castiñeira 1995). There are a series of studies of the formation processes of the archaeofaunistic records and their taphonomic histories (Beovide 1995, 2007; Beovide and Malán 2006; Moreno 2006, 2017; Moreno and Figueiro 2016, among others). With respect to the analysis of the formation processes of archaeological shell middens, the first works were of shell middens from the Atlantic coast of Uruguay (López et al. 1997, 2009). Posteriorly, the study of old shell middens formed by the same species that are living today on the coast of the Río de la Plata (ca. 3000–2000 years BP), involved the development of different approaches in order to discriminate between natural and anthropic accumulations (Beovide 2011, 2014; Beovide and Martínez 2014; Beovide et al. 2014, 2015, 2017). As part of this research program, the formation process of the current shell deposits in Penino Beach was studied in order to compare it with the formation processes of the archaeological shell middens of the area. The shell deposits are located on the coast of the Río de la Plata, at the mouth of the Santa Lucia River (UTM coordinates x: 56.766379 and y: 34.71167732), Department of San José (Fig. 9.1). They are in the margin of an extensive floodplain with a tidal and storm-wave influence, bordered by mixohaline wetlands. The objective of this work was to identify the natural and anthropic processes (n-transformer and c-transformer, according to Schiffer 1972, 1986) that influenced the formation of the current mollusk shell deposits on Penino Beach.

9.2 Methodology

In order to record the current formation processes, the topographical and stratigraphic characteristics of the deposit were first recorded. As a second step, different natural and anthropic events were observed and recorded between 2008 and 2010, and mollusk samples of the material were taken.

For the mesoscale analysis, the morphology and several stratigraphic aspects of the shell deposits were taken into account. The parameters to be evaluated were chosen based on the work of Henderson et al. (2002), Favier and Borella (2007), and Beovide (2011). The topography, geometry, and geological sections, considering the color, texture, structure, and thickness of the horizon, were surveyed. Mollusk shell concentrations were classified, taking into account the aspects mentioned in the works of Kidwell et al. (1986) and Henderson et al. (2002), namely: (1) metric data (length, width, volume); (2) predominant mollusk type (bivalves or gastropods); (3) disposition of the valves in the horizontal and vertical views, considering articulation, position, predominant orientation (chaotic, bimodal, unimodal, nests, stacked); and (4) form of the shell concentration observed in the vertical and horizontal views: lens: elliptical, spherical, curved or irregular and floor.

At the deposit scale, 4.5 l samples (15 × 15 × 20 cm) were collected every 10 cm, in addition to accounting for any stratigraphic discontinuities. The samples were sieved and the sediments were separated into two operational categories: "small clasts" (less than 4 mm) and "large clasts" (greater than 4 mm). In turn, "large clasts" were subdivided into: clasts A: 4–10 mm; clasts B: 10–20 mm, and clasts C: >20 mm. The "small clasts", "large clasts", mollusk shells, and archaeological material (ceramic sherds, charcoal/shell, lithic material, and non-malacological archaeofaunal remains) were weighed separately. The mollusks were identified for each sample and their relative weight with respect to the total sample was estimated.

Based on the work of Claassen (1998), Zuschin et al. (2003), Kotzian and Simões (2006) and Farinati et al. (2006), the following variables were considered for the analysis: height (ordered in frequencies derived from the study of living populations in Penino Beach conducted by Beovide 2011), preservation (degree of weathering, fragmentation, and abrasion), position and orientation, degree of articulation, and left/right valve proportion.

During the diagenetic and weathering processes, the original microscopic organic and inorganic components of the archaeological materials (in this case, valves) are separated and destroyed by physical and chemical agents (Dirrigl 1995; Claassen 1998, among others). Dirrigl (1995) categorized the deterioration of the valves into six grades. In this work, these categories were adapted to the valves of *Erodona mactroides*, ranging from E1 grade that corresponds to shells that still preserve the soft parts, to E6 grade that corresponds to shells heavily fissured, porous, and very fragile (see details in Beovide and Lorenzo 2011; Beovide et al. 2014, 2015). Fragmentation of *Erodona mactroides* was classified according to the remaining percentage of the shell: 0–<50, 50–80 and >80%. To describe the forms of the fractures on the valves, the types F0–F13 described in Beovide et al. (2014: 10, Fig. 2) were considered. A fragmentation index was also calculated: NR/NMI, where NR is the number of remains assigned to a species and NMI is the minimum number of individuals estimated, counting the articulated shells and the maximum number of right or left valves. The data were analyzed using cluster analysis and NMDS (non-metric multidimensional scaling) with the correlation coefficient association index, using the program PAST, version 2.17 (Hammer et al. 2001).

9.3 Characteristics of the Deposits

Near the mouth of the Santa Lucia River, the water level of the Río de la Plata varies throughout the day. These variations imply that the water of the river moves away from the shell deposits to a distance of more than 100 m, and, alternatively, it is close to or in some cases covers it (Figs. 9.2 and 9.3a, b). This variation depends on tide or storm events, and produces a broad sandy plain with very shallow bodies of water in some sectors (Fig. 9.2c, d). In this environment, several concentrations of *Erodona mactroides* and *Heleobia* sp. develop, since these species currently live in this area between the surface and 3 cm deep. The Penino Beach shell deposits are altered by

Fig. 9.2 **a**, **b** The shell deposits of Penino Beach. **c** Concentration of shells on the coastal line. **d** Concentration of shells after a storm event. **e** Anthropic impact. **f** Animal trampling on shell deposits

Fig. 9.3 Shell deposits of Penino Beach. **a** View of the deposits and sandy plain. **b** View of the floodplain after a storm. **c** Sand and shells overturned after a storm event. **d** Shells trapped by vegetation. **e** Shells trapped in depressions

anthropic actions, such as trampling, fires, and garbage resulting from the activity of the fishermen who sporadically camp in this location (Fig. 9.2e, f).

The topographical characteristics of the deposit are shown in Fig. 9.4 and Table 9.1. The deposits overlay the so-called "beach sediments" of the Villa Soriano Formation, a depositional product of the Pleistocene-Holocene sea level changes (Spoturno and Oyhantçabal 2004).

The beach deposits of the study area are generally composed of "fine to very thick and gravelly sand, yellowish white, composed of feldspar and quartz, regular to good sorted" (Spoturno and Oyhantçabal 2004: 58). The deposits are located at a height of +1.5 m over the present sea level, with an average thickness of 1 m and an approximate length of 400 m, and show different depression products of sand extraction for use in local building. Moreover, the deposit is affected by waves (Figs. 9.2, 9.3 and 9.4).

Three surveys were conducted on the Penino Beach deposit (Fig. 9.2d). Figure 9.5 shows the stratigraphy.

Table 9.1 Penino Beach shell deposit features

Deposit	Name	Geographical coordinates UTM 21 South, projection WGS84		Level	Length (max.) m	Width (max.) m	Area (m^2)	Perimeter (m)	Orientation	Thickness (m)
		x	y							
13	Penino beach	−56.766379	−34.71167732	(+) 1.5	443	32	8217.43	911	NWSE	1

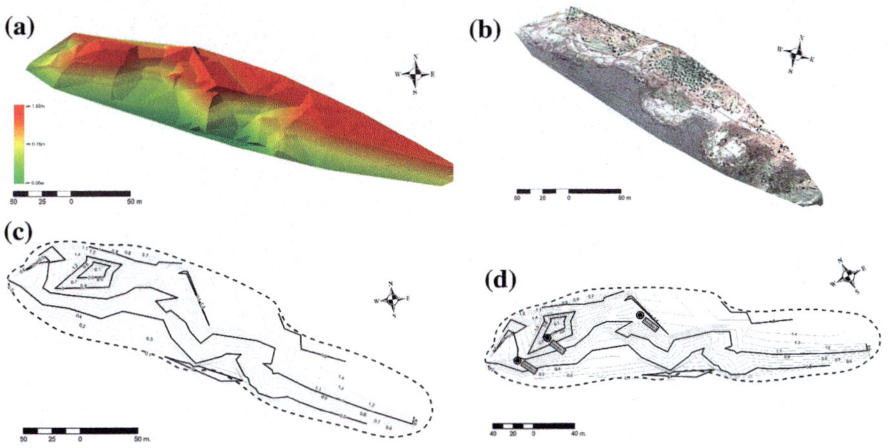

Fig. 9.4 **a–c** Topographic schemes of the shell deposits of Penino Beach. **d** Location of sampling holes 1–3

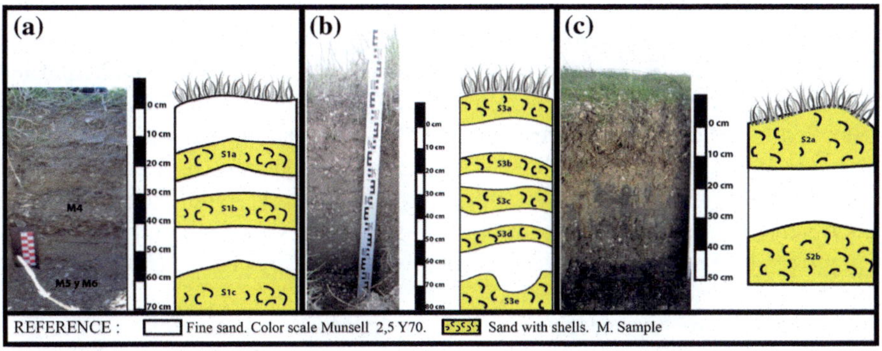

Fig. 9.5 Stratigraphy of the shell deposits. **a** Survey 1, **b** Survey 2, **c** Survey 3

9.4 Results

9.4.1 Analysis of the Samples

Survey 1 was conducted during 2009 and three concentrations of valves arranged in layers were identified; samples were taken from two of them, and labeled as M4, M5 and M6 (Fig. 9.5a). During survey 2, two concentrations of shells were identified (Fig. 9.5b). Survey 3 was conducted in the highest part of the deposit, and five concentrations of valves intercalated with fine sand layers were identified. Samples were also taken from a shell ridge that formed on the beach after storms (Fig. 9.4c). After these events, living mollusks accumulate in certain sectors, and articulated shells of mostly dead specimens accumulate on the coast line (see Figs. 9.4 and 9.6).

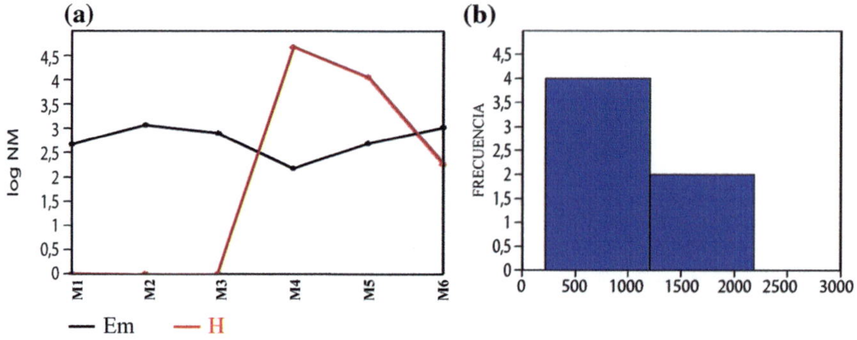

Fig. 9.6 **a** Minimum number of individuals (MNI) of *Erodona mactroides* (Em) and *Heleobia* sp. (H) per sample. **b** Frequency of remains in all samples

We collected samples after various events, as follows: shell concentration after a storm event (M1; Fig. 9.3c), accumulation of shells on the surface of the deposit (M2; Fig. 9.3d), and accumulation of shells on the surface of the deposit affected by the trampling of animals and humans (M3; Fig. 9.2e, f). The mollusk concentrations of the modern deposit were distributed in continuous layers, interspersed with sand and silt layers without mollusks (Fig. 9.5). The mollusks that predominated in the surveys of the Penino Beach deposits were the bivalves (*Erodona mactroides*). Gastropods (*Heleobia* sp.), when present in the samples, had an abundance of 30–1% with respect to the bivalves. All the shell concentrations had a "floor"-type geometry (sensu Kidwell et al. 1986). In the concentrations, disarticulated valves predominated (70%) over articulated valves (30%); their orientation was mostly chaotic (60%), over the convex up (35%) and convex down (5%) orientations.

As stated above, two species were clearly predominant: *Erodona mactroides*, identified in 100% of the samples, and *Heleobia* sp. The joint presence of *Erodona mactroides* and *Heleobia* sp. was recognized in three (50%) of the samples (M4–6). Table 9.2 summarizes the minimum number of individuals (MNI) and weight of the mollusk species in each of the six analyzed samples. It can be seen that the MNI of *Erodona mactroides* ranged from 152 to 1185 individuals. As shown in Fig. 9.6 and Table 9.2, the MNI of *Heleobia* sp. predominated in two of the edaphic samples (M4 and M5). The number of remains (NR) of *Erodona mactroides* was 300–1200 per sample (Fig. 9.6b). The samples that presented a lower NR (see Table 9.2) were those from the storm level (M1), together with the samples taken from the first levels of survey 1 (M4 and M5).

Regarding the relative weight among valves, small clasts and large clasts (Table 9.3), valves represented between 35 and 60%, small clasts between 39 and 63%, and large clasts between 0 and 8% (Fig. 9.7). The relationship among the concentrations of shells, small clasts, and large clasts was analyzed per sample on a logarithmic scale (Fig. 9.7a). Shells predominated in relation to the small and large clasts in the M1 (storm event), M2 (surface of the ridge), and M4 (first levels of sample 1) samples.

Table 9.2 Weight, minimum number of individuals (MNI), and number of remains (NR), in relation to the samples

Sample	Characteristics	*Erodona mactroides*	*Heleobia* sp.
M1	MNI	475	0
	NR	475	0
	Weight (g)	1.200	0
M2	MNI	1.185	0
	NR	2.185	0
	Weight (g)	1.455	0
M3	MNI	804	0
	NR	1.105	0
	Weight (g)	1.044	0
M4	MNI	152	47.502
	NR	217	47.502
	Weight (g)	478	522
M5	MNI	495	11.648
	NR	641	11.648
	Weight (g)	812	128
M6	MNI	1.067	182
	NR	1.953	182
	Weight (g)	887	2

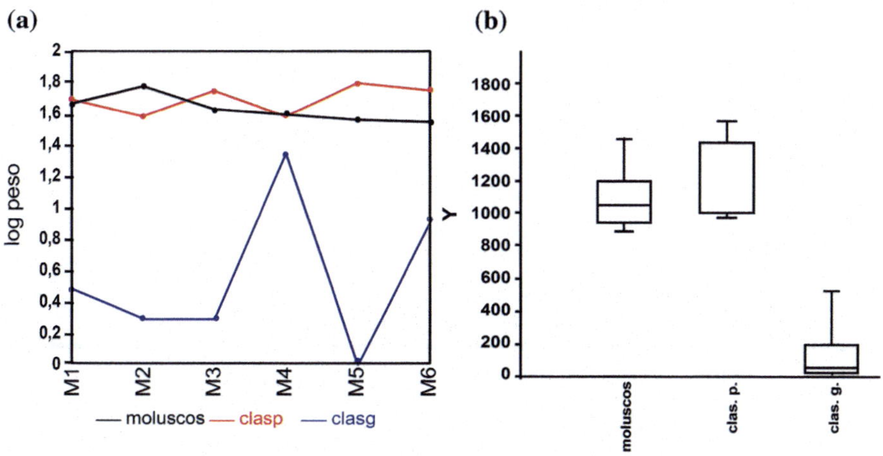

Fig. 9.7 **a** Weights of small clasts (clas. P.), large clasts (clas. G.), and mollusks per sample. **b** Mollusks, small clasts (clas. P.), and large clasts (clas. G.) in all samples

Table 9.3 Composition of the samples

	Species	Mollusk (g)	% mol-lusk	*Erodona mac-troides (g)*	% *Erodona mac-troides*	small clasts (g)	% small clasts	large clasts (g)	% large clasts	large clasts A (g)	% large clasts A	Large clasts B (g)	% large clasts B	large clasts C (g)	% large clasts C
	1	2	3	4	5	6	7	8	9	10	11	12	13	14	15
M1	1	1.200	48	1.200	100	1.250	50	50	2	0	0	50	100	0	0
M2	1	1.455	60	1.455	100	995	39	50	1	0	0	50	100	0	0
M3	1	1.044	43	1.044	100	1.431	56	25	1	0	0	0	0	25	100
M4	2	1.000	40	478	47.8	975	39	525	21	525	100	0	0	0	0
M5	2	940	37	812	32	1.560	63	0	0	0	0	0	0	0	0
M6	2	889	35.56	887	99	1.419	56.76	192	7.68	0	0	192	100	0	0

The presence of large clasts was an indicator of high energy. In fact, large B clasts (10–20 mm) prevailed in samples Ml and M2 (collected from the storm event and the deposit surface, respectively), and M6 (collected from the deepest level of survey 1) (Table 9.3). In sample M3 (collected from a shell concentration affected by trampling), large C clasts (greater than 20 mm in diameter) were found. In the sample M4 (sample of the first levels of survey 1), large A clasts were predominant (4–10 mm in diameter). The sample M5, which corresponds to the second level of mollusks in survey 1, did not include large clasts.

In Table 9.3, columns 1, 2, 4, 6, 8, 10, 12, 14 and 16 (log transformed + 1) were analyzed by means of a cluster analysis (Fig. 9.8) using the paired group (Manly 1994) and the correlation coefficient as the index of association. Three groups were defined: group 1, consisting of sample M4, which represents the most superficial level of survey 1; group 2, corresponding to the edaphic samples (M5 and M6) of survey 1 and those vinculated to the trampling event of the deposit surface (M3); group 3, corresponding to the shell sample products of a storm event (M1), and the sample of the ridge surface (M2).

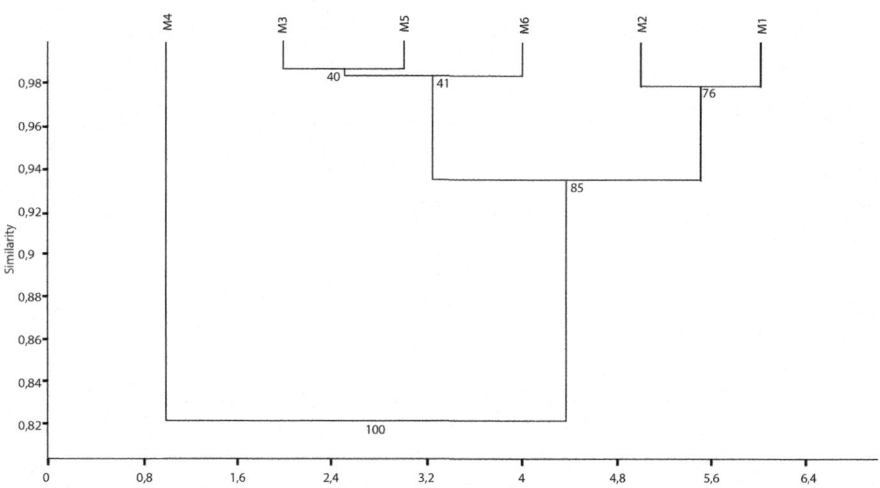

Fig. 9.8 Cluster analysis (correlation coefficient association index, paired group method)

9.4.2 *Analysis of the Attributes of the Set of Valves Per Sample*

9.4.2.1 Fragmentation

The least fragmented sample was Ml (Fig. 9.1), which corresponded to the mollusk strip deposited on the beach after a storm event. The most fragmented samples corresponded to the edaphic samples M4–M6, which were fragmented by 80% (Fig. 9.9a).

9.4.2.2 Disarticulation

In terms of numbers, right valves (n = 2851) predominated over left valves (n = 2795), but this difference was not significant (Mann–Whitney test: U = 16, z = −0.2402, $p = 0.8102 > 0.05$) (Fig. 9.9c). Articulated valves predominated in the storm sample M1.

9.4.2.3 Size

Considering all the samples from Penino Beach, small sizes predominated (T1 = 2697, T2 = 2351 and T3 = 1720) (Fig. 9.9c). However, these differences were not significant (Kruskal–Wallis test: Hc = 0.3629, $p = 0.8342 > 0.05$). In the storm event sample, large sizes (T1) predominated, as in the sample of the first 20 cm of survey 1 (M4).

9.4.2.4 Conservation Status

Globally, the E1 conservation type prevailed (3024 residues); this difference was significant ($p < 0.05$) among samples (Kruskal–Wallis test: Hc = 15.15, $p = 0.01865$; Fig. 9.9d). In the shells collected after a storm event (M1), as well as in the samples taken from the edaphic levels of survey 1 (M4 and M6), the E1 conservation type predominated. Regarding the samples of shells affected by trampling (M2 and M3), the first three conservation states (E1–E3, valves with incipient weathering) predominated. However, differences between the type of samples and conservation status were not significant ($p > 0.05$; Kruskal–Wallis test: p: Hc = 4.124, $p = 0.5947$; Fig. 9.9d).

9.4.2.5 Fractures

The predominant fracture type in sample M1 (product of a storm event) was the F3 form (valves fractured longitudinally). In the samples of the ridge surface (M2 and

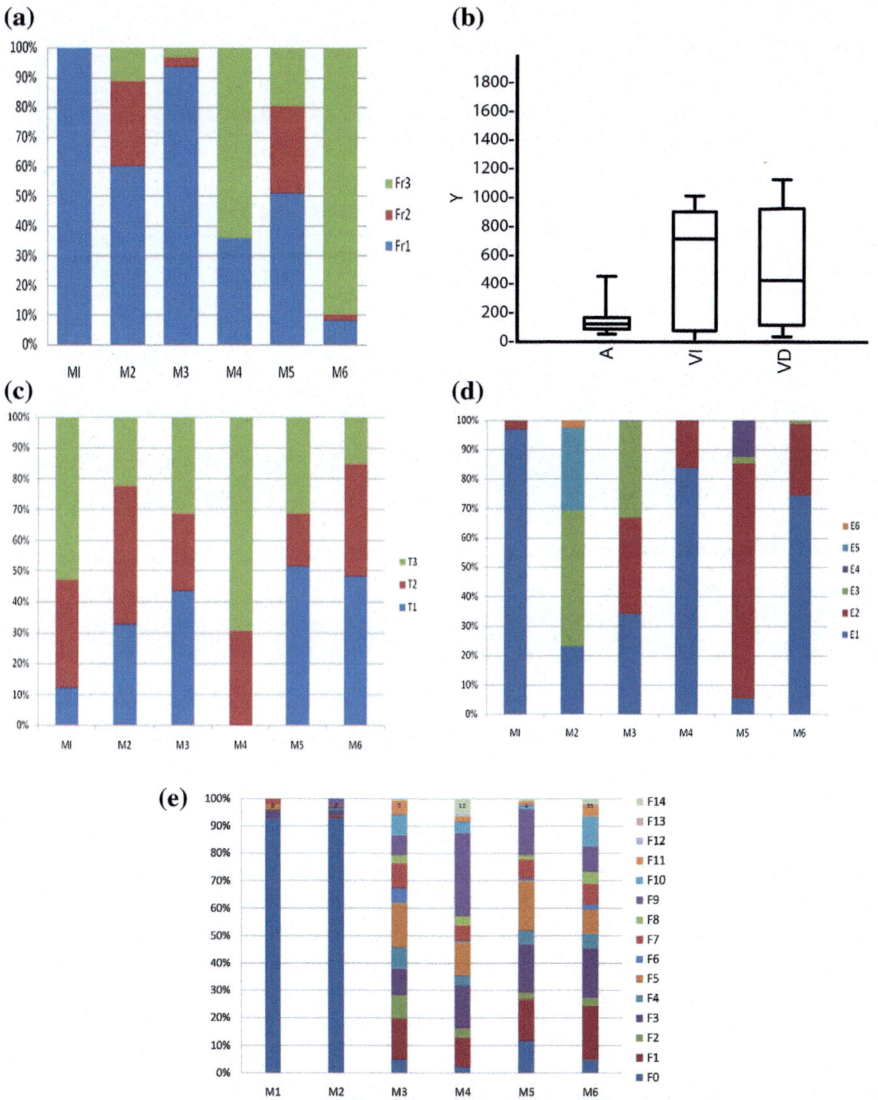

Fig. 9.9 Characteristics of the valves, by sample. **a** Fragmentation frequency of the valves per sample. **b** Frequencies of articulated valves (A), and disarticulated left (VI), and right (VD) valves. **c**. Sizes by sample. **d** Conservation stages by sample. **e** Frequency of shell type fractures per sample

Fig. 9.10 Formation processes of the shell deposits of Penino Beach. 1. Substrate of living *Erodona mactroides*. 2. Deposit of shells on a sandy ridge. 3. Valves retained by the vegetation. 4. Valves re-transported by the tide. 5. Trampling of animals and humans, and the presence of stoves. 6. Santa Lucia River

M3), the F3 and F9 (edge with V fracture) forms predominated, and in the edaphic samples (samples from survey 1, M4–M6), the F3, F5 (serrated edge), and F9 forms predominated (Fig. 9.9e). These differences were significant (Kruskal–Wallis test: $Hc = 35.9, p = 0.001372 < 0.05$). In the storm samples (M1) and shell ridge surface (M2), there was a high percentage of valves without modifications. This did not happen in the sample affected by trampling (M3), which had a great diversity of fractures. In general, form 13 (F13, valves without apex and anterior and posterior regions) was not identified in the samples from Penino Beach.

9.5 Discussion

The exploration of the formation and taphonomic processes, both regionally (Aliotta and Farinati 1990; Martínez 1990; Martínez et al. 2006; Farinati et al. 2008; Aliotta et al. 2013; Martínez and Rojas 2013, among others) and in particular environments (Beovide 2011), are complementary ways to understand the formation processes of the archaeological record (Borrero 1988, 2000, 2001). The particular exploration of the case of Penino Beach is a way to identify distinctive elements between the anthropic shell middens and the current, naturally-formed accumulations by the same species in the Río de la Plata estuary (Beovide 2011; Beovide et al. 2014, 2015).

From this analysis, it is clear that the Penino Beach deposits present most of the attributes proposed by Henderson et al. (2002), corresponding to a string of shells that fundamentally formed as a natural deposit, which distinguishes it from the archaeological shell middens (Beovide 2011; Beovide et al. 2014, 2015); these are mainly the intercalation of layers of shells and sediments, sharp contacts, and floor type as the predominant geometry.

In the studied case, the process can be summarized in the following scheme (Fig. 9.10): shelly deposits are formed by re-deposition in the sandy coastal ridges of the *Erodona mactroides* shells that live buried in the adjacent floodplain. During storm events, the valves are washed away by waves and re-deposited to different points of the ridge. Some valves are articulated and others are not. The valves are retained by the grass or by erosion hollows located at different points of the ridge. When no storm events occur, periodic changes (in the estuary water level) deposit the shells, forming a ridge. In some places, lenses of shells were identified as they were trapped in depressions. The valves, after being deposited in the ridge, suffer the action of different chemical and physical processes linked to edaphization. On these ridges, present human activities are developed over the valve concentrations.

9.6 Conclusions

The current deposits of shells in Penino Beach are mainly composed of two species: *Erodona mactroides* and *Heleobia* sp. Both live in the sandy sediments of the coastal plain adjacent to the studied deposit. Depending on storm or tidal events, the live mollusks, their shells, and sediments are dragged and deposited, forming the current studied shell ground. The deposits are presented in alternating sediment and mollusk shell layers, coinciding with previously recognized attributes of these deposits. The study of different samples of the sets of shells, obeying different natural processes (fluvial and aeolic sedimentation, storm events, and edaphization) and the use of space (trampling) by animals and humans, complements previous observations of the formation processes and makes it possible to acquire comparative elements for studies of the formation processes of the archaeological shell middens in the area.

Acknowledgements To Marco Lorenzo, for his collaboration in sampling and Javier Lemos, for helping with the topography of the deposit and graphics.

References

Aliotta S, Farinati E (1990) Stratigraphy of Holocene sand-shell ridges in the Bahia Blanca Estuary, Argentina. Mar Geol 94:353–360

Aliotta S, Ginsberg S, Spagnuolo J, Farinati E, Giagante D, Vecchi L (2013) Late Quaternary in a South Atlantic estuarine system: stratigraphic and paleontological indicators of coastal evolution. J South Am Earth Sci 45:194–208

Beovide L (1995) Análisis de los restos arqueofaunísticos de la "Casa del Diablo", San Miguel, Rocha. In: Consens M, López J, Curbelo C (eds) Arqueología en el Uruguay. Surcos, Montevideo, pp 54–65

Beovide L (2007) Animals, bones and Indians: patterns of butchering process in the indigenous economy from 16th to 18th centuries A.D., in the "Banda Oriental". In: Gutiérrez M, Miotti L, Barrientos G, Mengoni G, Salemme M (eds) Taphonomy and Zooarchaeology in Argentina. British Archaeol Rep, Oxford, pp 150–180

Beovide L (2011) Arqueozoología de los depósitos conchilíferos de la cuenca inferior del río Santa Lucía, Uruguay. Ph.D. Thesis, Universidad de la República, Montevideo, Uruguay

Beovide L (2014) Shell middens and the use of molluscs in the Late Middle Holocene in the Rio De La Plata: an ethnoarchaeological contribution. In: Szabó K, Dupont K, Dimitrijevi V, Gómez L, Serrand N (eds) Archaeomalacology: shells in the archaeological record. BAR International Series, Oxford, pp 111–121

Beovide L, Lorenzo M (2011) Huellas Antrópicas y Naturales en el material arqueomalacológico Platense. In: Vega-Centeno R, Olivera P, Petrick S (eds) II Congreso Latinoamericano de Arqueometría. Lima, pp 307–321

Beovide L, Malán M (2006) Procesos Posdepositacionales y dispersión espacial de los materiales arqueológicos del sitio Colonización, San José, Uruguay. In: Austral A, Tamagnini M (eds) Problemáticas de la Arqueología Contemporánea. Río Cuarto, pp 283–295

Beovide L, Martínez S (2014) Concheros Arqueológicos en la Costa Uruguaya: Revisión y Perspectivas. Rev Chil Antropol 29:26–32

Beovide L, Martínez S, Norbis W (2014) Etnobiología de *Erodona mactroides* (Mollusca, Bivalvia): análisis espacial y tafonómico de concheros actuales. Etnobiologia 12(2):5–20

Beovide L, Martínez S, Norbis W (2015) Discriminación entre acumulaciones de moluscos naturales, antrópicas modernas y arqueológicas, constituidas por las mismas especies. Cuad Inst Nac Antropol Pens Latinoam Ed Esp 4:56–70

Beovide L, Martínez S, Norbis W (2017) Space use patterns and resource exploitation of shell middens from the Rio de La Plata coast (ca. 6,000 to 2,000 Year BP), Uruguay. In: Mondini M, Muñoz S, Fernandez P (eds) Zooarchaeology in the Neotropics environmental diversity and human-animal interactions. Springer, Cham, pp 81–103

Binford L (1980) Willow smoke and dog's tails: hunter-gatherer settlement systems and archaeological site formation. Am Antiq 45:4–20

Binford L (2001) Constructing frames of reference: an analytical method for archaeological theory building using ethnographic and environmental data sets. University of California Press, Oakland

Borrero L (1988) Tafonomía regional. In: Ratto N, Haber H (eds) De procesos, contextos y otros huesos. ICA, Universidad de Buenos Aires, pp 9–15

Borrero L (2000) Ten years after: esquema para una Tafonomía regional de la Patagonia Meridional y norte de Tierra del Fuego. In: Desde el país de los gigantes. Perspectivas arqueológicas en Patagonia. Universidad Nacional de la Patagonia Austral, Rio Gallegos, pp 183–193

Borrero L (2001) Regional taphonomy: the scales of application to the archaeological record. In Buitenhuis H, Prummel W (eds) Animals and man in the past. Essays in honour of Dr. A. T. Clason. ARC-Pub 41, Groningen, pp 17–20

Bracco R, Ures C (1999) Ritmos y dinámicas constructivas de las estructuras monticulares. Sector sur de la cuenca de la Laguna Merin-Uruguay. In: López J, Sans, M (rds) Arqueología y Bioantropología en las Tierras Bajas. FHCE-UdelaR, Montevideo, pp 13–34

Bracco R, Montaña J, Nadal O, Gancio F (2000) Técnicas de construcción y estructuras monticulares. In: Durán A, Bracco R (eds) Simposio internacional de Arqueología de las Tierras Bajas. MEC, Montevideo, pp 285–299

Castiñeira C (1995) Batería Santa Ana: Una aproximación a los procesos de formación. Procesamiento de los datos: mapeo y animación. In: Consens M, López J, Curbelo M (eds) Arqueología en el Uruguay. Surcos, Montevideo, pp 287–293

Castiñeira C, Fernandez G, Céspedes C (2001) Procesos de formación del sitio cráneo marcado en el litoral de la laguna de Castillos (Departamento de Rocha, Uruguay). Una aproximación interdisciplinaria para su reconstrucción ambiental. In: Arqueología Uruguaya hacia el fin del Milenio, Fundación Fontaina Minelli. Asociación Uruguaya de Arqueología, Montevideo, pp 101–115

Claassen C (1998) Shells. Cambridge University Press, Cambridge

Dirrigl F (1995) Zoarchaeological and Taphonomic analysis of the gold krest site. In: Dlutowski E (ed) Proceedings of the annual ESAF Meeting Bulletin Eastern States Archeological Federation, vol 54, 11 pp

Farinati E, Spagnuolo J, Aliotta S (2006) Bioerosión en micromoluscos holocenos del estuario de Bahía Blanca, Argentina. Ameghiniana 43:45–54

Farinati E, Spagnuolo J, Aliotta S (2008) Tafonomía de bivalvos holocenos en la costa del estuario de Bahía Blanca, Argentina. Geobios 41:61–67

Favier C, Borella F (2007) Consideraciones acerca de los procesos de formación de concheros de la costa norte del Golfo San Matías, Río Negro. Cazadores Recolectores del Cono Sur 2:151–165

Gianotti C, Bonomo M (2013) De montículos a paisajes: procesos de transformación y construcción de paisajes en el sur de la cuenca del Plata. Comechingonia 17(2). http://www.scielo.org.ar/scielo.php?script=sci_arttext&pid=S1851-0272013000200006&lng=es&nrm=iso

Gianotti C, Criado F, Piñeiro G, Gazzán N, Capdepont I, Seoane Y, Cancela C (2009) Dinámica constructiva y formación de un asentamiento monumental en el Valle de Caraguatá, Tacuarembó. In: Excavaciones en el exterior 2008. Informes y Trabajos, IPCE, MCU, Madrid, pp 245–254

Hammer Ø, Harper D, Ryan P (2001) PAST: paleontological statistics software package for education and data analysis. Palaeontol Electron 4(1):9 pp. http://palaeo-electronica.org/2001_1/past/issue1_01.htm

Henderson W, Anderson L, Mcgimsey C (2002) Distinguishing natural and archaeological deposits: stratigraphy, taxonomy, and taphonomy of Holocene shell-rich accumulations from the Lousiana Chenier Plain. Palaios 17:192–205

Inda H, Del Puerto L, Capdepont I, Bracco R (2017) Formation processes of coastal archaeological sites: a changing prehistoric scenario on the Atlantic shore of Uruguay. Geoarchaeology 32:633–645

Iriarte J (2006) Landscape transformation, mounded villages and adopted cultigens: the rise of early Formative communities in south-eastern Uruguay. World Archaeol 38:644–663

Kidwell S, Fürsich F, Aigner T (1986) Conceptual framework for the analysis and classification of fossil concentrations. Palaios 1:228–238

Kotzian CB, Simões MG (2006) Taphonomic signatures of the recent freshwater mollusks, Touro Passo Stream, RS, Brazil. Rev Bras Paleontol 9:243–260

López J (2001) Las Estructuras tumulares (Cerritos) del litoral Atlántico uruguayo. Lat Am Antiq 12:231–255

López J, Piñeiro G, Castiñeira C, Gascue A (1997) Ocupaciones en el litoral Atlántico de Uruguay: Aproximación Paleoambiental al conocimiento de los sitios costeros: Sitio "La Esmeralda". Jorn Antropol Cuenca Del Plata 3:28–34

López J, Villarmarzo E, Brum L (2009) Análisis de las plantas arqueológicas del sitio La Esmeralda, (Rocha, Uruguay). In: Beovide L, Erchini C, Figueiro G (eds) La arqueología como profesión: los primeros 30 años. XI Congreso Nacional de Arqueología Uruguaya, pp 218–229

Manly B (1994) Multivariate statistical methods. Chapman & Hall, London

Martínez S (1990) Taphonomy and Paleoecology of Holocene mollusks from the western margin of the Merin Lagoon (Villa Soriano Fm., Uruguay). Quat S Am Antarct Pen 7:121–135

Martínez S, Rojas A (2013) Relative sea level during the Holocene in Uruguay. Palaeogeogr Palaeoclimatol Palaeoecol 374:123–131

Martínez S, Rojas A, Ubilla M, Verde M, Perea D, Piñeiro G (2006) Molluscan assemblages from the marine Holocene of Uruguay: composition, geochronology, and paleoenvironmental signals. Ameghiniana 43:385–397

Moreno F (2006) Arqueotafonomía costera: la conservación de restos óseos en yacimientos del litoral Atlántico Uruguayo. Rev Atlántica-Mediterr Prehist Arqueol Soc 8:71–85

Moreno F (2017) Modificaciones naturales y antrópicas en el conjunto zooarqueológico del sitio CH2D01, Excavación IA (Sudeste uruguayo): aportes a la discusión de los procesos de formación. Cad LEPAARQ 14:458–479

Moreno F, Figueiro G (2016) Metodologías de excavación y recuperación diferenciales en el sitio Ch2D01 IA (Rocha, Uruguay) y sus efectos en el registro arqueofaunístico. An Arqueol 3:5–48

Schiffer M (1972) Contexto Arqueológico y Contexto Sistémico. Am Antiq 2:156–165

Schiffer M (1986) El lugar de la Arqueología Conductual en la Teoría Arqueológica. In: Segundas Jornadas de Arqueología y Ciencia, Museo Nacional de Historia Natural, Santiago de Chile, pp 195–218

Spoturno J, Oyhantçabal P (2004) Mapas geológicos y de recursos minerales del Departamento de Montevideo a escala 1/50.000 y de Canelones y San José a escala 1/100.000. Facultad de Ciencias (UDELAR)-DINAMIGE, Uruguay

Suárez X, Gianotti C (2013) Earthen mound formation in the Uruguayan lowlands (South America): micromorphological analyses of the Pago Lindo archaeological complex. J Archaeol Sci 40:1093–1107

Zuschin M, Stachowitsch M, Stanton R (2003) Patterns and processes of shell fragmentation in modern and ancient marine environments. Earth-Sci Rev 63:33–82

Chapter 10
Actualistic Taphonomy in the Northeast of the Santa Cruz Plateau (Argentina). Advances and Archaeological Implications

Laura Marchionni, Darío O. Hermo, Bruno H. Mosquera, Lucía A. Magnín, Laura L. Miotti and Eloisa García Añino

Abstract This paper presents the first advances made in the framework of the actualistic taphonomy project that we recently started in the northeast of the Central Plateau of Santa Cruz, Patagonia Argentina. The main goal of this project is to strengthen, through this line of study, the archaeological interpretations of the área. Here we present the results corresponding to the field works developed between 2016 and 2017 in three archaeological localities, with the objective to assess the potential of the different characteristic environments of the area to accumulate, disperse, preserve, destroy and bury bone remains. The survey was carried out at a broad spatial scale and was based on systematic observations through transects in which taphonomic and environmental data were recorded. The results obtained offer the first approach to determine which are the environments with the greatest potential for the natural incorporation of bones into the archaeological record and the heterogeneity with which different taphonomic agents and processes operate in each one. This work provides new information for the Central Plateau region and reflects the development of a new line of research for the area, from which systematic data are provided that contribute not only to improve the interpretations of the archaeological assemblages from the region, but also favor the comparison with neighboring regions.

L. Marchionni (✉) · D. O. Hermo · B. H. Mosquera · L. A. Magnín · L. L. Miotti · E. G. Añino
CONICET/División Arqueología, Facultad de Ciencias Naturales y Museo, Universidad Nacional de La Plata, Paseo del Bosque s/n°, 1900 La Plata, Buenos Aires, Argentina
e-mail: lau_marchionni@yahoo.com.ar

D. O. Hermo
e-mail: dhermo@fcnym.unlp.edu.ar

B. H. Mosquera
e-mail: bruno_mosquera@hotmail.com

L. A. Magnín
e-mail: lumagnin@yahoo.com.ar

L. L. Miotti
e-mail: laura.miotti2@gmail.com

E. G. Añino
e-mail: eloisagargia89@yahoo.com.ar

S. Martínez et al. (eds.), *Actualistic Taphonomy in South America*,
Topics in Geobiology 48, https://doi.org/10.1007/978-3-030-20625-3_10

10.1 Introduction

The dynamics of the different sedimentary environments in a landscape play a fundamental role in the configuration of the formation processes of each archaeological site generating particular taphonomic histories (Butzer 1982; Waters and Kuehn 1996; Ravesloot and Waters 2004). Because the multiple processes that modify the regional and local faunal stages cannot be deduced simply from the observed results of the archaeological deposit itself, it is necessary to prepare designs of actualistic studies at different spatial scales (Pobiner and Braun 2005). The systematic study of assemblages of carcasses and skeletal remains in different disarticulation stage is a research line that allows us to identify the processes and agents that act in a specific environment, the type of modification that they produce and the intensity of the action (Abramova 1993; Brain 1981). Recently, driven by the need of this type of information for the northeastern of the Santa Cruz plateau (Argentina), we have begun the development of an actualistic taphonomy program for strengthening the archaeological interpretations of the area. From naturalistic observations at broad spatial scales (Marean 1995), we seek to understand how modern agents and processes generate different patterns of accumulation, dispersion and differential destruction of faunal remains. Taphonomic studies in large spaces, as the work we here proposed, have contributed enormously to archaeological interpretations from its beginnings in the late 1970s (Behrensmeyer 1978; Blumenschine 1989; Haynes 1980).

Although zooarchaeology has been has been strongly developed in the study area (Miotti 1998; Miotti and Salemme 2005; Miotti et al. 1999), taphonomy has had more discrete advances, which are systematized in recent years (Giardina et al. 2000; Marchionni 2013, 2016; Marchionni et al. 2012; Miotti and Marchionni 2011). In this context of research, studies from an actualistic taphonomy perspective at different spatial scales become relevant, since the archaeological sites we study present great variability and are distributed in an area with environmental heterogeneity, which translates into distinctive depositional environments (Gómez and Magnin 2008).

In particular, some of the archaeological sites that we study showed early chronologies, in which the human use of both extinct fauna and living species was recorded (Cardich 1977; Cardich et al. 1973; Miotti 1998; Miotti and Salemme 1999, 2005; Miotti et al. 1999, 2018). The strengthening of the taphonomic studies for the evaluation of these contextual associations are of paramount importance to advance in the study of the process of settlement of the American continent.

The progress in this perspective is also important to compare the results obtained with those from Pampa and Patagonia, where this type of research has been systematically developed for several years (Borella 2004; Borella and Borrero 2010; Borella and Muñoz 2006; Borrero 1988, 1990, 2000, 2001, 2007; Borrero and Martin 1996; Borrero et al. 2005; Cruz 2003, 2009; Cruz and Muñoz 2010; Fernández 2008; Gutiérrez et al. 2016; Kaufmann 2009; Martin 2008; Massigoge et al. 2015; Muñoz 2009, among others).

In this context, the aim of this paper is to present the first results obtained from systematic taphonomic transects in four different environments from the Northeast of the Central Santa Cruz Plateau. The transects were designed with the objective of recognizing the natural accumulation and scattering patterns (bones rain) and determining the potential of preservation, destruction, and burial of the bone remains that characterize each environment. This information is key to differentiate natural from cultural deposits and for evaluating the potential of each environment incorporating vertebrate remains to the archaeological record by natural processes. We assume each of them as a source of microenvironmental variability for the expression of the processes we seek to evaluate, and therefore, such variability has implications for the formation of the regional archaeological record. An important piece of information recorded in these transects was the location of vertebrate carcasses or parts of them and their taxonomic, anatomical, and taphonomic description. From the analysis of the distribution and density of the different types of findings and attributes of the modern bone record, we seek to characterize the taphonomic agents and processes that operate in the different environments in study area and to evaluate the potential of each one of them for the fossil record accumulation and preservation (Blumenschine et al. 1994). In this way, we hope to contribute to generate a new actualistic taphonomic model for this region. It allows us to account the variability of bone assemblages, according to the different geo-ecological area processes (Cruz 2003). From these results, expect reaching a better understanding of the formation processes of particular deposits and the way in which they affect the zooarchaeological record.

10.2 Study Area

The area studied lies between latitudes 47° 20′ and 47° 35′ South, and longitudes 68° 25′ and 68° 50′ West. It corresponds to the northeast of the Central Plateau of Santa Cruz, also known as Deseado Massif, Argentinian Patagonia (Fig. 10.1). This massif is geologically and physiographically characterized as a positive, rigid and stable geological structure (De Giusto et al. 1980), where the fluvial action is the main landscape geomorphic process, but also the volcanic activity, the action of the wind and the processes of mass removal (Panza 2001). The sector studied is mainly characterized by volcanic formations and a poorly integrated fluvial systems. The Zanjón Del Pescado is the most conspicuous temporary watercourse of this sector. As the other ephemeral courses and lagoons in the area, it presents intermittent regime, carrying water at the beginning of the spring, when the snow melts (Panza 1982, 2001).

This plateau corresponds to an arid, rigorous and windy environment, with a cold temperate climate (Paruelo et al. 2005). The annual temperatures average varies between 8 and 10 °C. The temperature variation is related to the presence of the Andes, the moderating action of the Atlantic Ocean, and the high latitude (Jobbágy et al. 1995). One of the most characteristic climatic feature of the region is the incidence of strong winds from the west, they are more frequent in spring-summer and can exceed 100 km/h. The Andean range acts as an orographic barrier generating a regional gradient of precipitations that decreases from west to east as a function of

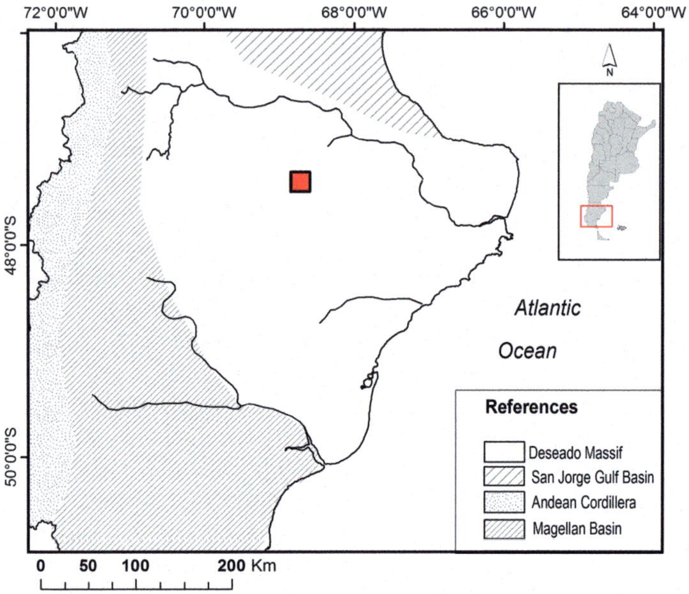

Fig. 10.1 Study area location in the Northeast of the Central Plateau (Santa Cruz, Argentina)

the wind direction (10.1). The average annual record of precipitations shows values of around 180–300 mm, with between 5 and 10 annual snowfalls with an average of 20 cm each.

From a biogeographical point of view, this area corresponds to the aridest sector of Patagonia and is located in the Neotropical Region, within the Dominion of the Patagonian steppe (Cabrera 1976). The characteristic vegetation belongs to the xerophilous type, represented by low, squat plants and shrubs that grow as plates adhered to the ground (Cabrera and Willink 1980). Among the animal species, *Lama guanicoe* (guanaco) stands out. It is a large wild ungulate that is widely distributed in South America (Raedecke 1978). It corresponds to the species most used by hunter-gatherer societies and, therefore, with the highest representation in the zooarchaeological contexts of the region. Other wild fauna includes several marsupials, as the Patagonian weasel (*Lestodelphis halli*), different species of bats (*Lasiurus* sp. and *Tadarida* sp.), the Patagonian hare or mara (*Dolichotis patagonum*), and a large rodent hilly vizcacha (*Lagidium* sp.). Among the common placental mammals, armadillos as the piche (*Zaedyus pichiy*) and rodents as the tuco-tuco (*Ctenomys*) are abundant along with other species that belong to the Cricetidae, and, less frequently, the Caviidae families. The carnivores are characterized by red fox (*Pseudalopex culpaeus*), gray fox (*Pseudalopex griseus*), puma (*Puma concolor*) which is the main predator in the region, the Patagonian ferret (*Lyncodon*) and the Patagonian hog-nosed skunk (*Conepatus humboldtii*). Among the birds, the most representative of the area is Darwin's rhea (*Rhea pennata*) or choique, after its Tehuelche name.

In addition to the autochthonous fauna that characterizes the region, the presence of domestic and wild introduced animals is also currently recorded within the study area. Among the first mentioned, and due to the livestock activity that characterizes economy of the region, there is a significative presence of sheep (*Ovis aries*), cows (*Bos Taurus*) and horses (*Equus caballus*); while one of the most frequent wild introduced species corresponds to the European hare (*Lepus capensis*).

10.2.1 Characterization of the Surveyed Sectors

La Huella (LH) sector corresponds to an open area (intermediate height pampa or plateau) located between 400 and 500 masl (Fig. 10.2), it is mainly characterized by a stony plain, constituted by outcrops of the La Matilde formation as well as the presence of ephemeral affluents and bodies of water (lagoons and swamps). La Matilde geological formation is composed mainly of primary or reworked tuffs, subordinate tuffites and, to a lesser extent, thin ignimbritic mantles, also of acidic composition (Panza 1982, 2001). Due to the erosive processes that act in this sector, raw materials of utility for knapping are distributed through screes, pediments and water courses. Recently, diverse archaeological evidence has been recorded including surface sites, artifacts concentrations and isolated finds (Hermo et al. 2017).

Los Toldos (LT) area includes the bottom, slope, and rim of a canyon and a top plateau area (over 800 masl) located to the North that presents a depression with a small lagoon (Fig. 10.2). The watercourse that flows through the canyon has a temporary regime and is tributary to the Zanjón Del Pescado. The canyon was formed in Chön Aike and La Matilde formations, composed by ignimbrites, tuffs and agglomerates, and its bottom is characterized by a substrate with a variable granulometry, mostly sandy loam with a contribution of gravel. In addition, the canyon is topographically sheltered from the winds, and numerous rock shelters and caves of different dimensions open along its slopes. These characteristics have been appreciated by hunter-gatherers in the past, as is shown by the consistent archaeological evidence known for this locality. In fact, Los Toldos archaeological locality is an important referent within the Patagonian archaeology since one of the most complete cultural sequences of the area was described here (Cardich 1977; Cardich et al. 1973). Researches carried out in these caves opened discussions about the Patagonian and American peopling at the end of pleistocenic times.

Finally, the Laguna Sierras Blancas sector (LSB) includes a temporary lagoon located 27 km to the NE of Los Toldos canyon. It is an endorheic system fed mainly by rain and snowfall, that has an extension of 2.43 km^2 (Fig. 10.2). This basin was formed on ignimbrites, tuffs and cinerits (Laguna Palacios and Chon Aike Formations) and the sediments deposited on its bottom include silts and claystones (Panza 2001). Surveys conducted by Miotti and collaborators in the ’90s resulted in the location of archaeological sites, which alerted about the potential of this locality.

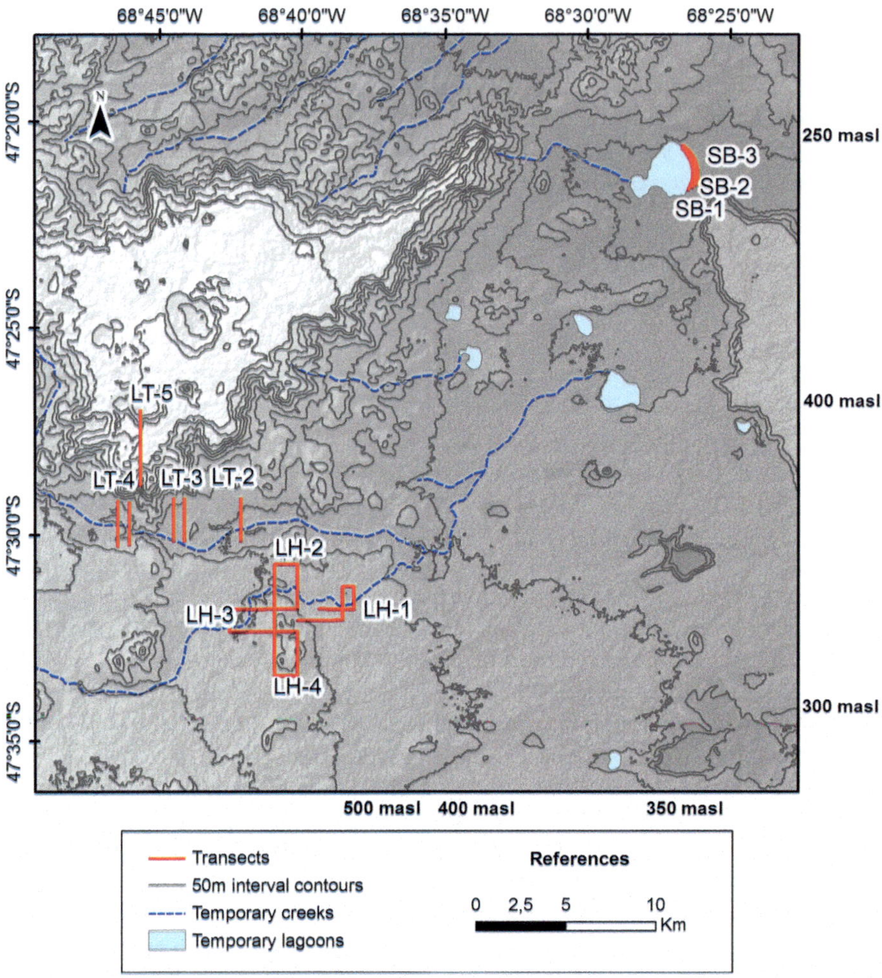

Fig. 10.2 Location of the survey transects on a digital elevation model (SRTM), the contour lines have equidistance of 50 m

10.3 Survey Design and Methodology

Four specific environments were surveyed during the fieldworks: the intermediate plateaus of LH, the canyon of LT, the high plateaus of LT and the edges lagoon of LSB (Fig. 10.3). The taphonomic and environmental information was registered based on surveys of systematic transects, following a naturalistic perspective (Marean 1995) and at a landscape scale (Borrero et al. 2005; Cruz 2009). The location of the transects were defined based on different proposals (Borrero et al. 1992; Cruz 2009; Foley 1981; Magnin 2010; Massigoge et al. 2015; Terranova 2013), considering the

Fig. 10.3 Examples of the different environments surveyed: **a** Intermediate Plateaus in La Huella; **b** Canyon area in Los Toldos; **c** High Plateus in Los Toldos; and **d** Edges of Lagoon in Laguna Sierras Blancas

particular characteristics of each sector to be surveyed, and seeking to cover large areas and according to the accessibility of each one.

As can be seen in Fig. 10.2 a system of 4 radial transects was used in LH (Fig. 10.2). In LT, six parallel transects crossed the canyon environment (Fig. 10.2, LT2, LT3 and LT4) and another transect was directed towards the high plateau (Fig. 10.2, LT5). Finally, in LSB, 3 subparallel transects were directed along the lagoon coastlines (Fig. 10.2).

Transects were walked by three observers, obtaining a visibility coverage of approximately 30 meters wide. Records were made each time a bone remain was found. Based on the proposal made by Massigoge et al. (2015), three categories of findings are used in this work: carcass, coded as (C): when more than 75% of the skeletal elements of the animal conjoined by soft tissue are present; occurrence of articulated bones (OAB): when two or more bone are joined by soft tissue and they represent less than 75% of the skeleton of the animal; and disarticulated bone (DB): when an individual bone element is found, spatially isolated. Each of these categories may consist of a scatter finding (SR) when there is a bone remain in less than 10 m of transect section or be found as part of concentrations (CO), when 2 or more bones are found in less than 10 m of transect. These concentrations can be also constituted by more than one type of finding (DB, OAB, C), and are assigned to combined concentrations (CO-COM), to differentiate them from those that present a unique type of finding. The categories of findings used here are also related to the different degrees of disarticulation that can occur in the carcasses, following Muñoz and Cruz

(2014), these can be: articulated (e.g. C), initial disarticulation (e.g. C and/or OAB), advanced disarticulation (e.g. OAB), or terminal/complete disarticulation (e.g. BD).

Both, bone remains and different features of taphonomic interest which we detailed below, were georeferenced by GPS. Each finding was analyzed and photographed in situ since the bone remains were not collected. All of them were left in their place of discovery available for longitudinal taphonomic studies, which imply a long-term series of return visit to the same site (Borella 2004; Borrero 2007; Haynes 1995; Otaola and Tripaldi 2016). The repetition of the observation, in the medium and long term, will allow us to achieve the general goals that we set ourselves here.

The detailed description of each bone finding was carried out in a systematic way by means of a sheet in which anatomical, taxonomic and taphonomic information was recorded. Taxonomic and anatomical abundance was estimated based on the NISP, %NISP, and MNI (Grayson 1984; Lyman 1994). Among natural modifications, we surveyed weathering (Behrensmeyer 1978; Borrero 2007; Massigoge et al. 2010); marks made by carnivores (Binford 1981; Borrero et al. 2005; Cruz and Muñoz 2010; Mondini 2002, 2003; Massigoge et al. 2014), rodent (Bocek 1986; Lyman 1994), root (Behrensmeyer 1978; Binford 1981); and trampling marks (Behrensmeyer 1978; Borrero 2007; Lyman 1994). Likewise, we took into account the presence of the different types of butchering marks (Binford 1981; Johnson 1985; Lyman 1994), and the relation between bone finding and substratum. The last variable was coded as buried (when more than 50% the bone surface is covered by sediment), half-buried (less than 50% is covered) and exposed (when there is no sediment covering it) (adapted from Behrensmeyer and Boaz 1980 in Massigoge et al. 2015). The survey also records environmental variables, such as the geomorphology and the depositional environments (slope, type of sediment, dominant process, burial potential, and the presence of water and vegetation). Finally, the vegetation was used as an indicator of archaeological visibility, following Massigoge et al. (2015) there are five possible categories: (1) excellent, when there is no plant coverage, (2) very good (between 1 and 25%), (3) good (26–50%), (4) regular (51–75%), and (5) poor (76–100%).

10.4 Results

A total area of 1.32 km^2 was covered by transects: 0.735 km^2 corresponds to LH, 0.405 km^2 to LT (canyon: 0.3 km^2; high plateau: 0.105 km^2), and 0.180 km^2 of land surface was transect covered at LSB.

10.4.1 Environmental Features

The intermediate plateaus sector of LH (Fig. 10.3a) registers slopes that vary between <3° and >14°. The more pronounced is in the central sector, from where transects are directed to the north, south, east and west. When moving away from that sector, the slopes are softened until they become low or null (<3°). The same happens with the type of substrate: while in the central sector is characterized by rocky outcrops, it turns to a sandier substrate in the margins of the area. Sectors lower and peripheral to the outcrops concentrate vegetation and therefore, and the visibility is reduced from very good to good-regular. The main process observed in these low areas is deposition, increasing the burial potential, constrasting with the registers in higher sectors, where erosion predominates. This tendency is more evident towards the west.

In LT canyon, significative variations in the slope of the terrain were observed. The bottom of the valley presents null to low slope values (< 3°) and its flanks, low and medium slopes (between 3° and 14°), reaching in some cases abrupt slope values (>14°) (Fig. 10.3b). The sediment is basically sandy, but it was also registered the presence of gravelly sand and silty sand. The visibility varies between regular and very good. The main bioturbation agents are shrub vegetation and, less frequently, tuco tuco (making caves) and other animals of greater size (such as horses, guanacos, and sheep) that caused trampling. Erosion is the dominant process of this environment, however, both deposition and geomorphological stability (evident where soil formation is present), were registered in punctual sectors. Due to this, although in general terms the burial potential is low, certain areas with a greater probability of materials burial were identified.

The second prospected environment in LT corresponds to a plain with sandy sediments where neither erosive nor deposition processes were identified (Fig. 10.3c). Rather it seems a stable environment, with <3° slopes, low burial potential, and regular archaeological visibility.

Finally, in LSB a series of subparallel lines of the eastern coast of the temporary lagoon were surveyed. This environment is characterized by slope values that range between null and low (<3°), the presence of not consolidated gravelly sand sediment, and a good to very good visibility. Deposition is the main process and the burial potential is very variable since it ranges widely from low to high. Currently, the lagoon is dry but in the center, there are sectors of humidity. The presence of horse tracks and the scarce vegetation cover (small shrubs) are the only observed agents capable to generate bioturbation in this area.

10.4.2 Information from Bone Materials

A total of 75 bone findings distributed in the three prospected localities were recorded (Table 10.1; Figs. 10.4, 10.5 and 10.6), from them, more than 82% are located in LT,

Table 10.1 Surface, absolute frequency, density and type of faunal findings by surveyed sector

		LH	LT Canyon	LT Plateau	LSB	Total
	Km^2	0.735	0.3	0.105	0.18	1.32
	Absolute frequency	5	50	12	8	75
	$Density/Km^2$	6.8	166.6	114.28	44.44	64.1
	C	1	1	1		3
SR	OAB		4		2	6
	DB		5	1	2	8
	C					0
CO	OAB	1	3	1		5
	DB		31	7	4	42
	COM	3	6	2		11

SR Scatter finding; *CO* Concentrations; *C* carcass; *OAB* occurrence of articulated bones; *DB* disarticulated bone; *COM* combined concentrations

the second higher frequency was found in LSB (10.6%) and the last 6.6% came from LH. To compare the register among environments, we estimated the find's density per km^2. It showed that LT has the higher density in connection with the other sectors. In relation to the types of findings recorded (Table 10.1), the concentrations (CO) are more numerous than scatter rests (SR). In general, the presence of carcasses (C) is scarce; either as a scatter record or as part of the concentrations (Fig. 10.7a, b). On the other hand, most of the samples corresponds to completely disarticulated bone assemblages, and concentrations of articulated portions or a combination of both types—BD, OBA and/or some C—are the least.

The taxa identified among different types of finding correspond both, to wild (guanaco, choique, piche, fox and hare) and domestic animals (sheep, cow, and horse). Table 10.2 presents the frequency of finds (n) of each species and the there taxonomic abundance (MNI) by surveyed sector. Guanaco and sheep are the only taxa registered in the three localities, although in LT high plateaus only sheep remains were registered. Regarding the diversity of identified species, the NTAXA values show that the LT canyon hosts the highest taxonomic richness (Table 10.2), and is significantly higher than the remaining sectors.

Table 10.2 also presents both the information corresponding to the skeleton zones recorded and the different types of tissues observed for each identified species. The species were identified from bone remains, and in the cases of *Rhea pennata* and *Zaedyus pichiy*, their recognition was also possible from eggshells fragments or carapace plates, respectively.

In relation to the spatial distribution of the disarticulation patterns we note that, while in the LT canyon and in LSB the presence of completely disarticulated remains is dominant and advanced and/or terminal disarticulation, are recorded to a lesser extent; in the LT high plateau the articulated carcasses or the cases showing an initial

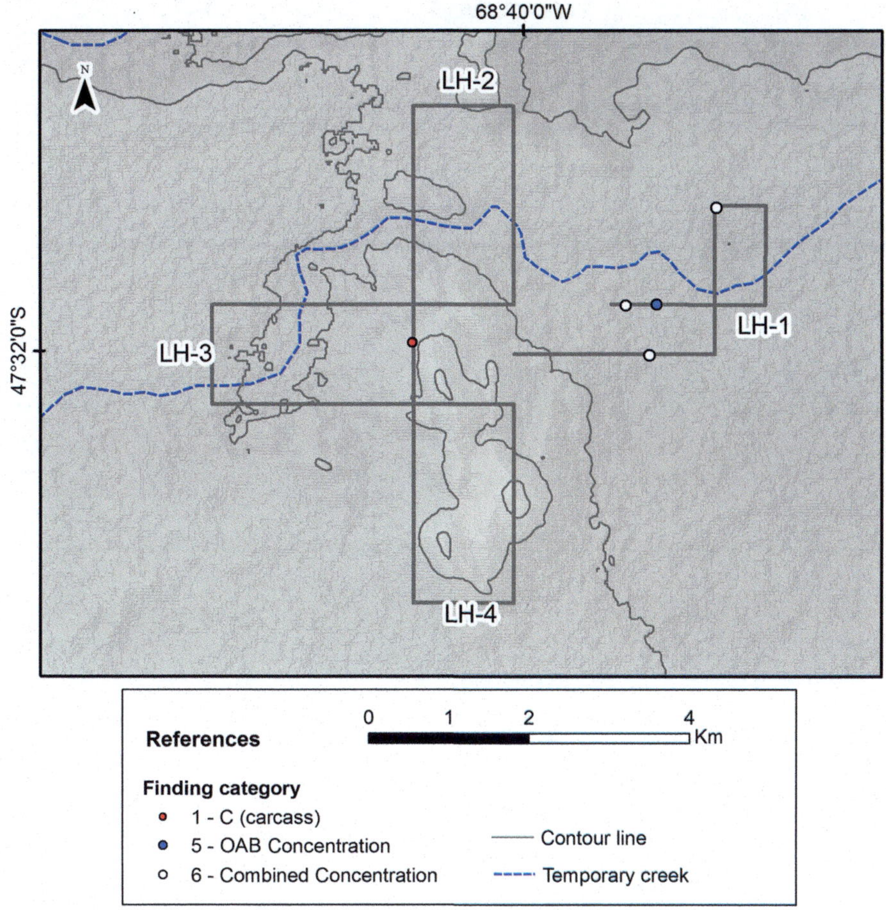

Fig. 10.4 Distribution of the different types of findings in LH

disarticulation are recurrent. The LH findings do not show a clear disarticulation pattern since there are finds in different stages.

The records made about the burying conditions show that the most of the time the bone is exposed –75.5% in LT canyon, 78% in LT hight plateau, 80% in LH and 75% in LSB. While LT canyon was the only sector where buried bones were found (in 6.12% of the cases, 50% of the bone were covered by sediments), the rest of findings are described as half-buried (18.36% of the cases registered in LT canyon, 22% in LT hight plateau, 20% in LH and 25% in LSB).

Regarding preservation indices, the guanaco and sheep profiles (the only two species with representation in all the studied environments) (Fig. 10.8), show the highest weathering values and the clearer evidence of loss of bone in the canyon and high plateau of LT. In this last sector, the greatest heterogeneity of weathering

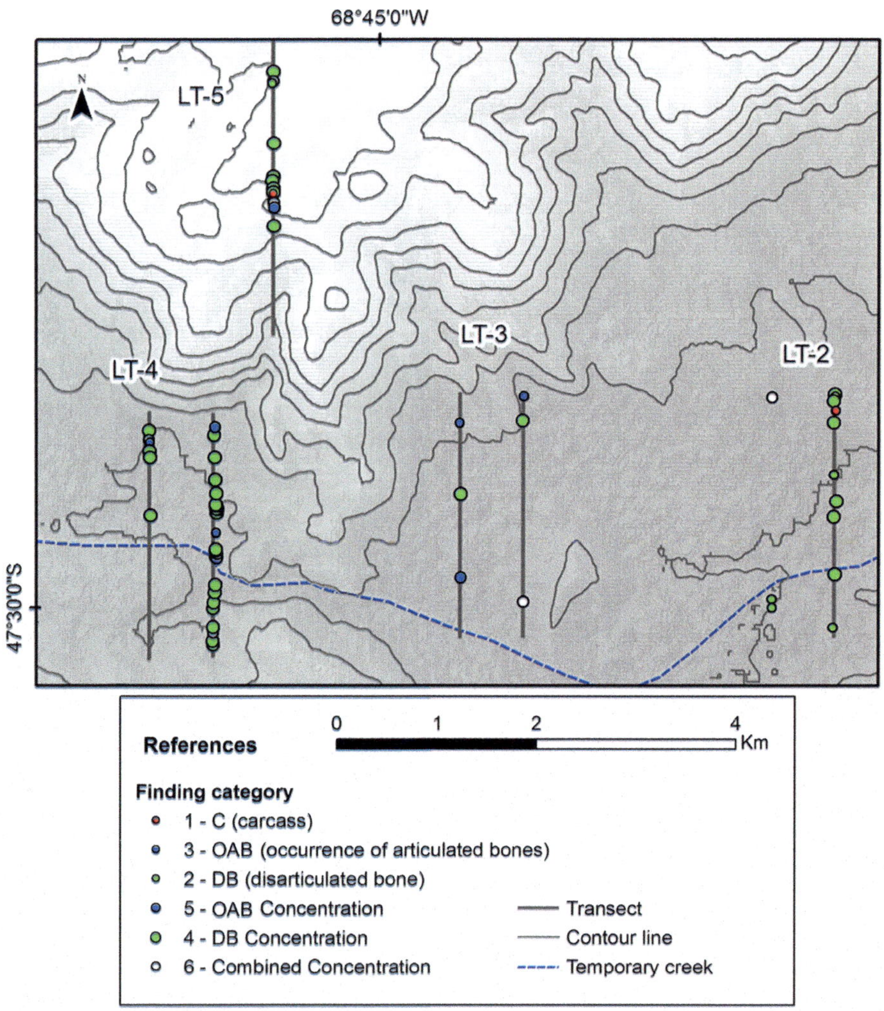

Fig. 10.5 Distribution of the different types of findings in LT

is recorded, with stages from 0 to 4 (Behrensmeyer 1978). Here, although the bone loss is evident, there is a clear low preservation tendency. While in LH there was a tendency similar to LT, but with a high representation of low stages; in LSB a different pattern of better bone preservation was observed.

With respect to the record of bone modifications, marks related to carnivore action (Fig. 10.7c, d), rodents and abrasion were observed, all of them in very low frequency. Abrasion was registered in a single guanaco specimen of LSB, and rodent marks were observed in a rib of this same species in LT canyon. Concerning carnivore marks, these were observed in three of the four sectors surveyed. The highest frequency of

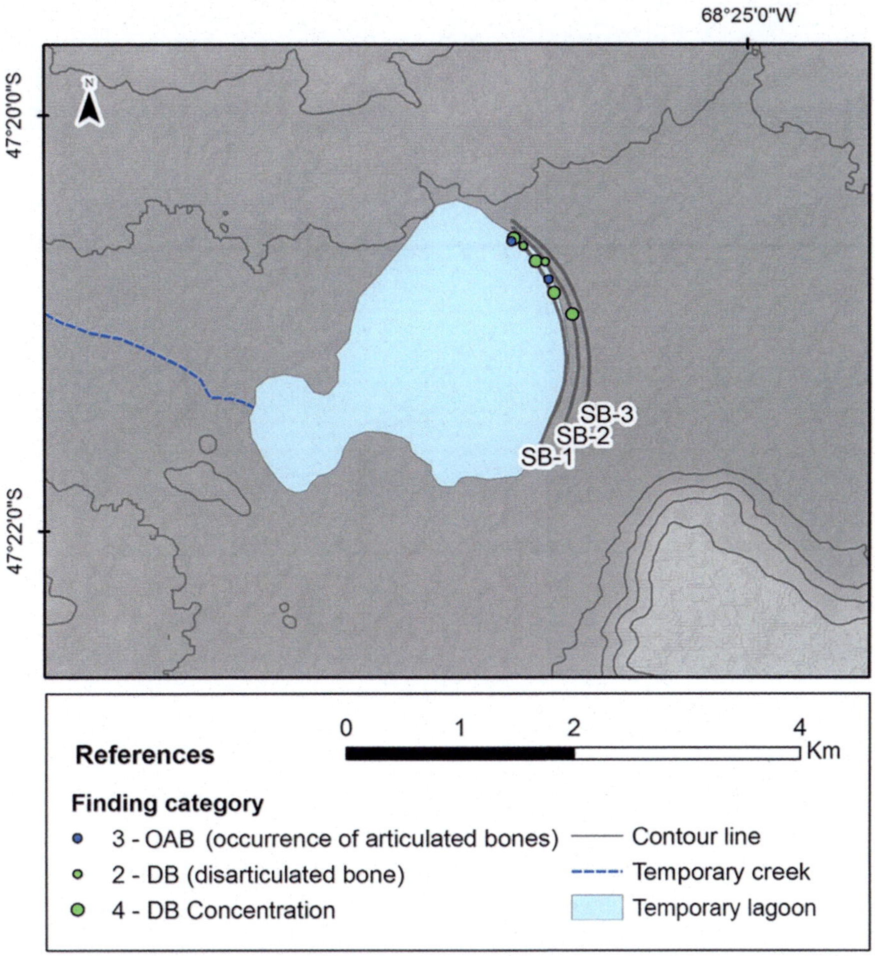

Fig. 10.6 Distribution of the different types of findings in LSB

this kind of marks was recorded in sheep carcasses in LT plateau (in 25% of the findings, n = 3) and in one case observed in LH (corresponding to 20%). Finally, 8% (n = 4) of the LT canyon findings recorded this modification pattern, which affects different species: guanaco, piche, medium bird and medium mammal.

Table 10.2 Taxonomic and anatomical representation by finding, and taxonomic richness by prospected environment

	LH						LT Canyon						LT Plateau						LSB					
	N	AX	AP	Both	Other	MNI	N	AX	AP	Both	Other	MNI	N	AX	AP	Both	Other	MNI	N	AX	AP	Both	Other	MNI
Medium bird							1																	
R. pennata							4	1	3	1	L	2	1				E		3	1			E	1
Medium mammal							2																	
Big mammal							3						1											
Zeadyus pichiy							7	1		4	A	6							1				A	1
Lepus sp.							3			3	S-L	1												
Lycalopex sp.							1		1			1												
Bos taurus							3			3		2												
Ovis aries	1			1	S-F-L	1	22	4	5	12	S-F-L	15	9		3	4	S-F-L	5	1		1			1
Lama guanicoe	5			5	S-F-L	6	6	1	2	2		3							2		2		L	1
Equus sp.							1				V	1	1		1			1	1			1		1
NTAXA	2						8						4						5					

N Frequency of findings; *AX* axial elements; *AP* appendicular elements; *Both* axial and appendicular elements; *Other* corresponds to the different types of tissues that were observed (*S* skin; *F* fleece; *L* ligament; *E* eggshell; *V* vessel; *A* armour)

Fig. 10.7 **a** Guanaco carcass in LH; **b** combined concentration in LT canyon; **c** detail of the evidence of damage caused by carnivores in the guanaco scapula found in LH; **d** damage by carnivores in sheep pelvis in LT plateau

10.5 Discussion and Conclusions

The concentrations of disarticulated bones (CO-DB) were notably the most frequent among the different types of findings in all the sectors surveyed. The only exception corresponds to LH, where the combined concentrations (CO-COM) were more abundant. This is because in LH the findings are characterized by concentrations in a better state of articulation, either present as occurrences of articulated bones (OBA) or as carcasses (C). In general terms, the advanced state of disarticulation in which the most of the registered materials are presented, provides information regarding that in these environments the burial processes would be very slow, favoring the disarticulation of the carcasses and therefore, increasing the probability of mixing and displacement of bones. A similar situation was registered on the Atlantic coast of the Buenos Aires province where the taphonomic studies carried out by Massigoge et al. (2015) with penguins, suggest that the presence of remains in different disarticulation and weathering states is an indicator of attritional assemblages. Considering this information, among the sectors that we prospected, the LH would be where the

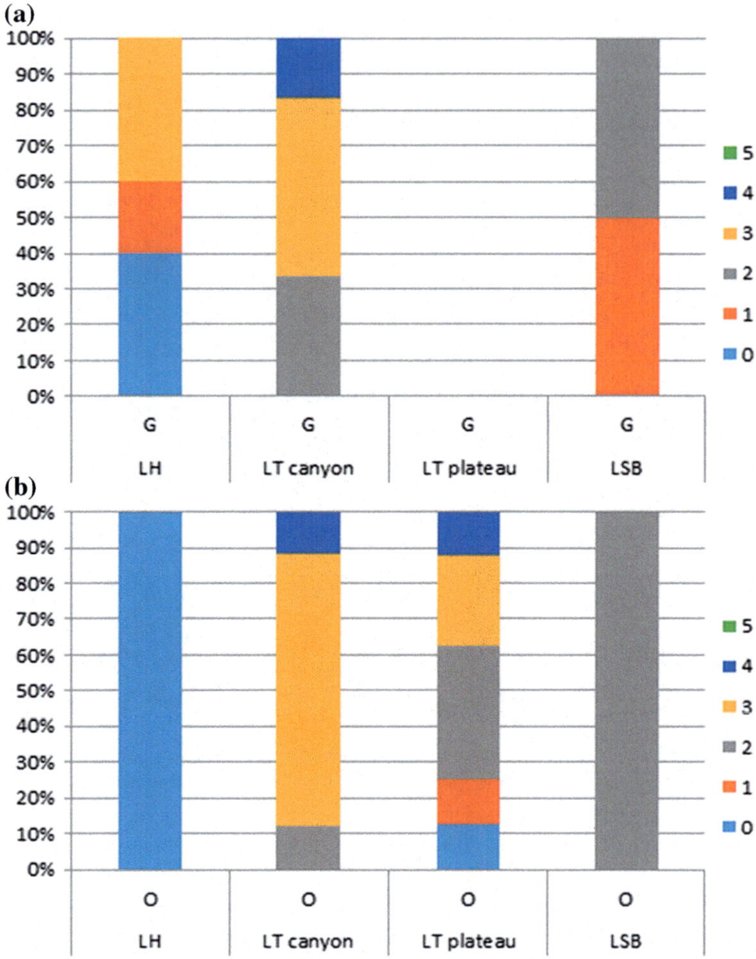

Fig. 10.8 Registered weathering profiles **a** in bone remains of guanaco (G) and **b** in bone sheep (O)

formation of this type of deposits is less evident, while both LT and LSB offer more prone scenarios for the formation of time-averaged assemblages. If we consider the relation of the findings with the substrate, more than 75% of the cases were recorded exposed on the surface and only a few findings were buried (covered with more than 50% by sediment) in the LT Canyon sector. Thus it is possible that the exposure of the materials could be one of the main processes responsible for the disarticulation. In order to evaluate the dynamics of these environments, we need to carry out programmed long-term studies to assess to what extent the mixture of materials is a direct result of the exposure of the carcasses or it is due to successive processes of sedimentary deposition and erosion that involve burial and exhibition cycles of the bones (Favier Dubois 2000). This last hypothesis is a better fit to explain the surveyed cases in LT Canyon where bones with heterogeneous weathering (Fig. 10.8) and different degrees of burial were recorded.

In relation to the distribution in the landscape, it is remarkable that LT canyon holds the largest number and variety of finding types. It is also where the greatest diversity and taxonomic abundance is recorded, and where the identified species acquire higher MNI frequencies (Table 10.2). As it was evidenced by the density estimates of findings/km^2, this unequal distribution of the findings is not related to the different size of the sampled areas (Table 10.1). Therefore, LT canyon would correspond to the sector that offers the best conditions for the deposition/preservation of materials. The available archaeological information shows that this sector has been inhabited by hunter-gatherer societies since the beginning of the settlement of the region (ca. 12,000 years BP) and throughout the Holocene, and that the caves located in this canyon have been recurrently used by the people (Cardich 1977; Cardich et al. 1973; Miotti 1998). This study allows us to suggest that there are other factors that may be conditioning the distribution of materials, for example, there are better conditions of general habitability in the LT canyon, due to the shelter it offers and the greater availability of water. These characteristics are surely also the ones that explain the greatest amount and variety of species, that favour a greater supply of animal resources. Also, among the surveyed sectors, LT canyon is the only one where currently the livestock activity takes place, and therefore, the existing biota is greater than that of the other sectors. However, considering only the wild species, LT still presents the highest taxonomic diversity. In this sense, both the density of findings and the registered NTAXA values would be realizing for the composition and relative abundance of species that inhabit each of these environments, among of which LT canyon is the sector that provides the greatest variety and supply of animal resources.

It is interesting to note that most of the carcasses were located in environments of pampas (intermediate and high plateaus), while the lower sectors (canyon and lagoon edge) are characterized by the more abundant presence of disarticulated materials or in advanced stages of disarticulation. Following Borrero (2001) this greater density of disarticulated bones could be indicative of a greater likelihood of mixing between naturally deposited bones and those that result from human activities, therefore, within the studied area it is expected that the low sectors are the most likely to record this situation. It is interesting to consider that, in the case of the LT canyon, there could also be a natural bone contribution by migration from slopes plateau by erosion, this aspect that should be evaluated in future observations.

The most variable weathering values observed in LT suggest that both canyon and high plateu are the environments with higher probabilities to accumulate bone remains and form palimpsests, while the weathering profiles observed in LSB show homogeneous and low stages of weathering values, suggesting more synchronic deposits. As already suggested, this variety of weathering stages may be related to the existence of a more dynamic environment, in which the probability of mixing increases due to successive cycles of materials exposure and burial. This aspect is relevant for the archaeology of the area since, within the sectors studied in this work, in LT canyon the known stratigraphic archaeological sites are located inside caves that are sedimentary traps (Farrand 1985; Waters 1992). On the other hand, we think that the variability observed in the weathering of LT high plateau responds to a constant exposure of the materials in an environment where the natural conditions impede the burial of bones. In the case of LH, an intermediate condition seems to be present, with areas of greater burial potential that could contribute to the preservation of bones, and other with low to null burial potential.

The bone records made in the field survey allow us to interpret the main causes of death for the animals as natural. They are mostly associated with the presence of wiring and ground accidents, although it cannot be discarded that some deaths could have been the result of poaching or climatic events, as it has already been registered in caves within the area (Marchionni et al. 2017). In connection with this point, it is important to note that the transect 4 at LT canyon (Fig. 10.5) coincides to a large extent with the wire that separates two farm properties, and, therefore, the great density of bone findings here could be due to this obstacle. In relation to the damage produced by carnivores, it was registered in few bone specimens, and this suggests moderate to low scavenging activities. The marks are few; they are localized and have not done serious damage on any segment surveyed. So, although local people believe that the puma is the main depredator of sheep livestock, there is no evidence pointing to carnivores as responsible for killing any of the individuals in the registered cases, nor is it expected that they have caused important disarticulation of carcasses. However, based on the records made, we are able to propose that the high plateau is the only sector where the action of predators contribute to the disarticulation and dispersion of the carcasses. Similar cases where carnivores have contributed to the disarticulation and dispersion of carcasses have been recorded in other areas of Patagonia (Borella 2004; Borrero 2007; Cruz and Muñoz 2010; Otaola and Tripaldi 2016, among others). Nevertheless, other naturalistic studies such as those carried out on the Atlantic coast of Buenos Aires, propose that carnivores would have significantly conditioned the representation of skeletal parts of vertebrates (Massigoge et al. 2015). The data collected during the surveys has allowed us to obtain a broader picture of the natural distribution of the bone records in an area of archaeological interest. This perspective is consistent with a current view of the way in which archaeological problems should be treated (Delcourt and Delcourt 1988; Favier Dubois 2000), that is, through the integration of the different spatial scales, from the site scale to the region. In this way, the results presented here are useful to assess how taphonomic processes operate and what is their variability range. The information coming from the actualistic taphonomy is still preliminary but it allows warning about the taphonomic dynamics of each of the environments studied, its potential for the natural accumulation of bones and their burying chances. In this sense, and based on the greater variety of agents and processes recorded here, LT canyon is the sector that seems to be the most dynamic in taphonomic terms, and where a lower integrity of the zooarchaeological record is expected (Binford 1981). However, despite these conditions, it is the sector with the highest density of bone materials naturally available and, therefore, with the greatest probability of natural incorporation of bones into the fossil record. The highest sedimentation recorded in some places within this sector, suggests burial processes that may favor the preservation of the bone registers. All this observations suggest a greater environmental variability than what was considered in the research so far.

This work provides relevant environmental and taphonomic information to the study area and contributes to the construction of archaeological expectations about the density, distribution, diversity, and preservation of the zooarchaeological record. It also contributes to start exploring the behavior of taphonomic agents and processes at more restricted spatial scales, which is part of the agenda of these investigations: the closer study of the environmental variability that has been just defined.

Acknowledgements We want to thank Sergio Martínez and all organizing committee of the first workshop "Tafonomía Actualista en América del Sur", who invited us to write this paper for a special volume. Grants from Agencia Nacional de Promoción Científica y Tecnológica (PICT 0176 and PICT 2015-1742), and University of La Plata (PI 665 and 805 FCNyM-UNLP). We would also like to thank the government authorities of Pico Truncado and the Echandi and Freile families that contributed to logistic tasks during our fieldwork. Also, we are grateful for the suggestions made by María Gutiérrez and people that collaborate with us in the fieldworks. Finally, we are grateful for the valuable comments made by the anonymous reviewer which undoubtedly contributed to improving this work.

References

Abramova Z (1993) Two examples of terminal Paleolithic adaptations. In: Soffer O, Praslov N (eds) From Kostenki to Clovis. Plenum Press, New York, pp 85–100

Behrensmeyer AK (1978) Taphonomic and ecologic information from bone weathering. Paleobiology 4:150–162

Behrensmeyer AK, Dechant-Boaz DE (1980) The recent bones of Amboseli Park, Kenya, in relation to east African Paleoecology. In: Behrensmeyer AK, Hill AP (eds) Fossils in the Making. Vertebrate taphonomy and paleoecology. University of Chicago Press, Chicago, pp 72–92

Binford L (1981) Bones: ancient men and modern myths. Academic Press, New York

Blumenschine RJ (1989) A landscape taphonomic model of the scale of prehistoric scavenging opportunities. J Hum Evol 18(4):345–371

Blumenschine RJ, Cavallo JA, Capaldo SD (1994) Competition for carcasses and early hominid behavioral ecology: a case study and conceptual framework. J Hum Evol 27:197–213

Bocek B (1986) Rodent ecology and burrowing behavior: predicted effects on archaeological site formation. Am Antiq 51(3):589–603

Borella F (2004) Tafonomía Regional y Estudios Arqueofaunísticos de Cetáceos en Tierra del Fuego y Patagonia Meridional. BAR International Series 1257, Archaeopress, England

Borella F, Muñoz S (2006) Observaciones tafonómicas sobre restos de pinnípedos en la costa norte fueguina (Argentina). Intersec Antropol 7:399–403

Borella F, Borrero LA (2010) Observaciones tafonómicas acerca de la desarticulación de carcasas de pinnípedos en ambientes litorales, el caso de Islote Lobos (Golfo San Matías, Río Negro). In: Gutiérrez MA, De Nigris M, Fernández PM, Giardina M, Gil A, Izeta A, Neme G, Yacobaccio H (eds) Zooarqueología a principios del siglo XXI: Aportes teóricos, metodológicos y casos de estudio. Ediciones del Espinillo, Buenos Aires, pp 371–380

Borrero LA (1988) Tafonomía Regional. De Procesos, Contextos y Otros Huesos. In: Ratto N, Haber A (eds) ICA. Sección Prehistoria, FFyL-UBA, Buenos Aires, pp 9–15

Borrero LA (1990) Taphonomy of guanaco bones in Tierra del Fuego. Quat Res 34:361–371

Borrero LA (2000) Ten years after: esquema para una tafonomía regional de la Patagonia meridional y norte de Tierra del Fuego. In: Belardi J, Carballo Marin, F., Espinosa, S (eds) Desde el país de los gigantes. Perspectivas arqueológicas en Patagonia, Tomo I. Universidad de la Patagonia Austral, Río Gallegos, pp 183–193
Borrero LA (2001) Regional taphonomy: The scales of application to the Archaeological Record. Animals and Man in the Past. Essay in honour of Dr. A. T. Clason emeritus professor of archaeozoology Rijksuuniversiteit Groningen, the Netherlands. In: Buitenhuis H, Prummel W (eds) ARC-Publicatie 41, Groningen, pp 17–20
Borrero LA (2007) Longitudinal taphonomy studies in Tierra del Fuego, Argentina. In: Gutierrez MA, Miotti L, Barrientos G, Mengoni Goñalons G, Salemme M (eds) Taphonomy and zooarchaeology in Argentina. BAR International Series, Oxford, pp 219–233
Borrero LA, Martin FM (1996) Tafonomía de carnívoros. Un enfoque regional. Arqueología. Sólo Patagonia. In: Gómez Otero J (ed) CENPAT-CONICET, Puerto Madryn, pp 189–198
Borrero LA, Lanata JL, Ventura BN (1992) Distribuciones de hallazgos aislados en Piedra del Águila. In: Borrero LA, Lanata JL (eds) Análisis espacial en la arqueología patagónica. Ayllu, Buenos Aires, pp 9–20
Borrero LA, Martin FM, Vargas J (2005) Tafonomía de la interacción entre pumas y guanacos en el Parque Nacional Torres del Paine, Chile. Magallania 33(1):95–114
Brain CK (1981) The hunters or the hunted? An introduction to African Cave Taphonomy. The University of Chicago Press, Chicago
Butzer K (1982) Arqueología, una ecología del hombre: Método y teoría para un enfoque contextual. Bellaterra, Barcelona
Cabrera AL (1976) Fitogeografía de la República Argentina. Bol Serie Botánica 14(1–2):1–42
Cabrera AL, Willink A (1980) Biogeografía de América latina. Programa regional de desarrollo científico y tecnológico. OEA, Washington
Cardich A (1977) Las culturas Pleistocénicas y post-pleistocénicas de Los Toldos y un bosquejo de la Prehistoria de Sudamérica. Obra Cincuecenten Mus La Plata II:149–172
Cardich A, Cardich L, Hadjuk A (1973) Secuencia arqueológica y cronología radiocarbónica de la cueva 3 de Los Toldos (Santa Cruz, Argentina). Relaciones de la Sociedad Argentina de Antropología VII:85–123
Cruz I (2003) Paisajes tafonómicos de restos de Aves en el sur de Patagonia continental. Aportes para la interpretación de conjuntos avifaunísticos en registros arqueológicos del Holoceno. Ph.D. thesis. Facultad de Filosofía y Letras-Universidad Nacional de Buenos Aires, Buenos Aires
Cruz I (2009) Tafonomía en escalas espaciales amplias: el registro óseo de las aves en el sur de Patagonia. In: Acosta D, Loponte A, Mucciolo L (eds) Temas de Arqueología 2: Estudios tafonómicos y zooarqueológicos. Instituto Nacional de Antropología y Pensamiento Latinoamericano, Buenos Aires, pp 15–34
Cruz I, Muñoz S (2010) Tafonomía comparativa: seguimiento de carcasas de mamíferos en Punta Entrada (Santa Cruz, Argentina). In: Gutiérrez M, De Nigris M, Fernández P, Giardina M, Gil A, Izeta A, Neme G, Yacobaccio H (eds) Zooarqueología a principios del siglo XXI. Aportes teóricos, metodológicos y casos de estudio. Ediciones del Espinillo, Buenos Aires, pp 387–396
De Giusto J, Di Persia C, Pezzi E (1980) Nesocratón del Deseado. In: Turner JCM (ed) Segundo Simposio de Geología Regional Argentina II:1390–1430. Academia Nacional de Ciencias, Córdoba
Delcourt HR, Delcourt PA (1988) Quaternary landscape ecology: relevant scales in space and time. Landscape Ecol 2(1):23–44
Farrand WR (1985) Rockshelters and cave sediments. In: Stein JK, Farrand WR (eds) Sediments in archaeological context. Center for the Study of Early Man, University of Maine, USA, pp 21–39
Favier Dubois CM (2000) La geoarqueología y los procesos de formación del registro. In: Nami HG (ed) La perspectiva interdisciplinaria en la arqueología contemporánea, vol 6. Arqueología Contemporánea pp 123–141
Fernández P (2008) Taphonomy and zooarchaeology in the Neotropics: a view from the northwestern Patagonian forest and steppe. Quat Int 180:63–74

Foley R (1981) A model of regional archaeological structure. Proc Prehist Soc 47:1–17
Giardina M, Hermo D, Miotti L, Vázquez M (2000) Resolución e integridad arqueológica del Componente inferior de Piedra Museo. In: Miotti L, Paunero R, Salemme M, Cattáneo R (eds) Guía de Campo de la visita a las Localidades arqueológicas. Taller Internacional "La colonización del Sur de América durante la transición Pleistoceno/Holoceno", Servicoop, La Plata, pp 79–83
Gómez JC, Magnin LA (2008) Cartografía geomorfológica aplicada a un sector de interés arqueológico en el Macizo del Deseado, Santa Cruz (Patagonia Argentina), Investigaciones Geográficas, Boletín del Instituto de Geografía, UNAM. Núm. 65, pp 22–37. ISSN 0188-4611. http://www.igeograf.unam.mx/instituto/publicaciones/boletin/bol65/bltn65art2.pdf
Grayson D (1984) Quantitative zooarchaeology. Academic Press, New York
Gutiérrez MA, Kaufmann CA, González ME, Scheifler NA, Rafuse DJ, Massigoge A, Álvarez MC (2016) The role of small carnivores in the movement of bones: implications for the Pampas archaeofaunal record, Argentina. Archaeol Anthrop Sci 8(2):257–276
Haynes G (1980) Evidence of carnivore gnawing on Pleistocene and Recent mammalian bones. Paleobiology 6(03):341–351
Hermo D, Marchioni L, Mosquera B (2017) Prospecciones arqueológicas en la cuenca del Zanjón El Pescado (Macizo del Deseado, Santa Cruz): los casos de Los Toldos y La Huella In: Abstracts of the X Jornadas de Arqueología de la Patagonia. Puerto Madryn. Chubut. 31 July–5 August 2017
Haynes G (1995) Clovis and pre-Clovis megamammals: a comparison of Carcass disturbance, age profiles, and other characteristics in light of recent actualistic studies. In: Johnson E (ed) Ancient peoples and landscapes. Museum of Texas Tech University, Lubbock, pp 9–27
Jobbágy EG, Paruelo JM, León R (1995) Estimación del régimen de precipitación a partir de la distancia a la cordillera en el noroeste de la Patagonia. Ecol Austral 5:47–53
Johnson E (1985) Current developments in bone technology. Adv Archaeol Meth Theory 8:157–235
Kaufmann CA (2009) Estructura de edad y sexo en guanaco: estudios actualísticos y arqueológicos en Pampa y Patagonia. Sociedad Argentina de Antropología, Buenos Aires
Lyman R (1994) Vertebrate taphonomy. Cambridge University Press, Cambridge
Magnin L (2010) Distribuciones arqueológicas en la meseta central de Santa Cruz: Implicancias para los estudios de uso del espacio y movilidad de sociedades cazadoras recolectoras. Ph.D. thesis. Facultad de Ciencias Naturales y Museo-Universidad Nacional de La Plata, La Plata. MS
Marchionni L (2013) Comparación de las distintas historias tafonómicas en conjuntos zooarqueológicos provenientes de la Meseta Central de la provincia de Santa Cruz. Ph.D. thesis. Facultad de Ciencias Naturales y Museo - Universidad Nacional de La Plata, La Plata. MS
Marchionni L (2016) Taphonomic study in Argentinian Patagonia: analysis of variability through time and space in the Central Plateau (Santa Cruz Province). Archaeol Anthropol Sci 8(2):241–255
Marchionni L, Mosquera B, Añino E (2012) Analysis of zooarchaelogical and taphonomical variability from Maripe Cave site, Santa Cruz province, Argentinian Patagonia. Quat Int 278:12–21
Marchionni L, García Añino E, Miotti L (2017) Actualistic study on a dense concentration of bone remains in the Central Plateau of Santa Cruz province (Argentina). J Taphonomy 15, special issue 1–3, pp. 29–44
Marean CW (1995) Of taphonomy and zooarchaeology. Evol Anthropol 4:64–72
Martin F (2008) Tafonomía y paleoecología de la transición Pleistoceno-Holoceno en Fuego-Patagonia: Interacción entre poblaciones humanas y de carnívoros y su importancia como agentes en la formación del registro fósil. Ph.D. thesis. Facultad de Ciencias Naturales y Museo-Universidad Nacional de La Plata. La Plata. MS
Massigoge A, González M, Kaufmann C, Gutiérrez M (2010) Observaciones actualísticas sobre meteorización ósea en restos esqueletales de guanaco. In: Berón M, Luna L, Bonomo M (eds) Mamül Mapu: pasado y presente desde la arqueología pampeana, Tomo I. Editorial Libros del Espinillo, Buenos Aires, pp 309–322

Massigoge A, Gutiérrez MA, Álvarez MC, Kaufmann CA, Rafuse DJ, González ME (2014) Estudio comparativo de las marcas de dientes producidas por dos pequeños carnívoros sudamericanos. Rev Chil Antropol 30:42–49

Massigoge A, Rafuse D, Álvarez MC, González ME, Gutiérrez MA, Kaufmann CA, Scheifler NA (2015) Beached penguins on the Atlantic Coast in the Pampas region of Argentina: Taphonomic analysis and implications for the archaeological record. Palaeogeogr Palaeoclim Palaeoecol 436:85–95

Miotti L (1998) Zooarqueología de la meseta central y costa de la provincia de Santa Cruz: un enfoque de las estrategias adaptativas aborígenes y los paleoambientes. Museo Municipal de Historia Natural de San Rafael, San Rafael

Miotti L, Marchionni L (2011) The study of archaeofauna at middle Holocene in AEP-1 rockshelter, Santa Cruz, Argentina: Taphonomic implications. Quat Int 245:148–158

Miotti L, Salemme M (1999) Biodiversity, taxonomic richness and specialists-generalists during Late Pleistocene/Early Holocene times in Pampa and Patagonia (Argentina, Southern South America). Quat Int 53–54:53–68

Miotti L, Salemme M (2005) Hunting and butchering events at the Pleistocene/Holocene transition in Piedra Museo: an example of adaptation strategies of the first colonizers of Patagonia. In: Bonnichsen R, Leppers B, Stanford D, Waters M, (eds) Paleoamerican origins: beyond Clovis. A&M University Press, Texas, pp. 209–220

Miotti L, Vázquez M, Hermo D (1999) Piedra Museo un Yamnagoo Pleistocénico en la colonización de la meseta de Santa Cruz. El estudio de la arqueofauna. In: Goñi R (ed) Soplando en el Viento. Universidad Nacional del Comahue, Instituto Nacional de Antropología y Pensamiento Latinoamericano, Neuquén- Buenos Aires, pp 113–136

Miotti L, Tonni E, Marchionni L (2018) What happened when the pleistocene megafauna became extinct? Quat Int 473:173–189

Mondini M (2002) Carnivore taphonomy and early human occupations in the Andes. J Archaeol Sci 29:791–801

Mondini M (2003) Formación del registro arqueofaunístico en abrigos rocosos de la Puna argentina. Tafonomía de carnívoros. Ph.D. Universidad de Buenos Aires, Buenos Aires

Muñoz S (2009) Los cánidos como agentes tafonómicos en los depósitos arqueológicos de la Isla Grande de Tierra del Fuego. In: Salemme M, Santiago F, Álvarez M, Piana E, Vázquez M, Mansur E (eds) Arqueología de Patagonia: una mirada desde el último confín. Editorial Utopías, Ushuaia, pp 799–812

Muñoz S, Cruz I (2014) Estado actual de las investigaciones Tafonómicas Naturalistas en Punta Entrada y Monte León (Patagonia Meridional). Rev Chil Antropol 29:102–110

Otaola C, Tripaldi A (2016) Longitudinal taphonomy studies of mammal carcasses from the Rio Salado Valley, Mendoza, Argentina. Ethnobiol Lett 7(1):1–13

Panza JL (1982) Descripción geológica de las Hojas 53d "Gobernador Moyano" y 54e "Cerro Vanguardia", provincia de Santa Cruz. Servicio Geológico Nacional (inédito), Buenos Aires

Panza JL (2001) Hoja Geológica 4769-IV Monumento Natural Bosques petrificados, Provincia de Santa Cruz. Boletín del SEGEMAR 258. Instituto de Geología y Recursos Minerales, Buenos Aires

Paruelo JR, Golluscio E, Jobbágy M, Canevari M, Aguiar M (2005) Situación ambiental en la estepa patagónica. In: Brown A, Martínez Ortiz U, Acerbi M, Corcuera J (eds) La situación ambiental argentina. Fundación Vida Silvestre Argentina, Buenos Aires, pp 303–313

Pobiner BL, Braun DR (2005) Applying actualism: considerations for future research. J Taphonomy 3(2):57–65

Raedecke K (1978) El guanaco de Magallanes, Chile. Su distribución y biología. Corporación Nacional Forestal. Publicación Técnica N° 4. Ministerio de Agricultura, Chile

Ravesloot J, Waters M (2004) Geoarchaeology and archaeological site patterning of the Middle Gila River, Arizona. J Field Archaeol 29:203–214

Terranova E (2013) Arqueología de la Cuenca del Arroyo Talagapa, Meseta de Somuncurá (Provincia de Río Negro). PhD, Facultad de Ciencias Naturales y Museo- Universidad Nacional de La Plata, La Plata. MS http://sedici.unlp.edu.ar/handle/10915/35034
Waters MR (1992) Principles of geoarchaeology. The University of Arizona Press, Tucson and London
Waters MR, Kuehn DD (1996) The geoarchaeology of place: the effect of geological processes on the preservation and interpretation of the archaeological record. Am Antiq 61(3):483–497

Chapter 11
Modern Bone Distribution in the Pampas of Argentina: Taphonomic Implications for the Regional Archaeological Record

Nahuel A. Scheifler, Agustina Massigoge, Cristian A. Kaufmann, Daniel J. Rafuse, Mariela E. González, María A. Gutiérrez and María C. Álvarez

Abstract In this paper, we present the results of naturalistic taphonomic observations made in different environmental contexts of the Pampas region. The objective of this study is to evaluate the formation processes of bone assemblages in differently sized vertebrates. In the hills, medium-sized vertebrates dominate the assemblages and the main accumulating agents are carnivores. In the shallow lakes and fluvial valleys, bones of large-sized vertebrates are predominant and accumulate mainly by natural death and discard of the carcasses by local farmers. The coast shows a more uniform representation of small, medium, and large-sized vertebrates, which is related to high biodiversity and good visibility of the bones deposited in this environment. Finally, the information from the plains indicates a similar representation of medium and large-sized vertebrates, both deposited by natural deaths. Results from the taphonomic analysis suggest higher destruction rates of bones of small-sized animals, and the likelihood of preservation and burial of animal bones varies in each of the environments. The mixing of bone remains with the archaeological deposits is dependent on these variables.

Keywords Naturalistic taphonomy · Vertebrate bone assemblages · Regional archaeology · Fossil record · Pampas region

---

Electronic supplementary material The online version of this chapter (https://doi.org/10.1007/978-3-030-20625-3_11) contains supplementary material, which is available to authorized users.

N. A. Scheifler (✉) · A. Massigoge · C. A. Kaufmann · D. J. Rafuse · M. E. González · M. A. Gutiérrez · M. C. Álvarez
Instituto de Investigaciones Arqueológicas y Paleontológicas del Cuaternario Pampeano (INCUAPA-CONICET), Facultad de Ciencias Sociales, Universidad Nacional del Centro de la Provincia de Buenos Aires, Av. del Valle 5737, B7400JWI Olavarría, Buenos Aires, Argentina
e-mail: nscheifler@soc.unicen.edu.ar

S. Martínez et al. (eds.), *Actualistic Taphonomy in South America*,
Topics in Geobiology 48, https://doi.org/10.1007/978-3-030-20625-3_11

11.1 Introduction

Taphonomic research at a regional scale in modern ecosystems has had an important development since the mid-1970s, mainly in African environments which were considered analogous to Plio-Pleistocene landscapes (Hill 1976; Behrensmeyer 1978, 1993; Tappen 1995; Behrensmeyer and Dechant Boaz 1980; Gifford 1984; Blumenschine 1988; Domínguez-Rodrigo 2001; Faith and Behrensmeyer 2006). One of the main objectives of these studies was to determine the natural processes that control the deposit, modification, and preservation of bones from recent vertebrates to improve the understanding of the ecological information contained in the fossil record of early hominid sites and its potential biases. These studies inspired some South American archaeologists, especially those working with hunter-gatherer sites from Patagonia. Here, a regional approach to taphonomy was encouraged by Borrero and colleagues (Borrero 1988, 1989, 1990, 2001; Borrero and Martin 1996; Borrero et al. 1991; Belardi and Carballo Marina 2003; Borella 2004; Cruz 2015). Instead of looking for past environmental analogous, these researchers proposed to carry out a taphonomic analysis in the same region where archaeological research was being undertaken. The basic goal of this approach is to understand the dynamics of the environment at a large spatial scale, particularly the processes responsible for bone deposit and preservation in different landscapes. Additionally, the researcher seeks to understand the impact of the natural "bone rain" (sensu Haynes 2018: 113) on the regional archaeological record (Borrero 1989, 2001, 2014).

In the Pampas region, our research team has recently begun to develop naturalistic taphonomy at a broad spatial scale (Massigoge et al. 2015; Gutiérrez et al. 2016, 2018). The late development of this perspective in the region is due in large part to the impact of agriculture and livestock on the natural environments. The significant differences between the modern and prehistoric environments have prevented researchers from taking a more comprehensive look at the analogous in the modern ecosystem. However, we have shown that valuable information can be obtained through the development of actualistic taphonomy in modern environments such as the Pampas. The main objective of our naturalistic research is to understand the formation of bone assemblages of different classes of vertebrates. This taphonomic knowledge will be useful to both understanding the biases and limitations of the faunal record when used for inferring past environments and live communities, as well as evaluating the potential of the faunal record to provide information on specific processes and ecological conditions. The objectives are to (a) determine the causal processes and agents of accumulation and modification of bone assemblages; (b) establish if differences exist in the rates of natural bone deposit ("bone rain"), in different environments; (c) evaluate the processes that lead to the preservation and burial of the bones; and (d) identify the potential for contamination of archaeological sites by naturally deposited bones in each environment.

11.2 Regional Setting

The Pampas region is a vast plain covered by grasslands (398,966 km^2), located in the central east of Argentina (Fig. 11.1). The southeast of the Pampas has a temperate-humid climate and contains a diversity of geomorphological landforms and environments, including two hill systems (Tandilia and Ventania). The general relief is flat, with a moderate slope towards the Atlantic coast. A complex drainage network, with headwaters in the plains and the hills, carves into Quaternary sediments, intensifying towards the ends of the mouth of the river's which flow into the ocean (Matteucci 2012). Also, there are numerous permanent and temporary shallow lakes. Major changes to the coastline occurred during the end of the Pleistocene (between 12,000 and 10,000 ^{14}C yrs BP) when the sea level was approximately 60 m below its current location (Ponce et al. 2011). During the Holocene, large sand dunes formed along the coast, and loess and loess-like sediments covered older eolian deposits in the plains (Fidalgo et al. 1991; Zárate and Rabassa 2005).

The vegetation of the Pampas grasslands was formerly dominated by tussock grasses and scattered shrubs (Soriano et al. 1992; Miñarro et al. 2008). Today, the most common genera of grasses are *Stipa*, *Piptochaetium*, *Paspalum*, and *Bothriochloa*

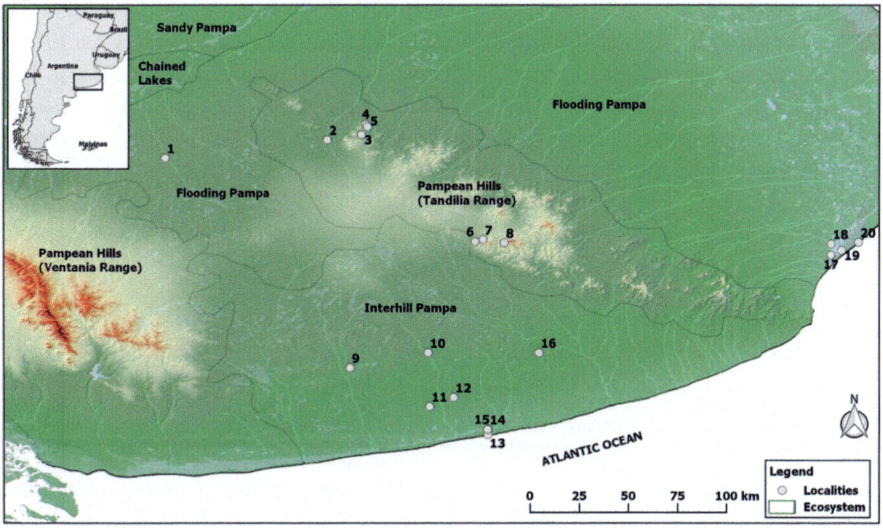

Fig. 11.1 Map of study area showing the location of transect localities. (1) General La Madrid, (2) Arroyo Tapalque, (3) Las Vertientes, (4) Cerro Largo, (5) Cerro San Cayetano, (6) Estancia La Tinta, (7) Cerro La Tinta, (8) La Juanita, (9) Arroyo Seco, (10) Arroyo Cristiano Chico, (11) Laguna Las Toscas, (12) Arroyo Cortaderas, (13) Balneario San Cayetano, (14) Laguna Salada San Cayetano, (15) San Cayetano vegetated dunes, (16) Laguna La Salada, (17) Recreo San Gabriel, (18) Estancia Nahuel Ruca, (19) Laguna Arsa, (20) Playa Puesto 1. Limits of the environmental ecosystems adapted from Matteucci (2012)

(Paruelo et al. 2007). Most natural vegetation has been replaced by modern-day agricultural crops (Bilenca and Miñarro 2004; Matteucci 2012).

From the end of the Pleistocene, the Pampas underwent a mass extinction of large mammals and a minor extinction and retraction of smaller sized taxa during the Holocene (Politis and Pedrotta 2006; Barnosky and Lindsey 2010; Prevosti et al. 2015). The introduction of livestock in the 16th century, and agriculture by the end of the 19th century, have intensely modified the original landscape and affected the vertebrate communities (Bilenca et al. 2017). Along with an array of livestock, Europeans also introduced wild species such as the European hare (*Lepus europaeus*) and Wild boar (*Sus scrofa*) (Soriano et al. 1992).

At present, wild large-sized vertebrates are scarce. The populations of the native ungulates that once constituted the main prey for Holocene hunter-gatherers (Guanaco—*Lama guanicoe*—and Pampas deer—*Ozotoceros bezoarticus*), are now reduced to small groups confined in reserves (Soriano et al. 1992; Martínez and Gutiérrez 2004). The small-sized vertebrate community however is abundant, including fish, amphibians, reptiles, birds, and mammals (Soriano et al. 1992; Matteucci 2012).

Carnivores represent an important agent in the accumulation and modification of the bone assemblages. Among the carnivorous mammals in the Pampas, there are canids, felids, mustelids, and mephitids. The canids include the Pampas fox (*Lycalopex gymnocercus*), Gray fox (*Lycalopex griseus*) and Dog (*Canis familiaris*) (Nowak 1991). The felids comprise the Mountain lion (*Puma concolor*), Eyra cat (*Puma yaguarondi*), Jaguar (*Panthera onca*), Geoffroy's cat (*Leopardus geoffroyi*) and Pampas cat (*Leopardus colocolo*) (Nowak 1991; Lucherini et al. 2004). The mustelids include the Neotropical otter (*Lontra longicaudis*), Lesser grison (*Galictis cuja*) and Patagonian weasel (*Lyncodon patagonicus*) (Nowak 1991). The Mephitidae Family encompasses the Molina's Hog-nosed skunk (*Conepatus chinga*). Except for the Domestic dog, these carnivores are identified in the fossil record since the Late Pleistocene.

11.3 Materials and Methods

Based on the diversity of environments in a relatively restricted area, we distributed taphonomic transects along five large environments: hills, shallow lakes, river valleys, plains, and coast (Fig. 11.2). The starting and ending position of each taphonomic transect was georeferenced in the field. Transects were walked by two individuals. All transects were 10 m wide. The length was variable depending on the landscape, and uninterrupted transects were finalized at 500 m. Each transect was partitioned in 50 m samples and the following variables were recorded: sediment type; slope; potential for burial of faunal material; type and distribution of vegetation; bioturbation; presence of living animals or modern human activity; archaeological materials; and visibility based on land cover (excellent visibility: without vegetation cover; very good visibility: 1–25%; good visibility: 26–50%; regular visibility: 51–75%; and poor visibility: 76–100%). Faunal remains were classified as follows: disarticulated bone (a specimen unrelated to another specimen by soft tissue); occurrence of articulated bones (two or more specimens joined by soft tissue, comprising less than 75% of the animal skeleton); and carcasses (specimens joined by soft tissue, comprising more than 75% of the animal skeleton). We also recorded if the bones were scattered or concentrated. A “concentration” is defined here as any assemblage of at least five remains (including disarticulated, occurrence of articulated bones, and carcasses) from the same or different individuals distributed in a small area (approximately 100 m^2) (Cruz 2007).

Except for domestic ungulate bones, which were analyzed in the field, all faunal remains were collected and analyzed in the laboratory. Each finding was recorded on a data sheet with specimen or carcass ID numbers and the following variables: taxon; presence of soft tissue; element; fusion state; completeness; articulation between elements; burial state (following Behrensmeyer and Dechant Boaz 1980); inclination; and taphonomic modifications, which included weathering stage; sedimentary abrasion; carnivore and rodent marks; root etching; manganese stains; and trampling (Behrensmeyer 1978; Haynes 1980; Binford 1981; Shipman 1981; Grayson 1984; Lyman 1984; Olsen and Shipman 1988; Behrensmeyer et al. 2003; Gutiérrez and Kaufmann 2007).

The bone assemblage was divided in three body size categories: small-sized vertebrates (<1 kg); medium-sized vertebrates (1–20 kg); and large-sized vertebrates (>20 kg). In the case of taphonomic analyses, we only studied those specimens that could be assigned to a taxonomic category. Finally, Ntaxa was obtained by counting the identified number of non-overlapping taxa (Grayson 1991).

11.4 Results

A total of 221 transects were conducted, covering an area of 1,046,130 m^2. The highest visibility was recorded along the coast, varying from poor to regular in the rest of the environments. Vertebrate remains were identified as disarticulated bones (n = 3055), occurrences of articulated bones (n = 76), and carcasses (n = 22). Burial was only identified in disarticulated bones (3.2%). The proportions of buried remains for the different environments are distributed as follows: river valleys (13.2%), shallow lakes (4.3%), plains (2.1%), hills (1.4%) and coast (1.4%). Shallow lakes contained the maximum density (0.0051/m^2) of remains and the plains the lowest density (0.0017/m^2) (Table 11.1). Different taxonomic classes were identified, among which the mammals (69.4%) dominate in all environments. Other categories identified in lower frequencies include birds (17.6%), fish (2%), reptiles (1.8%), and amphibians (0.1%). A total of 9.1% of the specimens could not be determined at any taxonomic level (Table 11.1). The total sample analyzed for taphonomic modifications include 2702 bone remains, distributed as follow: 352 small-sized vertebrates, 1224

Fig. 11.2 Examples of surveyed environments: **a** hills, **b** shallow lakes, **c** coast: backshore, **d** coast: active dunes, **e** plains, **f** river valleys

Table 11.1 Characteristics of the studied sample and taxonomic distribution of the bone assemblages by environment

Environments	Hills		Shallow lakes		Coast		Plains		River valleys		Total	
N transects	40		25		118		9		29		221	
Sampled area (m^2)	179,310		126,670		576,900		27,950		135,300		1,046,130	
Total number of remains	692		646		1465		48		302		3153	
Disarticulated bones	678		626		1414		46		291		3055	
Articulated bones	12		9		48		1		6		76	
Carcasses	2		11		3		1		5		22	
Density of remains	0.0039		0.0051		0.0025		0.0017		0.0022		0.0003	
Taxonomic category	N	%	N	%	N	%	N	%	N	%	N	%
Fish	0	0	15	2.3	48	3.3	0	0	1	0.3	64	2
Amphibians	0	0	1	0.2	1	0.1	1	2.1	1	0.3	4	0.1
Reptiles	53	7.7	0	0	3	0.2	0	0	0	0	56	1.8
Birds	28	4.0	51	7.9	472	32.2	0	0	4	1.3	555	17.6
Undetermined	1	0.1	40	6.2	47	3.2	0	0	1	0.3	89	2.8
Large (>20 kg)	0	0	0	0	13	0.9	0	0	1	0.3	14	0.4
Medium (1–20 kg)	17	2.5	2	0.3	261	17.8	0	0	0	0	280	8.9
Small (<1 kg)	10	1.4	9	1.4	151	10.3	0	0	2	0.7	172	5.5
Mammals	601	86.8	524	81.1	755	51.5	45	93.8	262	86.8	2187	69.4
Undetermined	0	1.3	4	0.6	12	0.8	0	0	0	0	25	0.8
Large (>20 kg)	183	26.3	429	66.4	295	20.1	20	41.7	182	60.3	1108	35.1
Medium (1–20 kg)	406	55.2	87	13.5	295	20.1	25	52.1	78	25.8	867	27.5
Small (<1 kg)	12	4.0	4	0.6	153	10.4	0	0	2	0.7	187	5.9
Undetermined	10	1.4	55	8.5	186	12.7	2	4.2	34	11.3	287	9.1
Total	**692**		**646**		**1465**		**48**		**302**		**3153**	**100**

medium-sized vertebrates, and 1126 large-sized vertebrates. The results of this study will be presented considering each environment: hills, shallow lakes, coast, plains, and river valleys. A brief geomorphological and ecological characterization based on published information of each environment is also included.

11.4.1 Hills

11.4.1.1 Environmental Characteristics

The hills landscape in which this research was developed belongs to the Tandilia Range. This environment presents a series of discontinuous elevations intersected by valleys, streams, semi-permanent watercourses and undulating plains. In general, the Tandilia Range contains low peaks (<520 masl), with extensive slopes which are more pronounced towards the top of the hills. The stratigraphy includes a lower portion crystalline base igneous and metamorphic rock, and an upper portion of sedimentary rocks (Dalla Salda 1999). The hills are surrounded and partially covered by late Cenozoic eolian sediments (Zárate and Rabassa 2005). Above these sediments, pedogenetic processes formed the current soils from the end of the Pleistocene and throughout the Holocene. These soils are fully developed in the surrounding plains and can be found with different degrees of development. Likewise, water erosion of rocks has generated various types of platforms, shallow openings (rock shelters), caves, and hollows (Flegenheimer and Zárate 1989; Martínez et al. 1999). Currently, the hills environment is dominated by fluvial and pluvial erosive processes that are more pronounced when vegetation is scarce or in the sectors where mining takes place. Runoff during periods of intense rains generates an important material drag and loss of soil (Campo de Ferreras and Piccolo 1999). Ecological records of the faunal communities in the central sectors of the Tandilia Range shows the presence of 224 vertebrate species: 187 birds, 21 mammals, 12 reptiles, 3 amphibians, and 1 fish (http://www.ecoregistros.org).

11.4.1.2 Taxonomic Diversity

Forty transects were conducted in the hills, covering a total of 179,310 m^2. The density of remains is 0.0039/m^2, and the visibility is regular. Vertebrate remains were classified as disarticulated bones (n = 678), occurrences of articulated bones (n = 12), and carcasses (n = 2) (Table 11.1). Taxonomic classes are distributed by mammals (86.8%), reptiles (7.7%), birds (4%), and undetermined taxa (1.4%) (Table 11.1). Medium-sized mammals are the most abundant group (Table 11.1; Fig. 11.3). European hare (*Lepus europaeus*) is the species with the highest frequency of bone remains recovered in this size category (79.1%) (Online Resource 1). The second most abundant taxonomic class is reptiles, exclusively assigned to Black and White tegu (*Tupinambis merianae*). Among the birds, medium-sized are predominant (60.7%), with the Chicken (*Gallus gallus*) being the most frequent (Table 11.1). Hills present an Ntaxa of 20, subdivided into the following classes: mammals (Ntaxa = 14), birds (Ntaxa = 5), and reptiles (Ntaxa = 1).

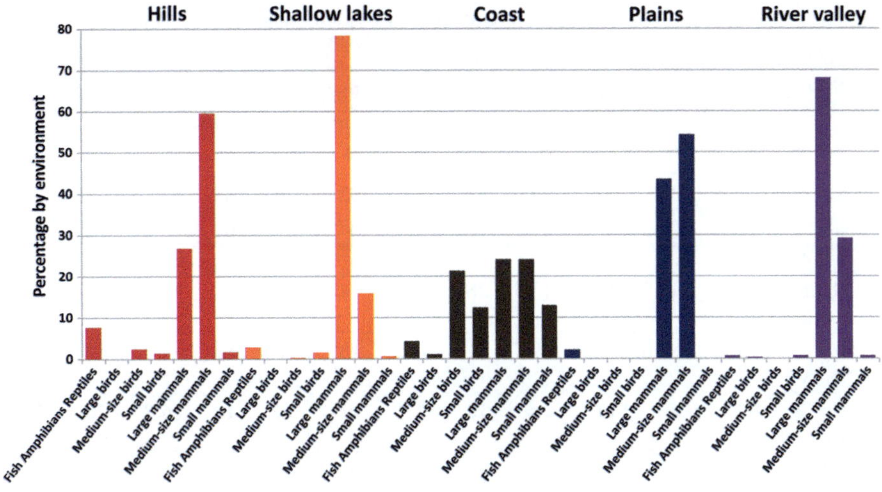

Fig. 11.3 Frequency of vertebrate classes in each environment

11.4.1.3 Taphonomic Modifications

A total of 681 bone remains were analyzed for taphonomic modifications, distributed as follows: 22 small-sized vertebrates, 476 medium-sized vertebrates, and 183 large-sized vertebrates. The bone assemblage recovered in the hills is characterized by a high frequency of carnivore marks (38%) (Table 11.2; Figs. 11.4 and 11.5d). Medium-sized vertebrates show the highest frequency (40.1%), followed by large-sized vertebrates (34.4%) (Table 11.2). Weathering (Fig. 11.5a) and fragmentation are taphonomic modifications that are well represented in this environment (34.9 and 40.5%, respectively). Bone assemblages of small-sized vertebrates do not show evidence of weathering. Medium-sized vertebrates present the highest percentage in Stage 0, although some remains reach as high as Stage 5 (Fig. 11.6). The weathering profile of large-sized vertebrates is more advanced, with all stages present. Despite its low total frequency (5.1%), the percentage of the chemical deterioration stands out in relation to other environments, mainly in small-sized vertebrates (22.7%) (Table 11.2).

11.4.2 Shallow Lakes

11.4.2.1 Environmental Characteristics

Geological evidence indicates that the current shallow lakes in the Pampas region are located in basins of varying ages, modeled by deflation and accumulation processes during dry and humid climates of the Late Pleistocene and Holocene (Frenguelli

Table 11.2 Distribution of taphonomic modifications by environment and taxonomic class size

Taphonomic modifications	Environments											
	Hills		Shallow lakes		Coast		Plains		River valleys		Total	
	N	%	N	%	N	%	N	%	N	%	N	%
Weathering	**238**	**34.9**	**222**	**34.4**	**617**	**52.4**	**32**	**69.6**	**86**	**32.3**	**1195**	**44.2**
Small vertebrate	0	0	0	0	125	40.3	0	0	0	0	125	35.5
Medium vertebrate	90	18.9	12	13.5	282	50.7	14	56.0	0	0	398	32.5
Large vertebrate	150	82.0	210	49.0	210	67.5	18	90.0	86	47	674	59.9
Root etching	**27**	**4**	**17**	**2.6**	**13**	**1.1**	**1**	**2.2**	**18**	**6.8**	**76**	**2.8**
Small vertebrate	0	0	0	0	3	1	0	0	0	0	3	0.9
Medium vertebrate	22	4.6	3	3.4	7	1.3	1	4	1	1.3	34	2.8
Large vertebrate	5	2.7	14	3.3	3	1	0	0	17	9.3	58	2.1
Carnivore marks	**259**	**38.0**	**81**	**12.5**	**135**	**11.5**	**6**	**13.0**	**30**	**11.3**	**511**	**18.9**
Small vertebrate	5	22.7	3	21.4	26	4.7	0	0	0	0	34	9.7
Medium vertebrate	191	40.1	14	15.7	71	12.8	3	12	13	16.7	292	23.9
Large vertebrate	63	34.4	64	14.9	38	12.2	3	15	17	9.3	185	16.4
Rodent marks	**3**	**0.4**	**1**	**0.2**	**1**	**0.1**	**0**	**0**	**0**	**0**	**5**	**0.2**
Small vertebrate	1	4.5	0	0	0	0	0	0	0	0	1	0.3
Medium vertebrate	2	0.4	0	0	0	0	0	0	0	0	2	0.2
Large vertebrate	0	0	1	0.2	1	0.3	0	0	0	0	2	0.2
Trampling	**9**	**1.3**	**18**	**2.8**	**2**	**0.2**	**0**	**0**	**6**	**2.3**	**35**	**1.3**
Small vertebrate	0	0	0	0	1	0.3	0	0	0	0	1	0.3
Medium vertebrate	3	0.6	1	1.1	0	0	0	0	0	0	4	0.3
Large vertebrate	6	3.3	17	4	1	0.3	0	0	6	3.3	30	2.7

(continued)

Table 11.2 (continued)

Taphonomic modifications	Environments											
	Hills		Shallow lakes		Coast		Plains		River valleys		Total	
	N	%	N	%	N	%	N	%	N	%	N	%
Chemical deterioration	**35**	**5.1**	**11**	**1.7**	**20**	**1.7**	**0**	**0**	**13**	**4.9**	**79**	**2.9**
Small vertebrate	5	22.7	0	0	3	1	0	0	0	0	9	2.6
Medium vertebrate	28	5.9	0	0	5	0.9	0	0	0	0	33	2.7
Large vertebrate	2	1.1	11	2.6	12	3.9	0	0	13	7.1	38	3.4
Thermal alteration	**1**	**0.1**	**0**	**0**	**4**	**0.3**	**0**	**0**	**1**	**0.4**	**6**	**0.2**
Small vertebrate	0	0	0	0	0	0	0	0	0	0	0	0
Medium vertebrate	1	0.2	0	0	0	0	0	0	0	0	1	0.1
Large vertebrate	0	0	0	0	4	1.3	0	0	1	0.5	5	0.4
Abrasion	**8**	**1.2**	**34**	**5.3**	**701**	**59.6**	**0**	**0**	**13**	**4.9**	**756**	**28.0**
Small vertebrate	0	0	5	35.7	216	18.4	0	0	0	0	221	62.8
Medium vertebrate	8	1.7	10	11.2	406	73	0	0	0	0	424	34.6
Large vertebrate	0	0	19	4.4	79	25.4	0	0	13	7.1	111	9.9
Manganese staining	**4**	**0.6**	**60**	**9.3**	**7**	**0.6**	**0**	**0**	**3**	**1.1**	**74**	**2.7**
Small vertebrate	0	0	0	0	1	0.3	0	0	0	0	1	0.3
Medium vertebrate	4	0.8	1	0.2	0	0	0	0	0	0	5	0.4
Large vertebrate	0	0	59	0.1	6	1.9	0	0	3	1.1	68	6.0
Calcium carbonate coating	**14**	**2.1**	**29**	**4.5**	**0**	**0**	**0**	**0**	**0**	**0**	**43**	**1.6**
Small vertebrate	0	0	0	0	0	0	0	0	0	0	0	0
Medium vertebrate	7	1.5	0	0	0	0	0	0	0	0	7	0.6

(continued)

Table 11.2 (continued)

Taphonomic modifications	Environments											
	Hills		Shallow lakes		Coast		Plains		River valleys		Total	
	N	%	N	%	N	%	N	%	N	%	N	%
Large vertebrate	7	3.8	29	6.8	0	0	0	0	0	0	36	3.2
Butchering marks	**9**	**1.3**	**14**	**2.2**	**1**	**0.1**	**1**	**2.2**	**9**	**3.4**	**34**	**1.3**
Small vertebrate	0	4.5	0	0	0	0	0	0	0	0	0	0
Medium vertebrate	0	0	0	0	0	0	0	0	0	0	0	0
Large vertebrate	8	4.4	14	3.3	1	0.3	1	5	13	7.1	37	3.3
Fracture	**276**	**40.5**	**196**	**30.3**	**430**	**36.5**	**33**	**71.7**	**143**	**53.8**	**1078**	**39.9**
Small vertebrate	12	54.5	2	14.3	135	43.5	0	0	0	0	149	42.3
Medium vertebrate	203	42.6	0	0	145	26.1	16	64.0	28	35.9	392	32.0
Large vertebrate	61	33.3	194	45.2	150	48.2	17	85.0	115	62.8	537	47.7

1956; Tricart 1973; Dangavs 2005). The size of the shallow lakes fluctuates, with dimensions ranging from <1 to 150 km^2. Small shallow lakes (<1 km^2) dominate the region (Dangavs 2005). The characteristics of this environment (shape, size, aggregation, distribution, hydro-chemical table, etc.) are highly dynamic and controlled by direct pluvial inputs, surface runoff and, above all, free groundwater (Quirós et al. 2002; Dangavs 2005; Grosman 2008; Grosman and Sanzano 2008). In the shallow lakes, aquatic birds such as herons, ducks, grebes, and coots are frequent (Darrieu and Bó 1992). A total of 53 native species of freshwater fish have been identified in the shallow lakes, including Pejerrey (*Odontesthes bonariensis*), Carp (*Cyprinus carpio*), Catfish (*Pimelodella laticeps*), and Tararira (*Hoplias malabaricus*) (Grosman 2008).

11.4.2.2 Taxonomic Diversity

Twenty-five transects were conducted in this environment, covering a total of 126,670 m^2. Visibility is regular and the density of remains is 0.0051/m^2. Vertebrate remains were classified as disarticulated bones (n = 626), carcasses (n = 11), and occurrences of articulated bones (n = 9) (Table 11.1). Taxonomic classes are distributed in mammals (81.1%), birds (7.9%), fish (2.3%), amphibians (0.2%), and indeterminate taxa (8.5%). Large-sized mammals are predominant (Table 11.1;

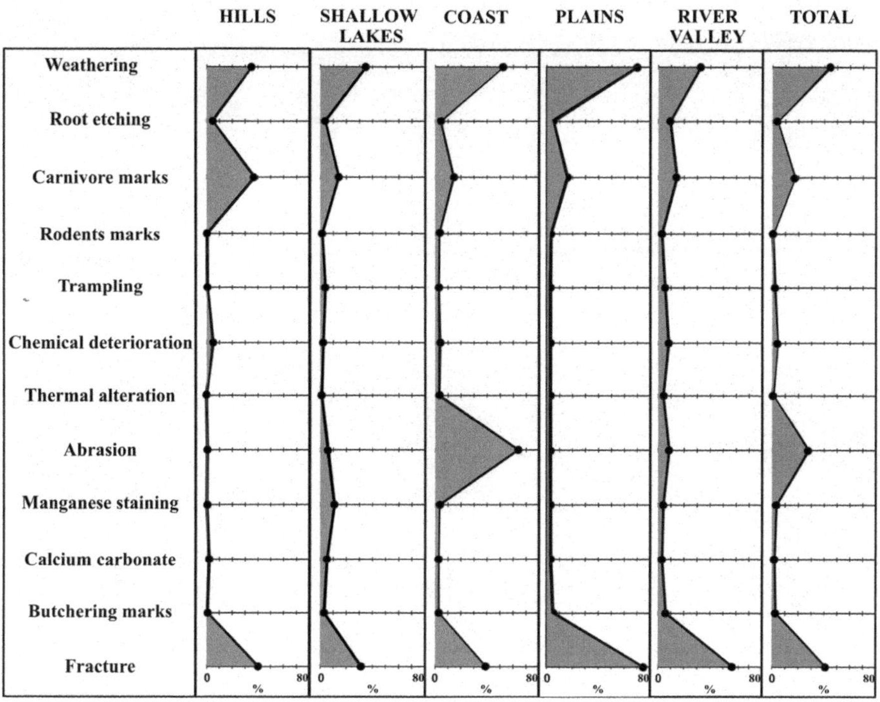

Fig. 11.4 Taphogram comparing taphonomic variables in the surveyed environments

Fig. 11.3). Undetermined domestic ungulates are the taxonomic category with the highest frequency of bone remains recovered within this size category (32.6%), followed by Cow (*Bos taurus*; 26.1%) (Online Resource 1). The second most abundant taxonomic class is birds, with a high percentage of indeterminate specimens (78.4%). Among the fish, the Order Siluriformes is predominant (66.7%). The total Ntaxa for the shallow lakes is 14, with a distribution by classes that shows the highest value for mammals (Ntaxa = 9), followed by birds (Ntaxa = 3), and then fish and amphibians (Ntaxa = 1).

11.4.2.3 Taphonomic Modifications

A total of 532 bone remains were analyzed for taphonomic modifications, distributed as follows: 14 small-sized vertebrates, 89 medium-sized vertebrates, and 429 large-sized vertebrates. Bone specimens mainly present modifications resulting from weathering (34.4%) (Table 11.2; Fig. 11.4). This process is manifested to a greater extent in large (49%) and medium (13.5%) vertebrates (Table 11.2). All small-sized vertebrates are in Stage 0 (Fig. 11.6). In the medium-sized category, the weathering profile only reaches Stages 1 and 2. For large-sized vertebrates,

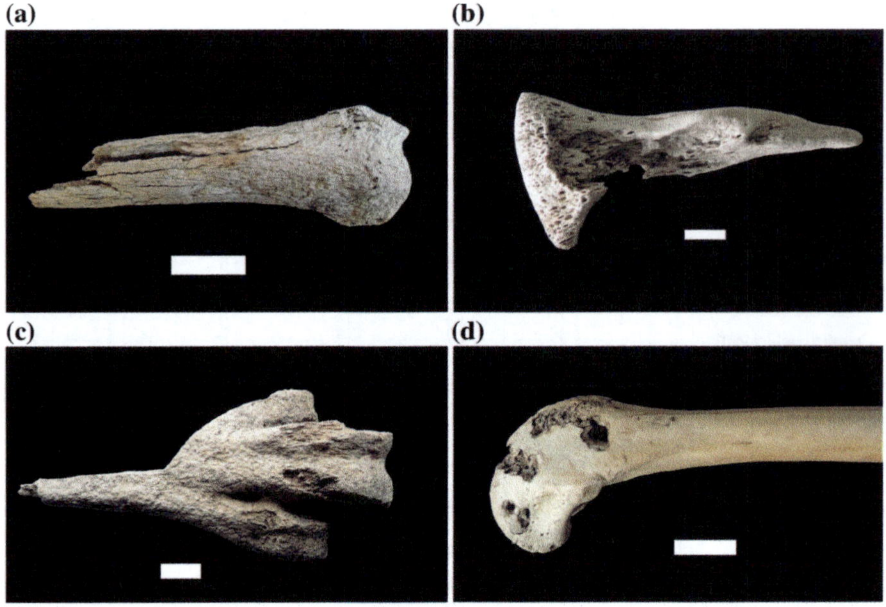

Fig. 11.5 Examples of taphonomic modifications. **a** Weathering on the distal radius of Canidae (coast); **b** abrasion on undetermined bone of large-sized mammal (fluvial valleys); **c** chemical deterioration on the distal tarsometatarsus of *Rhea Americana* (fluvial valleys), **d** carnivore punctures on the distal femur of *Lepus europaeus* (hills)

all stages are present, with a gradual decrease towards the more advanced stages. Other recurrent modifications are fragmentation (30.3%) and carnivore tooth marks (12.5%) (Table 11.2; Fig. 11.4). Fragmentation dominates in large-sized vertebrates (45.2%), while carnivore marks have similar frequencies in the three size categories (Table 11.2). With smaller frequency, but more important than in other environments, manganese (9.3%), abrasion (5.3%) (Fig. 11.5b), and calcium carbonate (4.5%) are observed.

11.4.3 Coast

11.4.3.1 Environmental Characteristics

The Pampas coast includes sandy beaches, cliffs, dune fields, and freshwater shallow lakes. During the Late Pleistocene and Holocene, the coastline has changed its position due to marine transgressions. Stabilization of the current environment occurred after the Middle Holocene (Ponce et al. 2011). The dune ridges are characterized by sand burial and wind erosion. Vegetation occurs only in the interdune areas, where

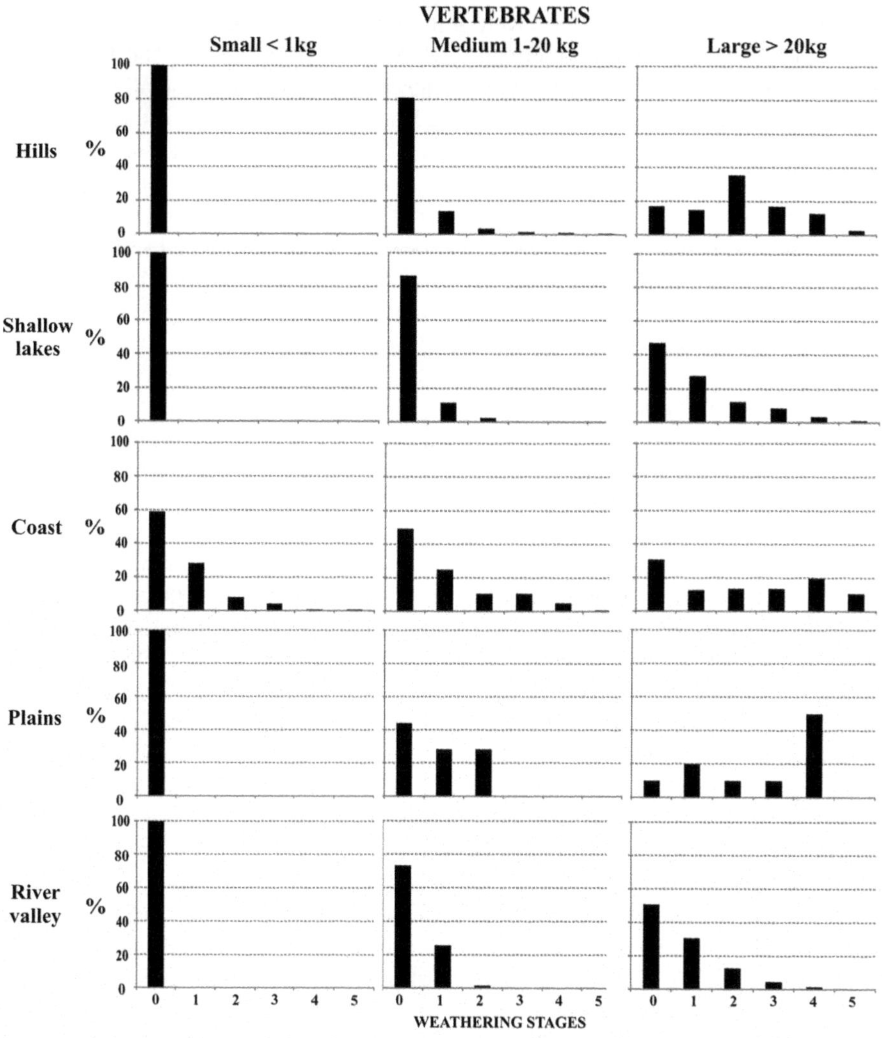

Fig. 11.6 Weathering profile per environment and class size

the freshwater table is at or just below the sand surface (Marcomini and López 2013). In the interdune areas, both in the active and stabilized dunes, the high level of the freshwater table and the obstruction of rainfall drainage by the dunes causes the formation of marshes and temporary shallow lakes, where abundant vegetation and fauna can be found. Behind the dune field, larger shallow lakes form, as the dunes impede the drainage of some watercourses from the plains (Frenguelli 1931; Bértola et al. 2009).

The environmental characteristics provide a nesting ground for numerous species of migratory birds (Darrieu and Camperi 2001). Among the marine species that breed in this coastal zone are the Kelp gull (*Larus dominicanus*), the Olrog's gull (*Larus belcheri*), the South American tern (*Sterna hirundinacea*), the Royal tern (*Sterna maxima*) and the Sandwich tern (*Sterna sandvicencis*). A small rodent that exclusively inhabits the coastal dunes of the Buenos Aires province is the Tuco-tuco (*Ctenomys australis*) (Celsi et al. 2010). The reptiles in the coastal dunes include lizards (e.g., *Liolaemus multimaculatus*, *L. wiegmanni*, *L. gracilis*), amphisbaena, snakes and vipers (e.g., *Liophis poecilogyrus* and *Philodryas aestivus*). Anurans (e.g., *Hyla pulchella*) inhabit the lower sectors where there is abundant vegetation and humidity. Among the marine fauna, Sea lions (*Otaria flavescens* and *Arctocephalus australis*) and Magallanic penguins (*Spheniscus magellanicus*) occasionally appear along the beach (Pütz et al. 2007). Currently, the Pampas coast is vulnerable to the erosive processes produced mainly by the circulation of off-road vehicles and the extraction of sand for urban construction.

11.4.3.2 Taxonomic Diversity

One hundred and eighteen transects were conducted in this environment, covering a total of 576,900 m^2. The density of remains is 0.0025/m^2, and there is excellent visibility. Vertebrate remains were classified as disarticulated bones (n = 1414), occurrences of articulated bones (n = 48), and carcasses (n = 3) (Table 11.1). Taxonomic classes are distributed by mammals (51.5%), birds (32.2%), fish (3.3%), reptiles (0.2%), amphibians (0.1%), and indeterminate taxa (12.7%) (Table 11.1). Mammals and birds are the most abundant taxonomic classes. Among mammals, large and medium-sized mammals share the same frequency (39.1%), including the Order Pinnipedia (38.6%) and European hare (30.5%), which are the best taxonomic categories represented, respectively (Table 11.1; Fig. 11.3; Online Resource 1). The Tuco-tuco (*Ctenomys* sp.) presents a high frequency of specimens and is the only small-sized mammal identified (Online Resource 1). The medium and small-sized categories of birds also presented similar percentages (Fig. 11.3). Magellanic penguin (*Spheniscus magellanicus*) is the most abundant identified species (85.1%). Fish is the third most abundant taxonomic class, with a high occurrence of bony fish (Actinopterygii; 47.9%) (Table 11.1). The total Ntaxa for the coast is 29, with the higher values for mammals (Ntaxa = 16), followed by birds (Ntaxa = 7), fish (Ntaxa = 4), and amphibians and reptiles (Ntaxa = 1 for both Classes).

11.4.3.3 Taphonomic Modifications

A total of 1177 bone remains were analyzed for taphonomic modifications, distributed as follows: 310 small-sized vertebrates, 556 medium-sized vertebrates, and 311 large-sized vertebrates. The main taphonomic modification in the coast is abrasion (59.6%) (Fig. 11.4), which occurs primarily in medium-sized vertebrates (73%) (Table 11.2). Another process that stands out is weathering (52.4%), with high frequencies for all three size categories (Table 11.2). Regarding the weathering profile, both the small and medium-sized vertebrates show a similar profile, with slightly more advanced stages in the medium-sized vertebrates. The weathering pattern for large-sized vertebrates is more homogenous, with similar frequencies for all the stages (Fig. 11.6). Bone fragmentation (36.5%) was identified in large (48.2%), small (43.5%), and medium-sized (26.1%) vertebrates. Finally, carnivore marks (11.5%) were equally represented in medium (12.8%) and large-sized (12.2%) vertebrates (Table 11.2).

11.4.4 Plains

11.4.4.1 Environmental Characteristics

This landscape consists of a flat plain with only slight undulations and a maximum elevation of around 200 masl (Fidalgo et al. 1991; Zárate and Rabassa 2005). The plains are composed of loess and loessoid sediments of Late Cenozoic age (Zárate 2005; Zárate and Rabassa 2005). Because the fine soil texture causes a low infiltration of water, during the periods of high water levels, surface laminar runoff occurs until it reaches the permanent river valleys (Munguía and Campo de Ferreras 2003). During events of excessive rainfall, the surface of the interfluvial plains environment is affected by water erosion. The soils are highly fertile, so almost all the landscape is exploited for agricultural and livestock purposes (Moscatelli and Puentes 2000). A census of birds and mammals in the Pampas grassland of the province of Buenos Aires showed the presence of 39 species of birds (e.g., *Rhea americana*, *Rhynchotus rufescens*, *Chloephaga rubidiceps*, *Pluvialis dominica*, *Sturnella defilippi*, *Vanellus chilensis*, *Milvago chimango*, *Anthus correndera*) and 15 of small mammals belonging to 9 Families (Didelphidae, Dasypodidae, Canidae, Mustelidae, Felidae, Cricetidae, Caviidae, Myocastoridae and Leporidae) (Comparatore et al. 1996).

11.4.4.2 Taxonomic Diversity

Nine transects were conducted in the plains, covering a total of 27,950 m^2. The density of remains is 0.0017/m^2, and there is very good to poor visibility. Faunal remains include disarticulated bones ($n = 46$), occurrences of articulated bones ($n = 1$), and carcasses ($n = 1$) (Table 11.1). The following taxonomic categories are present: mammals (93.8%), amphibians (2.1%), and undetermined *taxa* (4.2%) (Table 11.1). In the case of mammals, the medium and large-sized vertebrates are the most frequent (Table 11.1; Fig. 11.3). Plains vizcacha (*Lagostomus maximus*) and Cow (*Bos taurus*) are abundant (Online Resources 1). The total Ntaxa is 5, assigned to mammals (Ntaxa = 4) and amphibians (Ntaxa = 1).

11.4.4.3 Taphonomic Modifications

A total of 46 faunal remains were analyzed for taphonomic modifications, distributed as follows: 1 small-sized vertebrate, 25 medium-sized vertebrates, and 20 large-sized vertebrates. Fragmentation and weathering affected a high percentage of the specimens (71.7 and 69.6%, respectively) (Table 11.2; Fig. 11.4). Both processes are expressed with greater intensity in large-sized vertebrates (between 85 and 90%) (Table 11.2). Regarding the weathering profile, all small-sized vertebrate bones are in Stage 0 (Fig. 11.6). In medium-sized vertebrates, the absence of weathering predominates, and this process only reaches Stages 1 and 2. For large-sized vertebrates, the highest percentage of bones are in Stage 4 (Fig. 11.6). Again, the carnivore tooth marks are a frequent modification (13%), affecting to a greater extent the large (15%) and medium-sized (12%) vertebrates (Table 11.2).

11.4.5 River Valleys

11.4.5.1 Environmental Characteristics

The river valleys constitute early development watersheds integrated by a small number of irregular tributaries conditioned mostly by rainfall (Fidalgo et al. 1991; Fidalgo 1992). They do not present fluvial terraces but rather flood plains—poorly defined- that are only covered during maximum flood periods (Frenguelli 1950; Fidalgo et al. 1991). In the upper section of the river valleys, the sedimentary deposits are shallow, with ravine walls reaching a maximum of 2 m in height. In the lower basins, boxed meanders form deep ravines typically reach up to 8 m high (Campo de Ferreras and Piccolo 1999). The process of widening of the river valleys and the development of ravines can be intensified in some sectors by cattle trampling, which generates numerous gullies that are later eroded by surface runoff (Marini and Piccolo 2005). Unlike lentic environments (shallow lakes), lotic environments (rivers and streams) are characterized by basins of erosion and transport (Dangavs 2005). The freshwater species that inhabit this environment are the same as those of the shallow lakes. The rivers of the Pampas region work as ecological corridors for tropical species that expand their austral distribution in favor of increased humidity and temperature (Doumecq Milieu et al. 2012).

11.4.5.2 Taxonomic Diversity

Twenty-nine transects were conducted in this environment, covering a total of 135,300 m^2. The density of remains is 0.0022/m^2, and the visibility is regular. The vertebrate remains include disarticulated bones (n = 291), occurrences of articulated bones (n = 6), and carcasses (n = 5) (Table 11.1). Taxonomic classes include mammals (86.8%), birds (1.3%), fish (0.3%), amphibians (0.3%), and undetermined taxa

(11.3%) (Table 11.1). Mammals are the most abundant taxonomic classes (86.8%), being the large-sized mammals the category with the highest frequency (69.5%); however, the medium-sized category presents a high percentage (42.8%). Undetermined domestic ungulates are the mammals more represented within the large size category (43.4%), followed by Cow (*Bos taurus*; 21.4%), Sheep (*Ovis aries*; 12.63%) and Horse (*Equus caballus*; 2.7%). In the medium-sized category, Dog (*Canis familaris*; 44.9%) and Plains vizcacha (*Lagostomus maximus*; 33.3%) predominate (Table 11.1; Fig. 11.3; Online Resource 1).

11.4.5.3 Taphonomic Modifications

A total of 226 bones were analyzed for taphonomic modifications, distributed as follows: 183 large-sized vertebrates, 78 medium-sized vertebrates, and 5 small-sized vertebrates. The most frequent taphonomic modification is fractures (53.8%), affecting mainly large-sized vertebrates (62.8%) (Table 11.2; Fig. 11.4). Another process that shows a high frequency is weathering (32.3%); again, affecting both medium and large-sized vertebrates (Fig. 11.6). In medium-sized vertebrates, Stages 1 and 2 are recorded, but in low frequencies. For large-sized vertebrates, there is a gradual decrease of weathering from Stages 0 to 4 (Fig. 11.6). Carnivore tooth marks were recorded in 11.3% of the total remains, with the highest percentage in the medium-sized vertebrates (Table 11.2). Like in the hills, chemical deterioration (Fig. 11.5c) presents an important frequency (4.9%) but affects only large-sized vertebrates (Table 11.2).

11.5 Discussion

The main objective of this taphonomic research is to contribute to the understanding of the formation processes that configure the modern and fossil vertebrate bone assemblages in the Pampas region. To accomplish this objective, it is essential to estimate the natural "bone rain" and to measure the potential preservation of the remains of different vertebrates in distinctive environmental contexts (Behrensmeyer et al. 2000; Miller et al. 2014). Due to the characteristics of our work, we do not have direct information about the causes of animal deaths. However, considering ecological information and indirect data from our taphonomic modifications, in the case of large-sized vertebrates, which correspond mostly to livestock, the deaths are mainly due to diseases and environmental stress. Moreover, based on the record of butchering marks, killing for human consumption should also be considered as a cause of death for this size category. Regarding small and medium-sized vertebrates, natural deaths can include predation by small and medium-sized carnivores (Perovic and Pereira 2006).

In all environments, disarticulated bones (>95%) dominate over articulated ones, which suggest a low anatomical integrity of the carcasses. In addition to the microbial decomposition (DeVault et al. 2003), the early decay of carcasses may be accelerated by carnivore action, which is observed in all environmental contexts. Another taphonomic process that could favor the disintegration of carcasses is water transport, although this process is only prevalent in particular environments, such as the coast and the river valleys.

The comparison of the taxonomic richness (Ntaxa) contrasts with the current faunal biodiversity at a regional scale, which indicates a greater abundance of birds. In our sample, the mammals, especially large and medium-sized, are the more abundant in all the environments. There is also a low representation of fish and amphibians in the aquatic environments. Likewise, in the hills and coast, few reptiles were identified, which is also inconsistent with the current faunal biodiversity of these environments. The bias against smaller-sized taxa has also been noted in other naturalistic taphonomic studies, and it is mainly attributed to a greater bone destruction of these vertebrates (Behrensmeyer et al. 1978, 2003; Behrensmeyer and Dechant Boaz 1980; Borrero 2000; Miller et al. 2014). Different types of destruction processes where observed in our study, in particular, a high degree of weathering and carnivore action. The advanced stages of weathering observed in large-sized vertebrates (mainly mammals) suggest a prolonged surface exposure time related to a greater resistance of destruction processes. As other researchers have already noted, this taphonomic aspect establishes that large-sized vertebrate bone assemblages are more prone to time averaging (Behrensmeyer 1982; Borrero 2007; Cruz 2015). In relation to carnivore action, several ecological experiments point to a high effectiveness of scavenging on small-sized vertebrates in different types of environments (steppe, agricultural fields, temperate forests, tropical forests, tundra, grasslands, desert mountains); with a loss of 60–100% of the carcasses (DeVault et al. 2003; Schlacher et al. 2013).

In general, the density of vertebrate remains is low. This suggests that the chances of naturally deposited bones in the archaeological record at a regional scale are also low. However, there may be specific environments where "bone rain" is more likely, and in consequence, can have a higher impact on the archaeological record. To evaluate these specific situations; we will discuss the differences in the bone assemblages from each of the surveyed environments.

In the shallow lakes and river valleys, there is a greater frequency of large-sized mammal remains (mainly cow) and a low representation of birds and fish. Since animals are attracted to water sources, natural deaths due to etho-ecological reasons and carnivore predation can contribute to the deposition of vertebrate remains in these environments. However, in the case of livestock, one of the factors that explain its high representation is the common practice among farmers to discard the carcasses into bodies of water. This process establishes a high degree of uncertainty in the interpretation of the natural deposit rate of vertebrate remains in shallow lakes and river valleys.

As for the taphonomic processes acting in the shallow lakes and river valleys, weathering, carnivore action, and trampling are important. Although trampling is registered in a low percentage, its frequency is greater than in the other environments, which could be due to recurrent visits by livestock in search of water. In the case of shallow lakes, a higher frequency of manganese stains and calcium carbonate is observed. This is related to greater humidity and the susceptibility of these environments to periodic water fluctuations. On the other hand, in the fluvial valleys, there is a greater frequency of root etching and chemical deterioration. The presence of these modifications suggests the occurrence of re-exposure of skeletal remains, which is consistent with sedimentary erosion in these contexts. Likewise, in both aquatic environments, the highest percentage of buried bones was recorded, which could be the result of fluvial and lacustrine sedimentation and trampling.

Most archaeological sites from the Pampas region are found near fluvial valleys and shallow lakes. Carnivores or fluvial transport can destroy or transport bones from carcasses. At the same time, in these locations there are high chances of burial because of trampling and bioturbation. Common bioturbation processes include disturbance by plants and burrowing mammals that can favor the rapid burial of the bones and thus the contamination of archaeological deposits with modern remains. A clear example of this mixture is the presence of domestic ungulates remains in archaeological deposits dated to pre-Hispanic periods (e.g., Messineo et al. 2013; Scheifler and Messineo 2016). This process is particularly important for discussing topics in the regional archaeology record, for example, if Guanacos were available in the Pampas when the Europeans arrived (Tonni and Politis 1980; Silveira and Crivelli Montero 1982; Crivelli Montero et al. 1987–88). Our results indicate that contamination of archaeological sites with naturally deposited bones is expected in fluvial valleys and shallow lakes. The presence of introduced fauna in a pre-Hispanic site warns us of this mixture, and that it is important to consider that the same processes of contamination with native taxa can occur in the present and in the past.

As for the hills, there is a dominance of remains of medium-sized mammals (mainly European hare) followed by large-sized mammals (mainly domestic ungulates). The high frequency of carnivore marks on the different vertebrate size categories, in particular, the medium-sized mammals, suggests that this agent played an important role in the formation of bone assemblages in this environment. The accumulation of bones is related in part to the ethology of carnivores to transport their prey to rock shelters and cavities. Although we do not know the particular carnivore species responsible for the accumulation of remains, we consider that both medium and small-sized carnivores could be involved, probably canids, felines, mustelids, and mephitids (Gutiérrez et al. 2016). Taphonomic studies with camera traps in hill environments of South Africa have shown that the same rock shelter can be used in a short period of time by multiple animals (e.g., Brown hyaena, Wild boar, Badger, Jackal, Porcupine, Leopard), generating complex palimpsests (Bountalis and Kuhn 2014). The important taphonomic role of carnivores in the Pampas hills can explain in part the high density of remains, many of which appear as part of bone concentrations, and the high taxonomic richness. The taxonomic variability partly reflects the diet of these predators. Likewise, previous studies have shown that along the hills, some of the remains of domestic fauna are transported directly from nearby farm houses located along the base of the hills. These locations act as resource catchment areas for carnivores (Gutiérrez et al. 2016: 267).

Another modification registered in the hills is weathering, which is more frequent and shows more advanced stages in large-sized vertebrates. The differences in the weathering profiles among the size categories may be due to their differences in resistance to destructive processes, an aspect that was discussed earlier. However, the characteristics of the hills may also condition the intensity of weathering. Smaller-sized vertebrates come mainly from sheltered locations, such as the base of rock walls or cavity interiors, while the remains of larger vertebrates were more frequent along the hilltops (where livestock tend to graze). Another prominent aspect of this environment that conditions bone preservation is pluvial erosion (Campo de Ferreras and Piccolo 1999), which may help explain in part the low percentage of buried bones.

Regarding the possibilities of contamination of archaeological sites in the hills, the transport of prey by carnivores will enhance the possibilities of mixing bones deposited naturally with archaeological remains, since these are also favorable places for human occupations (Mazzanti 2006). Even though our results indicate low burial possibilities, the recurrence in the use of these spaces by carnivores over time suggests that, in the long term, the impact of the natural bone rain on the archaeological record may be significant, especially when considering the shallow deposits in this environment (Martínez and Mazzanti 2017).

With respect to the coast, the taxonomic richness and the more even representation of vertebrates of different body sizes better reflects the natural diversity of living fauna. This could be in part because of the excellent visibility (Massigoge et al. 2015). However, sampling bias cannot be ruled out, since this environment was the largest surveyed area, which could have also increased the richness values in the sample. Much of the bone assemblage from the coast is dominated by penguins. As discussed in a previous paper from this environment (Massigoge et al. 2015), the abundance of penguin remains is related to the beaching of this seabird during seasonal migration. Other well-represented taxa are the Tuco-tuco and Sea lions. The Tuco-tuco is a small endemic rodent, with high population densities, that builds underground galleries in the dunes near the coast (Mora et al. 2006). The high representation of this rodent differs from that recorded in other naturalistic studies, where there is a bias against burrowing animals (Miller et al. 2014). In our case, we consider that the high frequency of Tuco-tuco remains in the coastal transects is because eolian action along the active dunes destroys their burrow systems and exposes the skeletal remains of individuals who died inside the burrows.

After the deposit of carcasses in the coastal environment, different natural processes contributed to soft tissue decay, disarticulation, spatial dispersal of bones, and finally, bone destruction. Taphonomic information points to an important role of abrasion caused by sand particles transported by water and wind. This suggests relatively long exposure times to atmospheric agents, coinciding with advanced weathering profiles for all size categories of vertebrates. Regarding the modifications by carnivores, the coast presents similar percentages to the rest of the environments, with the exception of the hills. The low percentage of carnivore marks in these types of open environments could be related to the fact that transects did not intercept carnivore dens. A prominent effect identified in previous studies, however, suggests that carnivore action along the coast is an important variable, especially when considering the transport of bones from the beach to the interior sites (Massigoge et al. 2015; Gutiérrez et al. 2016). Comparatively, ecological information on the East coast of Australia shows that scavengers (carnivores, reptiles, and birds) act as important biological vectors among different environments, taking advantage of the carrion deposited on the beach and moving the carcasses or depositing their scats or pellets in terrestrial context (Schlacher et al. 2013).

Archaeological sites from the Pampas coast mainly consist of surface assemblages composed almost exclusively of lithic artifacts in the blowout depressions of active dunes (Bonomo 2005). Our results show that there are high chances of the natural incorporation of modern bones into these assemblages, especially in the sites located close to the shoreline (Massigoge et al. 2015). In some transects along the coast, we recorded both lithic artifacts and modern faunal remains (Massigoge et al. 2015). This mixture could have also occurred in the past if some of the environmental conditions remained the same.

Finally, in the plains, a very low diversity of species is observed, with a clear predominance of medium-sized vertebrates (mainly Plains vizcacha) and large-sized vertebrates (primarily Cow). It is important to mention that most of the remains of Plains vizcacha were recorded just in a few transects that crossed an area with a high concentration of their burrows (Rafuse et al. 2019). Due to this situation, and the small size of the sampled area in the plains environment, it is not possible to ensure that the taxonomic representation registered so far is representative of the plains. As for the acting taphonomic processes, it is important to recognize the percentage and high frequency in advanced stages of weathered bones. This process may help to explain in part the high fragmentation of the sample. Weathering data, together with a low percentage of buried bones, suggests prolonged exposure times. Although the results of the plains are preliminary, the chances of bone preservation and burial appear to be lower in this environment.

11.6 Conclusions

The naturalistic taphonomic research in the Pampas region presented here is designed to help better understand the natural formation processes of the fossil record, particularly those processes affecting different classes of vertebrates. While sampling is still needed, particularly in some environmental contexts, our results show some interesting trends. First, Pampean carnivores have an important role in the accumulation, dispersal, and destruction of bones, especially smaller-sized vertebrates (<20 kg). Second, rock shelters and cavities of the hills present a higher rate of natural accumulation of bones, since they are places commonly occupied by carnivores. Third, the chances of burial of modern bones are higher in the river valleys and shallow lakes and lower for the rest of the environmental contexts. Forth, weathering contributes to bone destruction in all the environmental contexts but appears more frequently in the coast and the plains. Fifth, more advanced weathering stages in large-sized vertebrate indicate the importance of time-averaging on surface assemblages of these animals.

Sixth, bone destruction, primarily by carnivores and weathering, is more intense for small-sized vertebrates. Lastly, in general, the impact of the natural "bone rain" would be more substantial in the hills and shallow lakes. So far, our results suggest that taxonomic representation in the archaeofaunal assemblages from the Pampas region can only be confidently used for subsistence and paleoenvironmental interpretations after careful consideration of taphonomic biases, particularly those related to body size and mixing of cultural and natural bones. Future studies should focus on testing these general trends, especially through longitudinal taphonomic studies of carcasses of different-sized vertebrates, combined with controlled experiments on particular processes and agents.

References

Barnosky AD, Lindsey EL (2010) Timing of quaternary megafaunal extinction in South America in relation to human arrival and climate change. Quat Int 217:10–29. https://doi.org/10.1016/j.quaint.2009.11.017

Behrensmeyer AK (1993) The bones of Amboseli: bone assemblages and ecological change in a modern African ecosystem. Natl Geogr Res 9:402–421

Behrensmeyer AK (1982) Time resolution in fluvial vertebrate assemblages. Paleobiology 8:211–227

Behrensmeyer AK (1978) Taphonomic and ecologic information from bone weathering. Paleobiology 4:150–162. https://doi.org/10.2307/2400283

Behrensmeyer AK, Dechant Boaz DE (1980) The recent bones of Amboseli Park, Kenya, in relation to East African paleoecology. In: Behrensmeyer AK, Hill AP (eds) Fossils in the making. University of Chicago Press, Chicago, pp 72–92

Behrensmeyer AK, Kidwell SM, Gastaldo RA (2000) Taphonomy and paleobiology. In: Erwin DH, Wing SL (eds) Deep time: paleobiology's perspective. The Paleontological Society, Lawrence, KS, pp 103–147

Behrensmeyer AK, Stayton CT, Chapman RE (2003) Taphonomy and ecology of modern avifaunal remains from Amboseli Park, Kenya. Peleobiology 29:52–70

Behrensmeyer AK, Western D, Dechant Boaz DE (1978) New perspectives in paleoecology from a recent bone assemblage, Amboseli Park, Kenya. Paleobiology 5:12–21

Belardi JB, Carballo Marina FC (2003) Tafonomía regional en la cuenca media del río Coyle (Santa Cruz, Patagonia Argentina). Intersecc en Antropol 4:59–73

Bértola GR, Cortizo LC, Isla FI (2009) Dinámica litoral de la costa de Tres Arroyos y San Cayetano, Buenos Aires. Rev la Asoc Geol Argentina 64:657–671

Bilenca D, Miñarro F (2004) Identificación de áreas valiosas de pastizal (AVPs) en las pampas y campos de Argentina, Uruguay y sur de Brasil. Fundación Vida Silvestre Argentina, Buenos Aires

Bilenca DN, Abba AM, Corriale MJ et al (2017) De venados, armadillos y coipos: los mamíferos autóctonos frente a los cambios en el uso del suelo, los manejos agropecuarios y la presencia de nuevos elementos en el paisaje rural. Mastozool Neotrop 24(2):277–287

Binford LR (1981) Bones: ancient men and modern myths. Academic Press, New York, United States

Blumenschine RJ (1988) An experimental model of the timing of hominind and carnivore influence on archaeological bone assemblages. J Archaeol Sci 15:483–502

Bonomo M (2005) Costeando las llanuras: arqueología del litorial marítimo pampeano. Sociedad Argentina de Antropología, Buenos Aires

Borella F (2004) Tafonomía Regional y Estudios Arqueofaunísticos de Cetáceos en Tierra del Fuego y Patagonia Meridional. B.A.R. International Series, 1257. Archaeopress, Oxford
Borrero LA (1988) Estudios tafonómicos en Tierra del Fuego: su relevancia para entender procesos de formación del registro arqueológico. In: Yacobaccio HD (ed) Arqueología Contemporánea Argentina. Ediciones Búsqueda, Buenos Aires, pp 13–32
Borrero LA (1989) Sites in action: the meaning of guanaco bones in Fueguian archaeological sites. ArchaeoZoologia III:9–24
Borrero LA (1990) Taphonomy of guanaco bones in Tierra del Fuego. Quat Res 34:361–371. https://doi.org/10.1016/0033-5894(90)90047-O
Borrero LA (2000) Ten years after: esquema para una tafonomía regional de la Patagonia meridional y norte de Tierra del Fuego. In: Desde el país de los gigantes : perspectivas arqueológicas en Patagonia : jornadas de Arqueología de la Patagonia, pp 183–193
Borrero LA (2001) Regional taphonomy: background noise and the integrity of the archaeological record. In: Kuznar LA (ed) Ethnoarchaeology of Andean South America. Contributions to archaeological method and theory. International Monographs in Prehistory, Ann Arbor, Michigan, pp 243–254
Borrero LA (2007) Longitudinal taphonomic studies in Tierra Del Fuego, Argentina. In: Gutiérrez MA, Miotti L, Barrientos G et al (eds) Taphonomy and Zooarchaeology in Argentina. BAR International Series 1601. Archaeopress, Oxford, pp 219–233
Borrero LA (2014) Taphonomy, regional. In: Smith C (ed) Encyclopedia of global archaeology SE-832. Springer, New York, pp 7232–7235
Borrero LA, Lanata JL, Cárdenas P (1991) Re-estudiando cuevas: Nuevas Excavaciones en Ultima Esperanza. An del Inst la Patagon Ser Ciencias Soc 20:101–110
Borrero LA, Martin FM (1996) Tafonomía de carnívoros: un enfoque regional. In: Otero G (ed) Arqueología. Sólo Patagonia, CENPAT-CONICET, Puerto Madryn, pp 189–198
Bountalis AC, Kuhn BF (2014) Cave usage by multiple taphonomic agents: issues towards interpreting the fossil bearing cave deposits in South Africa. Am J Zool Res 2:55–61
Campo de Ferreras AY, Piccolo C (1999) Hidrogeomorfología de la cuenca del Río Quequén Grande, Argentina. Papeles Geogr Univ Murcia 29:35–46
Celsi CE, Mac-Lean HD, Yezzi A, Triches M (2010) Dunas Costeras de la Pampa Austral. Biodiversidad, ecología y conservación entre el río Quequén Salado y el balneario Pehuen-có. Buenos Aires. Cintia Celsi, Buenos Aires
Comparatore VM, Martínez MM, Vassallo AI et al (1996) Abundancia y relaciones con el hábitat de aves y mamíferos en pastizales de *Paspalum quadrifarium* (Paja Colorada) Manejados con fuego (Provincia de Buenos Aires, Argentina). Interciencia 21(4):228–237
Crivelli Montero E, Eugenio E, Silveira M (1987–88) El sitio Fortín Necochea (partido de General La Madrid, provincia de Buenos Aires). El material de superficie. Paleoetnológica 4:7–37
Cruz I (2007) Avian taphonomy: observations at two Magellanic penguin (*Spheniscus magellanicus*) breeding colonies and their implications for the fossil record. J Archaeol Sci 34:1252–1261. https://doi.org/10.1016/j.jas.2006.10.016
Cruz I (2015) Las investigaciones sobre preservación de huesos de aves y mamíferos grandes en Patagonia (Argentina). Archaeofauna 24:209–224
Dalla Salda L (1999) Cratón del Río de la Plata. El basamento granítico-metamórfico de Tandilia y Martin García. An Geol Argentina Subsecr Minería 29:97–106
Dangavs N (2005) Los ambientes acuáticos de la Provincia de Buenos Aires. In: de Barrio RE, Etcheverry RO, Caballé MF, Llambías E (eds) Geología y recursos minerales de la provincia de Buenos Aires. 16° Congreso Geológico Argentine. Relatorio, La Plata, pp 219–236
Darrieu CA, Bó NA (1992) Nuevos aportes a la distribución de las razas de Zenaida auriculata en la República Argentina (Aves, Columbidae). Rev del Mus La Plata, nueva Ser 14:137–146
Darrieu CA, Camperi AR (2001) Nueva lista sistemática de las aves de la provincia de Buenos Aires. Serie CO-BIOBO 3/PROBIOTA 2. Secretaria de Política Ambiental (FCNyM, UNLP), La Plata

DeVault TL, Rhodes JROE, Shivik JA (2003) Scavenging by vertebrates: behavioral, ecological, and evolutionary perspectives on an important energy transfer pathway in terrestrial ecosystems. Oikos 102:225–234. https://doi.org/10.1034/j.1600-0706.2003.12378.x

Domínguez-Rodrigo M (2001) A study of carnivore competition in riparian and open habitats of modern savannas and its implications for hominid behavioral modelling. J Hum Evol 40:77–98

Doumecq Milieu RE, Morici A, Nigro NA (2012) Ampliación de la distribución austral del carpincho (*Hydrochoerus hydrochaeris*) en la provincia de Buenos Aires, Argentina. Nótulas Faunísticas, Segunda Ser 92:1–10

Faith JT, Behrensmeyer AK (2006) Changing patterns of carnivore modification in a landscape bone assemblage, Amboseli Park. J Archaeol Sci 33:1718–1733

Fidalgo F (1992) Provincia de Buenos Aires Continental. In: Iriondo M (ed) El Holoceno en Argentina. CADINQUA, pp 23–28

Fidalgo F, Riggi JC, Gentile R et al (1991) Los 'sedimentos postpampeanos' continentales en el a´mbito sur bonaerense. Rev la Asoc Geológica Argentina 46:239–256

Flegenheimer N, Zárate MA (1989) Paleoindian occupation at Cerro El Sombrero locality, Buenos Aires Province, Argentina. Curr Res Pleistocene 6:12–13

Frenguelli J (1931) Nomenclatura estratigráfica patagónica. An la Soc Ciencias St Fe 1–115

Frenguelli J (1950) Rasgos Generales de la Morfología y la Geología de la Provincia de Buenos Aires. LEMIT 2:72

Frenguelli J (1956) Rasgos generales de la hidrografía de la Provincia de Buenos Aires. LEMIT Serie II:1–19

Gifford D (1984) Taphonomic Specimens, Lake Turkana. Washington, D.C.

Grayson DK (1984) Quantitative zooarchaeology: topics in the analysis of archaeological faunas. Academic Press, Orlando

Grayson DK (1991) Alpine faunas from the White Mountains, California: adaptative change in the late prehistoric Great Basin? J Archaeol Sci 18:483–506

Grosman F (2008) Espejos en la llanura. Nuestras lagunas de la región pampeana. Universidad Nacional del Centro de la provincia de Benos Aires, Tandil

Grosman F, Sanzano P (2008) Las lagunas seleccionadas y rutinas de muestreo. In: Grosman F (ed) Espejos en la llanura. Universidad Nacional del Centra de la provincia de Buenos Aires, Tandil, Nuestras lagunas de la región Pampeana, pp 39–46

Gutiérrez MA, Kaufmann CA (2007) Criteria for the identification of formation processes in guanaco (*Lama guanicoe*) bone assemblages in fluvial-lacustrine environments. J Taphon 5:151–175

Gutiérrez MA, Kaufmann CA, González ME et al (2016) The role of small carnivores in the movement of bones: implications for the Pampas archaeofaunal record, Argentina. Archaeol Anthropol Sci 8:257–276. https://doi.org/10.1007/s12520-015-0272-1

Gutiérrez MA, Rafuse DJ, Álvarez MC et al (2018) Ten years of actualistic taphonomic research in the Pampas region of Argentina: contributions to regional archaeology. Quat Int. 492:40–52. https://doi.org/10.1016/j.quaint.2017.09.025

Haynes G (1980) Evidence of carniore gnawing on Pleistocene and recent mamalian bones. Paleobiology 6:341–351

Haynes G (2018) Raining more than cats and dogs: looking back at field studies of noncultural animal-bone occurrences. Quat Int 466:113–130

Hill AP (1976) On carnivore and weathering damage to bone. Curr Anthropol 17:335–336

Lucherini M, Soler L, Luengos Vidal EM (2004) A preliminary revision of knowledge status of felids in Argentina. J Neotrop Mammal 11:7–17

Lyman RL (1984) Bone density and differential survivorship of fossil classes. J Anthropol Archaeol 3:259–299. https://doi.org/10.1016/0278-4165(84)90004-7

Marcomini SC, López RA (2013) Erosion and management in coastal dunes. In: Finkl CW (ed) Coastal hazards. Springer, Netherlands, Dordrecht, pp 511–553

Marini F, Piccolo MC (2005) Hidrogeomorfología de la cuenca del río Quequén Salado, Argentina. Investig. Geográficas 37:59–71

Martínez G, Gutiérrez MA (2004) Tendencias en la explotación humana de la fauna durante el Pleistoceno final y Holoceno en la Región Pampeana (Argentina). In: Mengoni Goñalons G (ed) Zooarchaeology of South America. BAR International Series 1298, Oxford, pp 81–98

Martínez GA, Mazzanti DL (2017) Evidencia geoarqueológica de la transición Pleistoceno-Holoceno en reparos rocosos de Tandilia Oriental (Provincia de Buenos Aires). Relac la Soc Argentina Antropol XLII(1):83–106

Martínez GA, Osterrieth M, Mazzanti D (1999) Estratigrafía de sitios arqueológicos en reparos rocosos en las sierras de La Vigilancia y Valdés, sistema de Tandilia, provincia de Buenos Aires. In: Actas del XII Congreso Nacional Arqueología Argentina, pp 139–144

Massigoge A, Rafuse DJ, Álvarez MC et al (2015) Beached penguins on the Atlantic Coast in the Pampas region of Argentina: taphonomic analysis and implications for the archaeological record. Palaeogeogr Palaeoclimatol Palaeoecol. https://doi.org/10.1016/j.palaeo.2015.06.045

Matteucci S (2012) Ecorregión Pampa. In: Morello J, Matteucci S, Rodríguez A, Silva M (eds) Ecorregiones y complejos ecosistémicos argentinos. Orientación Gráfica, Buenos Aires, pp 391–446

Mazzanti D (2006) La constitución de territorios sociales durante el Holoceno Tardío. El caso de las sierras orientales de Tandilia, Argentina. Relac la Soc Argentina Antropol XXXI:277–300

Messineo P, Álvarez MC, Favier Dubois C et al (2013) Estado de avance de las investigaciones arqueológicas en el sitio Empalme Querandíes 1 (centro de la subregión Pampa Húmeda, provincia de Buenos Aires). Comechingonia 17:123–148

Miller JH, Behrensmeyer AK, Du A et al (2014) Ecological fidelity of functional traits based on species presence—absence in a modern mammalian bone assemblage (Amboseli). Paleobiology 40:560–583. https://doi.org/10.1666/13062

Miñarro FO, Ortiz UM, Bilenca DN, Olmos F (2008) Río de la Plata Grasslands or Pampas & Campos (Argentina, Uruguay and Brazil). In: Michelson A (ed) Temperate grasslands of South America. World Temperate Grasslands Conservation Initiative Workshop, Hohhot-China, pp 24–33

Mora MS, Lessa EP, Kittlein MJ et al (2006) Phylogeography of the subterranean rodent Ctenomys australis in sand-dune habitats: evidence of population expansion. J Mammal 87(6):1192–1203. https://doi.org/10.1644/05-MAMM-A-399R1.1

Moscatelli G, Puentes MI (2000) Atlas de Suelos. In: Durán D, Casas R (eds) La Argentina Ambiental II. Lugar Editora, Buenos Aires

Munguía S, Campo de Ferreras AM (2003) Características Hidro-Geomorfológicas De La Cuenca Del Arroyo Pescado Castigado. Buenos Aires. Argentina. Papeles Geogr 38:137–150

Nowak RM (1991) Walker's Mammals of the World, fifth. The John Hopkins University Press, Baltimore

Olsen SL, Shipman P (1988) Surface modification on bone: trampling versus butchery. J Archaeol Sci 15:535–553. https://doi.org/10.1016/0305-4403(88)90081-7

Paruelo JM, Jobbágy E, Oesterheld M et al (2007) In: Paruelo, JM; M Bertiller; T Schlichter, FR Coronato, Veblen T, Young K, Orme A (eds) The physical geography of South America. The Oxford Regional Environments Series. Oxford University Press, Oxford, pp 232–248

Perovic P, Pereira JA (2006) Felidae. In: Bárquez R, Díaz M, Ojeda R (eds) Mamíferos de la Argentina. Sistemática y distribución. Sociedad Argentina para el Estudio de los Mamíferos (SAREM), Mendoza, pp 93–100

Politis GG, Pedrotta V (2006) Recursos faunísticos y estrategias de subsistencia en el este de la región pampeana durante el Holoceno tardío: el caso del guanaco (*Lama guanicoe*). Relac la Soc Argentina Antropol XXXI:301–336

Ponce JF, Rabassa J, Coronato A, Borromei AM (2011) Palaeogeographical evolution of the Atlantic coast of Pampa and Patagonia from the Last Glacial Maximum to the Middle Holocene. Biol J Linnean Soc 103:363–379. https://doi.org/10.1111/j.1095-8312.2011.01653.x

Prevosti FJ, Ramírez MA, Schiaffini M et al (2015) Extinctions in near time: new radiocarbon dates point to a very recent disappearance of the South American fox Dusicyon avus (Carnivora: Canidae). Biol J Linnean Soc 116(3):704–720. https://doi.org/10.1111/bij.12625

Pütz K, Schiavini A, Raya Rey A et al (2007) Winter migration of Magellanic penguins (*Spheniscus magellanicus*) from the southernmost distributional range. Mar Biol 152:1227–1235

Quirós R, Rennella A, Boveri M et al (2002) Factores que afectan la estructura y el funcionamiento de las lagunas pampeanas. Ecol Austral 12:175–185

Rafuse DJ, Kaufmann CA, Gutiérrez MA et al (2019) Taphonomy of modern communal burrow systems of the Plains vizcacha (*Lagostomus maximus*, Chinchillidae) in the Pampas region of Argentina: implications for the fossil record. Hist Biol 35:517–534. https://doi.org/10.1080/08912963.2017.1374957

Scheifler NA, Messineo PG (2016) Exploitation of faunal resources by hunter-gatherers in the center of the pampa grasslands during the Holocene. The archaeofauna of the Laguna Cabeza de Buey 2 site (San Carlos de Bolivar, Buenos Aires, Argentina). Quat Int 391:61–73. https://doi.org/10.1016/j.quaint.2015.08.078

Schlacher TA, Strydom S, Connolly RM, Schoeman D (2013) Donor-control of scavenging food webs at the land-ocean interface. PLoS One 8. https://doi.org/10.1371/journal.pone.0068221

Shipman P (1981) Life history of a fossil. An introduction to taphonomy and paleoecology. Harvard University Press, Cambridge

Silveira M, Crivelli Montero E (1982) El Sitio Fortín Marías II. Informe preliminar. Actas del VII Congr Nac Arqueol 128–135

Soriano A, León RJC, Sala OE et al (1992) Rio de la Plata grasslands. In: Coupland RT (ed) Natural grasslands. Elsevier; Ecosystems of the World, vol 8A, pp 367–407

Tappen M (1995) Savanna ecology and natural bone deposition: implications for early hominid site formation, hunting, and scavenging. Curr Anthropol 36:223–260. https://doi.org/10.1086/204353

Tonni EP, Politis GG (1980) La distribución del guanaco (MAMMALIA, CAMELIDAE) en la provincia de Buenos Aires durante el Pleistoceno Tardío y Holoceno. Los factores climáticos como causas de su retracción. Ameghiniana XVII:53–66

Tricart J (1973) Geomorfologia de la pampa deprimida, XII. Colecci6n Científica. INTA, Buenos Aires

Zárate M (2005) El Cenozoico tardío continental de la provincia de Buenos Aires. In: de Barrio RE, Etcheverry RO, Caballé MF, Llambías E (eds) *Geología y Recursos Minerales de la Provincia de Buenos Aires. Relatorio del XVI Congreso Geológico Argentino*. Ministerio de la Producción Gobierno de la Provincia de Buenos Aires/Asociación Geológica Argentina, La Plata, pp 139–158

Zárate M, Rabassa J (2005) Geomorfología de la Provincia de Buenos Aires. In: de Barrio RE, Etcheverry RO, Caballé MF, Llambias E (eds) Geología y Recursos Minerales de la Provincia de Buenos Aires, XVI Congreso Geológico Argentino. La Plata, Argentina, pp 119–138

Chapter 12
Expanding the Scope of Actualistic Taphonomy in Archaeological Research

Karen Borrazzo

Abstract This chapter presents the application of actualistic taphonomy to the study of one of the inorganic remains produced by hominins since 3 million year BP up to historical times: lithic artifacts. As rocks are among the most durable raw materials employed by modern humans and their ancestors, differential preservation has conferred a leading role in archaeological research upon lithic artifacts. Indeed, lithics—flaked artifacts in particular—are the *proxy* for culture or anthropic presence most commonly used by scholars all over the world. This artifact-human relationship promoted actualistic research on flintknapping in archaeology but no similar effort was devoted to assessing alternative non-cultural (i.e. taphonomic) sources for flaked stone objects. Even though actualistic studies have already shown that taphonomic processes may produce lithic pseudomorphs, this fact is only rarely considered in archaeological practice and research design. Furthermore, it is commonly assumed that human products are different enough from any natural specimen to be detected by lithic analysts. However, the current lack of knowledge on non-cultural flaking processes and their byproducts prevents their identification in the archaeological record, thus undermining the accuracy and reliability of archaeological interpretations. This paper illustrates the contribution of actualistic taphonomy to study the inorganic remains of the archaeological record and its critical role in assessing the cultural versus natural origin of lithic specimens in Fuego-Patagonia (South America). Naturalistic and experimental research on rockfall and trampling presented here suggests that the effects of these taphonomic processes result in pseudoartifacts that progressively incorporate to the regional archaeological record.

Keywords Lithic taphonomy · Pseudoartifacts · Middle-range research · Experimental archaeology · Fuego-Patagonia

---

K. Borrazzo (✉)
CONICET, Instituto Multidisciplinario de Historia y Ciencias Humanas (CONICET-IMHICIHU), Buenos Aires, Argentina
e-mail: kborrazzo@yahoo.com.ar

Facultad de Filosofía y Letras, Universidad de Buenos Aires (FFyL-UBA), Buenos Aires, Argentina

S. Martínez et al. (eds.), *Actualistic Taphonomy in South America*, Topics in Geobiology 48, https://doi.org/10.1007/978-3-030-20625-3_12

12.1 Introduction

Although the study of the fossil record nowadays encompasses dealing with numerous topics and methods to understand the history of its formation, taphonomy is unquestioningly a mandatory constituent of any paleo-research (e.g. Behrensmeyer et al. 1992; Behrensmeyer et al. 2000; Kowalewski and Labarbera 2004; Domínguez-Rodrigo et al. 2011). In archaeology, there is currently a general agreement among scholars dealing with a zooarchaeological and bioanthropological record on the necessity and benefits of incorporating taphonomy as a regular component of research (Gifford 1981, Lyman 1994; Pobiner and Brown 2005; Gutiérrez et al. 2007). More recently, several researchers highlighted the contribution of a taphonomic perspective to the study of other archaeological remains (Hiscock 1985; Valin et al. 2001; Barton et al. 2002; Borrazzo 2006; Mallol and Bertran 2010; Thiébaut et al. 2010; Domínguez-Rodrigo et al. 2011; Borrazzo and Weitzel 2014; Yeshurun et al. 2014, among others). Thus, theoretical and methodological approaches of taphonomy are becoming progressively (and decidedly) an integral part of archaeological research on the formation history of all the components of the record. As Borrero put it, archaeology should pursue an unrestricted or multi-service taphonomy to benefit from the integration (comparison and contrast) of taphonomic signatures on different types of remains (Borrero 2011, 2014).

In archaeology as in paleontology, actualistic research focuses on the study of present-day patterns and processes in contemporary settings to learn about the relationship between processes and effects. These lessons from the present aid and guide the interpretations of fossil data on the unobservable past processes that formed the historical records (Binford 1977, 1981; Behrensmeyer and Kidwell 1985; Gifford-Gonzalez 1989; Kowalewski 1999; Kowalewski and Labarbera 2004; Pobiner and Brown 2005). However, the recent character of present-day observations may limit their relevance to explain some features of the fossil record (Kowalewski 1999). Nevertheless, actualistic approach is always informative since it provides access to material situations in which the linkage between the responsible actor (taphonomic agent) and a trace (taphonomic effect) is unequivocal (Marean 1995; Pobiner and Brown 2005; Lin et al. 2017). The material expectations of models and hypothesis derived from actualistic research can be tested against the fossil record; differences and similarities emerging from the comparison of fossil data against expectations may suggest future directions of research (Lin et al. 2017).

The protocol outlined by Marean (1995) has been widely adopted as a general model for actualistic taphonomy in archaeology (e.g. Pobiner and Brown 2005; Álvarez and Alunni 2017). He differentiates two components in actualistic research: naturalistic and experimental studies. The former includes the direct observations of natural situations and sets the agenda for experiments; the latter improves the knowledge of trace-actor linkage as the analyst controls some parameters of the process (Marean 1995). Kowalewski and Labarbera (2004) make further distinction within actualistic research by identifying three strategies to acquire data, an observational approach (direct field observation and sample collection in modern settings) and two experimental approaches (field experiments and data collected in laboratory settings). According to Lin et al. (2017), experimental processes in archaeology include pilot studies and second-generation experiments. While the former detects potential material relationships, the latter verify the existence of those linkages by following a protocol to ensure repeatability and allow quantifiable results (Lin et al. 2017).

Due to the increasing recognition of the key role of taphonomy in the study of fossil records and the regional availability of "natural taphonomic laboratories", actualistic taphonomy has experienced a significant growth and diversification in South America (Ritter et al. 2016). In spite of the large contribution of actualistic taphonomy to archaeological research in the region (e.g. Cruz 2007; Gutiérrez et al. 2007; Massigoge and González 2012; Álvarez and Alunni 2017), its application to the study of lithicartifacts—usually the most frequent remains in the record—is still scarce (Borrazzo 2011a, 2013; Balirán 2014; Weitzel et al. 2014; Méndez Muñoz 2015; Carranza 2017; Carranza Elola and Méndez 2017).

This chapter aims to illustrate the contributions of expanding the scope of actualistic taphonomy into current topics of South American archaeology. It presents an actualistic research—that includes both naturalistic and experimental studies—focused on the taphonomic production of flaked stone objects (named pseudoartifacts, mimics or pseudomorphs) and its identification in the archaeological record. More specifically, the research summarized in this paper aims for contributing to two topics in archaeology. Firstly, while the early concern with the recognition of the complete repertoire of human flaked artifacts has prompted systematic archaeological and actualistic research in flintknapping (e.g. Johnson et al. 1978 and references therein) no comparable effort was devoted to natural or accidental flaking process (but see Warren 1914; Mason 1965; Nash 1993; Hosfield and Chambers 2003; Lopinot and Ray 2007; Demeter et al. 2010; Carranza Elola 2015, among others). Hence, the uneven knowledge on human and taphonomic flaking processes and byproducts prevents archaeology from achieving a more comprehensive and accurate understanding of the patterns displayed by the lithic record. Secondly, pseudoartifact is a topic relevant for current discussions in South American archaeology among which the lack of a taphonomic research program undermines argumentation (Aschero et al. 2017; Prentiss et al. 2015, 2016; Garvey and Mena 2016; Boëda et al. 2014, 2016; Aimola et al. 2014; Lahaye et al. 2013; Parenti 2015; Fiedel 2017; Fariña et al. 2014; Suárez et al. 2014; see Borrero 2015, 2016 for further discussion).

The results presented here are part of a larger lithic taphonomy research program conducted in Fuego-Patagonia (southern South America) for the last 15 years. Here I focus on the study of Casa de Piedra Roselló (CP), an archaeological site located in Chubut Province, Argentina (Castro Esnal et al. 2017a, b) (Fig. 12.1). Departing from naturalistic regional observations, two sets of experimental studies using the raw material available at the site were designed to assess the pseudoartifact component (or local taphonomic background noise, Borrero 2001, 2015). Actualistic data is applied to the analysis of an archaeological lithic sample collected at CP talus (Fig. 12.1).

(a) (b)

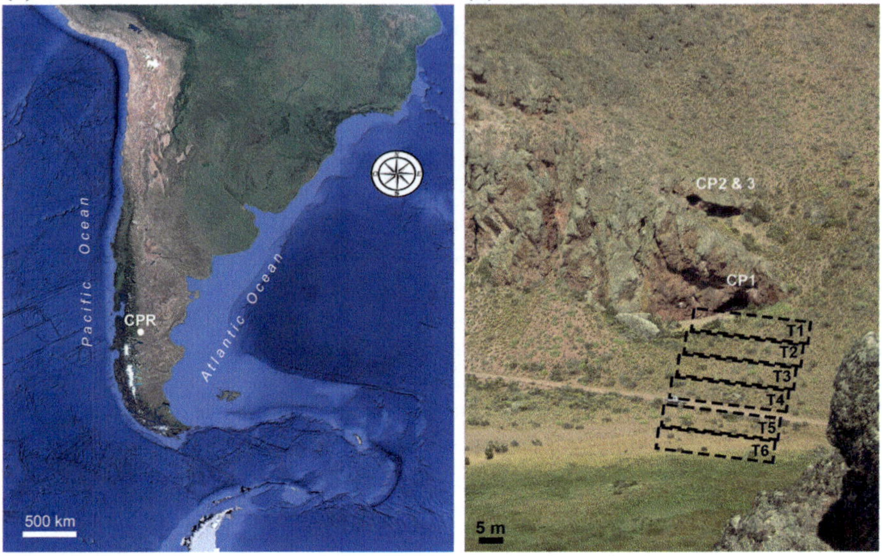

Fig. 12.1 **a** Location of the study area. **b** Casa de Piedra Roselló archaeological site and talus surface survey

12.2 The Role of Actualistic Taphonomy in the Archaeological Research on Pseudoartifacts

From a geological point of view, artifacts can be defined as a subset of rock fragments whose morphological properties were transformed by manufacture or human use. Thus, considering the morphological spectrum represented within the rock fragments universe, artifacts can usually be distinguished from natural clasts by displaying several unique traits that result from human agency and that are not replicated by natural or accidental processes (i.e. taphonomic processes). Nevertheless, there exists a morphological space shared by taphonomic as well as cultural lithic products. Pseudoartifacts, lithic mimics, or pseudomorphs are the natural lithic pieces that resemble artifacts (Breuil and Lantier 1965; Haynes 1973; Gillespie et al. 2004), while geofacts is the term reserved for geogenic mimics.

Among pseudoartifacts, the degree of similarity of their form (i.e. shape and size) with stone artifacts is also variable. In some cases, their non-cultural origin is easily established after performing a regular techno-morphological analysis. That is the case of the fragments produced by thermal fatigue ("cup-like" scars) or haloclasty (Warren 1914; Barnes 1939; Breuil and Lantier 1965; Andrews et al. 2004). But still remains a smaller portion of the morphological spectrum that can result either from cultural or taphonomic processes and that may be not detected by regular lithic analysis only (e.g. Mason 1965; Nash 1993; Borrazzo 2011b). Furthermore, from a technological perspective, the morphology of those pseudoartifacts may be interpreted as informal tools, by-products of rudimentary, simple technologies or resulting from the application of an expedient strategy within a context plenty of lithic raw material (e.g. Chlachula and Le Blanc 1996; Gillespie et al. 2004). Lyman (1984) addressed similar issues on pseudo-bone tools study.

A detailed examination of the natural and cultural pieces that look identical shows that, in spite of their different origins, the physical mechanisms operating during their production are the same: pressure and percussion (Warren 1914; Mason 1965; Cotterell and Kamminga 1987).

This chapter addresses the study of those common morphologies produced on stone by percussion and pressure as a result of either cultural or taphonomic processes (Barnes 1939; Mason 1965; Duvall and Venner 1979; Gillespie et al. 2004; Andrefsky 2014; Wiśniewski et al. 2014). I propose that before advancing behavioral interpretations, archaeologists need to undertake the taphonomic study of lithic assemblages and their context in order to assess the potential contribution of pseudoartifacts to the archaeological record (Mason 1965). Indeed, this is especially important for any archaeological context where raw material is locally available and not only a requirement for sites exhibiting old dates or exclusively "simple or expedient technologies".

Today, several issues condition the advance of pseudomorph research in archaeology. Firstly, the available taphonomic frame of reference is too general, ambiguous, and incomplete. We need detailed information on the effects that different processes acting upon different raw materials produce (e.g. Nash 1993; Luedtke 1986; Balirán 2014). Secondly, several discussions on pseudoartifacts have stressed the differences and overlooked the similarities exhibited by naturefacts (e.g. Boëda et al. 2014, 2016; Prentiss et al. 2015). Also, they have focused on mimics produced by mechanisms that do not operate in the flintknapping process (such as thermal stress or frost weathering, Andrews et al. 2004), which are relatively easy to detect with current lithic analysis protocols because their morphological attributes differ significantly from those of man-made artifacts. Note that the latter does not include the worrisome application of authoritarian criteria based on analyst expertise and knapping skill level to define the anthropogenic or non-anthropogenic origin of lithics (Mason 1965; Garvey and Mena 2016; Prentiss et al. 2016; Lin et al. 2017). Finally, we currently lack the knowledge on the lithic taphonomic background noise (Borrero 2001, 2015) of each study region, i.e. the aspect of rock fragments resulting from past and present taphonomic processes operating on locally available lithic raw materials.

Since archaeological knowledge about natural or taphonomic flaking is very limited, an actualistic taphonomy approach to pseudoartifacts research is adequate to provide information on the morphological effects of percussion and pressure on different lithic raw materials under non-technological situations.

12.3 Case Study

Casa de Piedra Roselló archaeological site (CP) (45.3°S, 71.2°W) is located within a ravine in the forest-steppe ecotone of Chubut Province (Castro Esnal et al. 2017a, b) (Argentina, Fig. 12.1a). It consists of three rock shelters (CP1, 2 and 3) eroded in the ignimbrite of Carrenleufú Formation. The cavities are located ca. 12 m above the bottom of the gully; both spaces are connected by a ca. 30 m long talus whose slope is 0–10° (Fig. 12.1b). The discontinuous vegetation cover of the talus includes bushes (*Berberis* sp. and *Senecio* sp.) and grasses (*Festuca* sp.). The substrate is fine sediment with gravel primarily derived from in situ weathering of the ignimbrite outcrop.

Cattle from the Roselló Ranch and wild animals frequent the CP ignimbrite outcrop looking for shelter. Dung, tracks, footprints, and carcasses record their presence at the site.

Excavations undertook at the main cave (CP1, Fig. 12.1b) provided evidence for human occupations between ca. 9000 yr BP and historical times. The stratigraphic sequence of CP1 has yielded an assemblage of lithic artifacts adequate for the study of changes and continuities in lithic technology in the area throughout the Holocene (Castro Esnal et al. 2017a, b).

All layers in CP1 stratigraphy included debris and tools on obsidian—largely from Pampa del Asador source, 270 km south of the site (Castro Esnal et al. 2017b). However, the predominant lithic raw material in the stratigraphic sequence since the early occupations is a local microcrystalline silicate that appears in veins within the Carrenleufú Formation (Castro Esnal et al. 2017a). Indeed, good flaking quality veins were observed on blocks as well as the walls of rock shelters 2 and 3 of CP (Fig. 12.1b). Nevertheless, current information suggests that artifacts made on CP chert were not deposited beyond the site.

12.4 Materials and Methods

12.4.1 Actualistic Observations

Pseudoartifacts research at Casa de Piedra Roselló implemented the two components of actualistic taphonomy. Naturalistic studies consisted of field observations of archaeological and non-archaeological loci within different geomorphic contexts of Mendoza, Neuquén, Río Negro, Chubut, Santa Cruz and Tierra del Fuego Provinces (Argentina) (Borrazzo 2006, 2011a, b, 2016; Borrazzo and Borrero 2015). They also included the collection of natural specimens to build taphonomic reference assemblage. Along with pseudoartifact sampling, this exploration of non-cultural lithic morphologies offered a general image on the range of taphonomic processes regionally available and their effects on different lithic raw materials. It is worth mentioning that taphonomic specimens collected for reference fulfilled several requirements (such as in situ refitting and the presence of detached fragments in anatomical position, Borrazzo 2016). In addition, detailed environmental information on the context of recovery of each sample was recorded. The information gathered through naturalistic studies guided the subsequent experimental research reported here.

Naturalistic data suggested that rockfall and trampling are two primary taphonomic processes with the energy required to change lithic morphology at a rock shelter context such as CP. Also, taphonomic specimens collected underscored the variability displayed by different raw materials subjected to the same taphonomic process. Departing from these general observations, experimental research was conducted to assess (1) the capacity of rockfall and trampling to produce pseudoartifacts on local CP chert nodules and, if that was the case, (2) providing a general description of the morphological attributes exhibited by those mimics. Together these data would allow estimating the contribution of taphonomic processes to local lithic assemblages. Overall, experiments seek to improve the accuracy of the link between processes and their effects on local chert at CP.

12.4.2 Experimental Research

Unlike rockfall effects, the modifications induced by trampling on the archaeological record are one of the topics more abundantly addressed in formation processes research (see Eren et al. 2010; Weitzel et al. 2014 and references therein). Notwithstanding the availability of experimental information on these taphonomic processes (especially for trampling), the variability recorded on naturalistic regional data indicated that it is imperative to conduct experiments using local raw materials to improve the conditions for taphonomic analysis. Therefore, two sets of experiments were designed to approach the formation processes of lithic assemblages at CP. All experimental pieces were painted with water-based paint to improve the visibility of modifications after rockfall and trampling processes in the subsequent analyses (Fig. 12.2).

In rockfall experiments, natural clasts of chert collected at CP talus (N = 22, Table 12.1) were deposited on a flat, dry loamy soil and a nodule of the same chert (initial weight: 222.4 g) was dropped (freefall[1] mode of motion, see Dorren 2003) from 1, 2, and 3 m high. The effects of rockfall impacts were recorded on the specimens deposited on the substrate. The freefalling rock (percussor) used during all rockfall experimental series was the same because neither its shape nor its weight underwent significant changes after rockfalls (final weight: 208.7 g). Chert pebbles and slabs used in the experiment have a fairly fresh surface. They are mostly angular, spherical to plate-like in shape what agrees with the scarce to no transport suffered by the nodules derived from in situ weathering.

[1]Although at least two other modes of motion take place on a talus after freefalling rockfalls (bouncing and rolling, Dorren 2003), experiments reported here focused on freefalling byproducts.

Fig. 12.2 Rockfall experiment. **a**, **b** Chert nodules before and after their preparation for rockfall experiments. **c–h** Specimen number 11 from its preparation and after receiving twelve rockfall impacts from 1 m high. **i** Specimen collected in Transect 2 on CP talus surface

Pieces in the 1 m height rockfall were subject to impacts individually until the specimen broke. As it was almost impossible to hit with the hammer an individual piece from 2 and 3 m high, pieces were deposited in groups of six and the hammer was dropped a hundred times per group (only effective hits were counted; when the hammer did not impact any specimen in the group, the thrown was repeated until it did).

Trampling was performed by a 60 kg individual wearing leather sole shoes. Experimental pieces (N = 42 flakes, Table 12.1) were manufactured by flintknapping from CP chert nodules. Two plots were set on different substrates: a soft substrate (dry loamy soil) and a hard substrate (paving stone). Each experiment included four series of 10 min trampling (between 70 and 80 passes).

Table 12.1 Descriptive statistics of CP surface sample and experimental materials

	CP Surface assemblage		Experimental assemblage	
	Weight (g)	Size (mm)	Weight (g)	Size (mm)
N	151	151	64	64
Min.	0.05	10	0.05	15
Max.	169	80	166.2	75
Sum.	4172.2	6030	1663.4	2270
Mean	27.63046	39.933770	25.99062	35.46875
Std. error	2.799017	1.214325	5.07707	2.26231
Variance	1183.009	222.662300	1649.70600	327.55460
Stand. dev.	34.394900	14.921870	40.61658	18.09847
Median	13.2	40.0	3.3500	30.0
25 prcntil	5.3	30.0	1.0500	20.0
75 prcntil	40.1	50.0	41.6250	50.0
Skewness	2.114434	0.447631	1.85229	0.73236
Kurtosis	4.648714	−0.357027	2.90474	−0.70989
Geom. mean	12.274010	37.071640	5.27611	31.30994
Coeff. var.	124.481800	37.366540	156.27400	51.02652

12.4.3 Archaeological Sample

A surface survey was conducted on the main talus of CP (Fig. 12.1b). It included six parallel 25 by 5 m transects placed perpendicular to the talus main slope. Sampling units 1–6 (top to bottom of the slope) recorded variations in slope gradient (mean transect values: 4, 6, 8, 10, 5 and 0°, respectively). A gravel road separates transects 5 and 6 from the rest of the units (Fig. 12.1b). The sample obtained (N = 151, Table 12.1) was analyzed from a taphonomic and technological perspective (e.g. Borrazzo 2006, 2011a, b).

12.5 Results

12.5.1 Experimental Studies

12.5.1.1 Rockfalls

During the rockfall experiments, freefalls from all three heights produced crushes and retouches (i.e. small flake removals) on deposited chert pebbles and slabs. Impacts from 1, 2, and 3 m also detached larger flakes, although they seem to occur more often in higher rockfalls.

Table 12.2 Technological description of rockfall experimental specimens larger than 5 mm

Lithic class	1 m	2 m	3 m	Total
Debitage	14	25	23	62 (66.67%)
Tools	5	5	0	10 (10.75%)
Cores	1	3	2	6 (6.45%)
Unmodified pebble/slab	6	5	4	15 (16.13%)
Total	**26**	**38**	**29**	**93**
Debitage	**14**	**25**	**23**	**62**
Number of dorsal flake scars				
0	9	17	16	42 (67.74%)
1	4	7	7	18 (29.03%)
2	1	0	0	1 (1.61%)
3	0	1	0	1 (1.61%)
Flaked stone tools	**5**	**5**	**0**	**10**
Tool types ($N = 13$)				
Indet. retouched edge	3	4	0	7 (58.85%)
Notch	2	0	0	2 (15.39%)
Knife	0	1	0	1 (7.69%)
Denticulate	1	0	0	1 (7.69%)
Sidescraper	0	1	0	1 (7.69%)
Endscraper	1	0	0	1 (7.69%)
Cores	**1**	**3**	**2**	**6**
Number of blows				
1	1	0	0	1 (16.67%)
2	0	2	1	3 (50.00%)
3	0	1	1	2 (33.33%)

Note that the sum of tool types is larger than the count of 'flaked stone tools' because several specimens exhibit more than one tool type on their edges

After 483 freefalls, the original sample (22 nodules) resulted in 98 specimens. However, five small flakes (~5 mm) are excluded from the following analysis in order to maximize the comparability of experimental results with CP talus surface sample. No specimen smaller than 10 mm was recovered in the surface collection and its absence may be due to visibility conditions during the survey. Therefore, the experimental rockfall sample analyzed here onwards is composed of 93 pieces (Table 12.2). Based on their morphological features, the 16.13% of the experimental sample remained classified as pebble/slab without artifact attribute after techno-morphological analysis (Table 12.2).

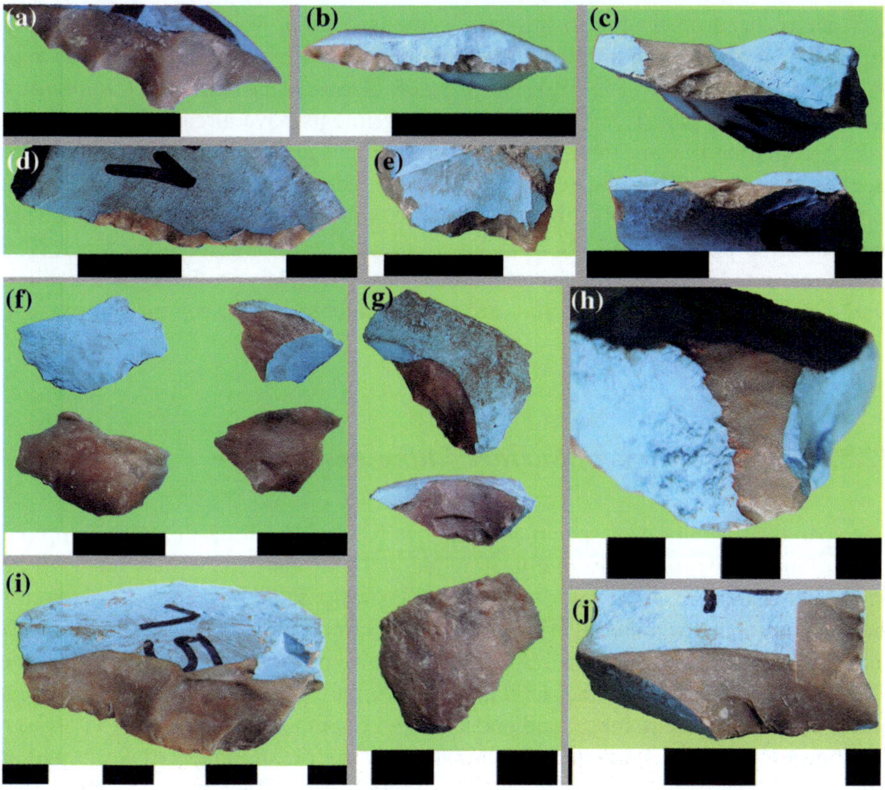

Fig. 12.3 Experimental pieces after trampling (**a**–**e**) and rockfalls (**f**–**j**)

Debitage (flakes, angular shatters, and debris) and tools comprise ca. 80% of the experimental rockfall sample (Table 12.2; Fig. 12.3). Cores represent 6.45% of the assemblage. Among debitage, pieces with none to one dorsal flake scar comprise 96.77% of the sample, although specimens without flake removals are the most frequent (Table 12.2). Only 1 and 2 m high rockfalls produced stone tools, which are dominated by nonspecific retouched edges (ca. 60%, Table 12.2, Fig. 12.3). However, several typological groups were recorded (notch, denticulate, knife, sidescraper and endscraper). Experimental specimens resembling cores exhibit only a few isolated blows; therefore, no formal core categories were identified in the experimental sample.

12.5.1.2 Trampling

After trampling experimental flakes, mimics of stone tools were recorded. They exhibited several techno-typological groups (e.g. Bordes 1961) that include different kinds of continuous retouched edges, serrated edges, retouched points, notches, and denticulates.

Fracture occurrence due to trampling process was high in both plots (55% in soft and 63.64% in the hard plot). However, the frequency of pseudotools was high on hard substrate only (45.45%). No pseudotool was identified after trampling in the soft substrate plot. Even though other experiments have already informed that flaking and fractures are more common when treadage occurs on a hard substrate (see Weitzel et al. 2014 and references therein), additional studies are required to further test our result on soft substrate. The tool types obtained after trampling on the hard substrate were notches (N = 4), long retouched edges (knife/sidescraper, N = 3) and restricted retouched edges (e.g. cutter, N = 5). Flakes produced by trampling were small (≤5 mm) and therefore are not considered in subsequent comparative analyses.

12.5.2 The Analysis of Surface Lithic Assemblage

The spatial distribution of specimens was heterogeneous along the talus sampling units (Table 12.3), probably due to gravitational and visibility issues that will be addressed elsewhere. As it was observed in CP stratigraphic sequence, local chert is the most frequent raw material in the talus surface sample as well (83.44%). Non-chert rocks are especially frequent in transect 4 which also offered the largest lithic sample (Table 12.3). Tools (42%) and debitage (38.46%) are the most represented artifact classes in the non-chert subset and endscraper is the dominant tool type (54.55%).

Within local chert specimens in the CP talus sample, debitage (flakes, angular shatters, and debris) and tools are the most represented artifact classes (Table 12.4). Among the former, pieces with none to two dorsal flake scars comprise ca. 70% of

Table 12.3 Lithic raw material composition of CP talus surface collection

Raw material	T1	T2	T3	T4	T5	T6	Total
CP chert	16	17	13	34	29	17	126
Non-local chert	0	0	0	5	1	0	6
Rhyolite	0	0	0	1	2	1	4
Chalcedony	0	0	0	3	1	0	4
Shale	0	1	0	1	0	0	2
Basalt	0	0	0	1	0	0	1
Jasper	0	0	0	1	0	0	1
Obsidian	0	1	0	0	0	0	1
Opal	0	0	0	1	0	0	1
Quartz	0	0	0	1	0	0	1
Ignimbrite bedrock	1	0	0	0	0	0	1
Indet. rock	0	1	0	1	0	1	3
Total	**17**	**20**	**13**	**49**	**33**	**19**	**151**

Table 12.4 Technological description of CP chert specimens in talus surface sample

Lithic class	T1	T2	T3	T4	T5	T6	Total
Debitage	12	10	8	20	15	7	72 (57.14%)
Tools	3	6	4	7	13	10	43 (34.13%)
Cores	1	1	1	4	1	0	8 (6.35%)
Indet. modified edges	0	0	0	3	0	0	3 (2.38%)
Total	**16**	**17**	**13**	**34**	**29**	**17**	**126**
Debitage	**12**	**10**	**8**	**20**	**15**	**7**	**72**
Number of dorsal flake scars							
0	2	2	0	4	3	1	12 (16.67%)
1	3	1	1	6	5	2	18 (25.00%)
2	5	1	3	6	4	1	20 (27.78%)
3	1	3	2	0	1	1	8 (11.11%)
4	1	2	2	1	2	0	8 (11.11%)
5	0	1	0	1	0	1	3 (4.17%)
6	0	0	0	0	0	1	1 (1.38%)
Indet	0	0	0	2	0	0	2 (2.78%)
Flaked stone tools	**3**	**6**	**4**	**7**	**13**	**10**	**43**
Tool type ($N = 58$)							
Sidescraper	0	2	0	1	7	4	14 (24.14%)
Notch	1	2	1	2	3	4	13 (22.42%)
Endscraper	0	2	0	2	2	1	7 (12.07%)
Indet. retouched edges	0	2	0	0	1	3	6 (10.35%)
Knife	1	0	1	1	2	0	5 (8.62%)
Woodscraper	0	0	1	1	2	0	4 (6.90%)
Denticulate	1	0	0	0	1	1	3 (5.17%)
Cutter	0	0	1	0	0	0	1 (1.72%)
Point	0	0	1	0	0	0	1 (1.72%)
RBO (restricted endscraper)	1	0	0	0	0	0	1 (1.72%)
Indet.	2	0	0	0	1	0	3 (5.17%)
Cores	**1**	**1**	**1**	**4**	**1**	**0**	**8**
Number of blows							
1	0	0	0	1	0	0	1 (12.50%)
2	1	0	0	2	0	0	3 (37.5%)
3	0	0	0	0	0	0	0 (0.00%)
4	0	0	1	0	1	0	2 (22.00%)
5	0	0	0	1	0	0	1 (12.50%)
6	0	1	0	0	0	0	1 (12.50%)

Note that the sum of tool types is larger than the count of 'flaked stone tools' because several specimens exhibit more than one tool type on their edges

Fig. 12.4 Chert specimens collected from the surface of Casa de Piedra Roselló talus. **a**, **b** Wood-scraper, note the less weathered flake scar on **b**. **c** Endscraper. **d** Notch. **e**, **i** Sidescraper. **f**, **g** and **k**, **l** Dorsal and ventral side of flakes. **h** Core. **j** Dorsal face of a flake with plunging termination. **m** Retouched point

the sample, although specimens with one and two scars are the most frequent. Cores represent 6.35% of the sample (Fig. 12.4).

12.6 Discussion

The results of rockfall and trampling experiments show that at least two of the taphonomic processes available at CP may produce pseudoartifacts when act upon the local chert. Indeed, pseudomorphs obtained in rockfall experiments included (pseudo) blanks, tools, and cores (Table 12.2) while treadage on chert flakes produced fundamentally (pseudo) tools. The following sections compare and discuss the morphological traits of experimental and talus lithic assemblages.

12.6.1 Debitage

A first issue highlighted by the comparison of data in Tables 12.2 and 12.4 is the higher number of blows recorded by talus flakes and cores. Rockfall experimental flakes exhibit lower dorsal scar counts; only the specimens in talus sample recorded four to six scars. Experimental flakes exhibit smaller values for length (mean: 14.56 vs. 36.18 mm; median: 12.15 vs. 34.75 mm) and width (mean: 16.35 vs. 31.61 mm; median: 13.75 vs. 30 mm) than complete flakes collected on CP talus surface. Also, the full range of length and width values recorded on the experimental specimens are contained within talus sample values.

As Nash (1993) noted, experimental flakes produced by rockfall are relatively short and wide (see also Luedtke 1986). A comparison of the length/width index values exhibited by complete flakes from both assemblages suggests that talus flakes are more elongated (L/W mean values 1.24 vs. 1.11), but the variance of both populations are significantly different ($f = 26.652$, $p < .001$), being higher in the experimental assemblage (L/W index values: talus = 0.5–3.5 vs. experimental = 0.4–5). A Mann-Whitney U test shows that there are no statistically significant differences in L/W median values between the samples ($U = 2048$, $p = .8511$).

Experimental flakes exhibit feathered (63.16%), plunging (10.53%), and hinge (7.02%) terminations. Flakes in talus assemblage also show feathered terminations as the most frequent type (69.39%), followed by hinge (12.24%), step (12.20%) and plunging (6.12%) terminations. The high frequency of feathered terminations in the present experimental sample contrasts with patterns observed in other rockfall studies (Nash 1993).

Cortical and flat striking platforms are the most frequent types among flakes of both experimental (61.90 and 28.57%) and talus (35.71 and 35.71%) samples. Crushed platforms are equally represented within both assemblages (ca. 5%). Dihedral striking platforms show different frequencies in talus and experimental samples: while they are well represented in the former (16.07%), their presence is scarce in the latter (2.38%). Filiform platforms are scarce in both talus and rockfall assemblages (5.36 and 2.38%, respectively).

These results suggest that taphonomic flakes like the ones produced during our rockfall experiments do not exhibit differential attributes that allow their identification within CP talus surface assemblage. Moreover, as it was already pointed out by other researchers (e.g. Mason 1965; Duvall and Venner 1979), the present study suggests that the differences between experimental and talus flakes are more quantitative than qualitative in nature. Further contextual studies need to explore the occurrence of larger and thus heavier freefall rocks at CP as well as if a larger number of rockfall events may have affected chert nodules deposited at the site. It is worth mentioning that during the talus survey, the presence of chert nodules and blocks larger than the ones used in our rockfall experiments were observed, thus suggesting that higher energy rockfall processes have taken place at the site. Therefore, their frequency and effects need further consideration.

12.6.2 Tools

As shown in Figs. 12.2, 12.3 and 12.4, several stone tools collected from the surface of CP talus display morphologies similar to those exhibited by the experimental samples after rockfalls and trampling. Indeed, the experiments produced almost all of the typological groups recorded within talus sample tools (Tables 12.2 and 12.3, see Sect. 12.5.1.2). It is worth mentioning that no formal tools on chert were recovered during the talus survey.

Mean value for tool size is 43 mm (Min.: 20 mm; Max.: 80 mm) in talus sample, 46 mm (Min.: 20 mm; Max.: 60 mm) in rockfall sample, and 26.5 mm (Min.: 20 mm; Max.: 45 mm) in trampling sample. Tool sizes recorded in trampling experimental assemblage are statistically different from talus (Mann-Whitney U $= 84$, $p < .01$) and rockfall (Mann-Whitney U $= 18.5$, $p < .01$) samples, but no statistical difference was recorded between rockfall and talus tool sizes (Mann-Whitney U $= 202.5$, $p = .5607$). Mean value for retouched edge angle in talus sample (N $= 58$ edges) is 70.8° (Min.: 30°; Max.: 105°), while rockfall sample (N $= 13$ edges) value is 76.9° (Min.: 45°; Max.: 110°) and trampling sample (N $= 12$ edges) value is 54.6° (Min.: 40°; Max.: 75°). Tool angles recorded in trampling experimental assemblage are statistically different from talus (Mann-Whitney U $= 24$, $p < .01$) and rockfall (Mann-Whitney U $= 107$, $p < .01$) samples, but no statistical difference was recorded between rockfall and talus tool angles (Mann-Whitney U $= 314.5$, $p = .2744$). Thus, talus and rockfall sample exhibit virtually identical mean values for tool size and retouched edge angle. Differences exhibited by trampling specimens are probably due to the small size of the blanks (flakes) used in the experiment.

It should be considered that the current and past presence of larger trampling agent at CP, such as guanaco (*Lama guanicoe*) and puma (*Felis concolor*), as well as introduced European livestock (cattle, sheep, and horses) indicates that trampling experimental results should be considered minimum values for the expected effects of local larger trampling agents.

12.6.3 Cores

The relative frequency of 'cores' is virtually identical in both experimental and talus samples (6.45 vs. 6.35%, Tables 12.2 and 12.4). Mean number of blows recorded on specimens from surface assemblage is larger than values exhibited by experimental pieces that resemble cores (3.7 vs. 2.2 blows per piece). However, as the rockfall rate averaged by CP talus surface sample is unknown, this difference between the assemblages may be indicating that the number of experimental rockfall events was below the mean number of impacts the specimens from talus collection experienced in average throughout their depositional history. A comparison of flake scars length on cores from both assemblages shows no statistically significant differences between their medians (Mann-Whitney U $= 29$, $p = .724$); however, flake scar width means are statistically different between the talus and experimental core samples (Mann-Whitney U = 11.5, p = .034). This latter result together with data provided by L/W rate on complete flakes suggests that talus assemblage averages conditions not present in the experiments conducted so far.

In sum, this research shows that taphonomic processes produce lithic specimens indistinguishable from artifacts. The comparison of experimental rockfall assemblage with specimens collected at CP talus highlighted the existence of morphological similarities suggesting that lithic assemblages from CP may include a taphonomic component (pseudoartifact) as well. As Mason (1965) earlier observation made it clear, an evaluation on the genesis of fractured stone can detect quantitative differences between natural versus artificial objects if it is conducted in an aggregate manner, that is to say, if both natural and artificial objects are considered as groups and not as single specimens. Moreover, "Simple examination of isolated specimens on the bases of personal opinion is unlikely to give a valid conclusion. Each object should be considered in terms of its context and as part of a series large enough for random or deliberate agencies to express themselves in comparison with all the relevant natural fractures" (Mason 1965: 3).

Current experimental data suggests that flakes equal or smaller than 30 mm in length and 54 mm in width, with three or less dorsal flake scars and cortical, flat, dihedral, filiform or crushed platform are likely to be produced by ~200 g freefalling rocks. When we apply these criteria to assess the potential taphonomic contribution to CP talus debitage sample, we find that only 21 out of 66 complete chert flakes can be disregarded as unlikely taphonomic specimens.

12.7 Conclusions

This chapter explored the potential effect of two taphonomic processes (rockfall and trampling) in the formation (and transformation) of the lithic record of Casa de Piedra Roselló archaeological site, a complex of rock shelters located in Patagonia (South América) with evidence of hunter-gatherer occupations since Early Holocene. Chert of good flaking quality is available at the geological formation where the rock shelters were formed and human groups used that raw material during the entire occupation span. Rockfall on chert pebbles and slabs takes place inside the shelters and on the talus as fragments removed from inner walls or outer cliff faces due to weathering fall downslope. The second process explored here was trampling, which is probably the ubiquitous taphonomic process for Fuego-Patagonian surface assemblages (Balirán 2014; Weitzel et al. 2014).

Considering that the two taphonomic processes explore here have been operating at the site since before the arrival of humans, I conclude that at least special attention should be given to the analysis of those pieces manufactured on local chert at Casa de Piedra Roselló. Moreover, the results presented emphasize the need for further experiments with larger samples adequate for statistical analysis in order to test and expand the pool of patterns recorded here. Field experiments (Kowalewski and

Labarbera 2004) at CP and second generation experiments (Lin et al. 2017) emerge as mandatory next steps in our taphonomy actualistic research.

In addition, this Patagonian case study shows that the application of a taphonomic perspective to the analysis of lithic assemblages is always informative even though unequivocal evidences of human agency are present at the archaeological contexts under study. In the latter case, this will allow assessing the human and taphonomic contributions to the lithic record. Consequently, a general outline in pseudoartifact research is needed to evaluate the taphonomic component within any lithic assemblage. Based on the present study, several of its primary constituents and aims can be advanced. First, research should focus on characterizing the natural availability of the lithic raw materials represented in the assemblage under study and the regional taphonomic background noise. Second, it needs to identify the local taphonomic agents and processes available at the site today and in the past, paying particular attention to their spatial range (that is to know the local context and its dynamic). Third, it is necessary to undertake naturalistic observations and experimental research using local raw materials. Besides mechanics (Cotterell and Kamminga 1987), several studies have underscored that variations in the physical properties of different lithic raw materials can condition the morphological patterns exhibited by flaked specimens (Goodman 1944; Nash 1993; Amick and Mauldin 1997, McBrearty et al. 1998; Gillespie et al. 2004). Therefore, pseudoartifact evaluation needs to be raw material-specific at some point of the actualistic research. In addition, further studies on rockfall need to assess the variability introduced by bouncing and rolling, the modes of motion taking place on the talus after freefalling (Dorren 2003). Finally, the comparison of local technological patterns against the regional technological background will highlight morphological continuities or discontinuities that may deserve further research (Mason 1965; Borrazzo 2011b).

Overall, actualistic taphonomy research indicates that we need to learn from non-archaeological contexts to get a more comprehensive understanding of the archaeological record.

Acknowledgements I am grateful to Sergio Martínez, Alejandra Rojas, Mariano Verde and Fernanda Cabrera for organizing the first *Workshop on Actualistic Taphonomy in South America* (Montevideo, October 9–11, 2017) and inviting me to participate in this volume. Analía Castro Esnal and Cecilia Perez de Micou directed fieldwork at CP and provided CP photograph. Thanks to María Laura Casanueva, Florencia Ronco, Lucía Gutiérrez, Cecilia Gutiérrez, and Esteban Ali Brouchoud who collaborated in field work. To the Roselló, Solsona, and Pérez del Barrio families for their kind and permanent support to archaeological studies. Luis Alberto Borrero encouraged and made insightful comments throughout all stages of this taphonomic study. This paper was draft as part of a postdoctoral study at Facultad de Filosofía y Letras, Universidad de Buenos Aires, Argentina. This research was funded by the Agencia Nacional de Promoción Científica y Tecnológica [PICT2015-2141 "Estudios de tecnología y tafonomía lítica en el sitio Casa de Piedra (Aldea Beleiro, Sudoeste de Chubut)"] and CONICET (República Argentina).

References

Aimola G, Andrade C, Mota L, Parenti F (2014) Final Pleistocene and Early Holocene at Sitio do Meio, Piauí, Brazil: stratigraphy and comparison with Pedra Furada. J Lithic Stud 1(2):5–24
Álvarez MC, Alunni D (ed) (2017) New perspectives in actualistic taphonomy in Argentina: limitations, contributions, and archaeological implications. J Taph 15(1–3)
Amick DS, Mauldin RP (1997) Effects of raw material on flake breakage patterns. Lithic Technol 22(1):18–32
Andrefsky W (2014) Fingerprinting flake production and damage processes: toward identifying human artifact characteristics. In: Graf K, Waters M, Geobel T (eds) Paleoamerican odyssey. Texas A&M Press, College Station, pp 415–428
Andrews BW, Murtha TM, Scheetz B (2004) Approaching the Hatch Jasper Quarry from a technological perspective: a study of prehistoric stone tool production in Central Pennsylvania. Midcont J Archaeol 29(1):63–101
Aschero C, Faundes Catalán W, Bobillo F (2017) Cacao 1: lithic evidence and mobility ranges during the Pleistocene in the Atacama Puna (Antofagasta de la Sierra, Catamarca, Argentina). In: Alberti J, Borrazzo K, Buscaglia S, Castro Esnal A, Elías A, Franco N (eds) 11th international symposium on knappable materials "From tools tone to stone tools", Book of abstracts, IMHICIHU-CONICET, Buenos Aires, p 116
Balirán C (2014) Trampling, taphonomy, and experiments with lithic artifacts in the Southeastern Baguales Range (Santa Cruz, Argentina). Intersec Antropol 3:85–95
Barnes A (1939) The differences between natural and human flaking on prehistoric flint implements. Am Anthropol 41(1):99–112
Barton CM, Bernabeu J, Maura JE, García O, La Roca N (2002) Dynamic landscapes, artifact taphonomy and landuse modeling in the Western Mediterranean. Geoarchaeology 17(2):155–190
Behrensmeyer AK, Damuth JD, DiMichele WA, Potts R, Sues H-D (eds) (1992) Terrestrial ecosystems through time. The University of Chicago Press, Chicago
Behrensmeyer AK, Kidwell SM (1985) Taphonomy's contributions to paleobiology. Paleobiology 11(1):105–119
Behrensmeyer AK, Kidwell SM, Gastaldo RA (2000) Taphonomy and paleobiology. Paleobiology 26(4):103–147
Binford LR (1977) General introduction. In: Binford LR (ed) For theory building in archaeology. Academic Press, New York, pp 1–13
Binford LR (1981) Bones: ancient men and modern myths. Academic Press, New York
Boëda E, Clemente-Conte I, Fontugne M, Lahaye C, Pino M, Daltrini Felice G, Guidon N, Hoeltz S, Lourdeau A, Pagli M, Pessis AM, Viana S, Da Costa A, Douville E (2014) A new Late Pleistocene archaeological sequence in South America: the Vale da Pedra Furada (Piauí, Brazil). Antiquity 88:927–955
Boëda E, Rocca R, Da Costa A, Fontugne M, Hatté C, Clemente-Conte I, Santos JC, Lucas L, Felice G, Lourdeau A, Villagran X, Gluchy M, Ramos MP, Viana S, Lahaye C, Guidon N, Griggo C, Pino M, Pessis A-M, Borges C, Gato B (2016) New data on a Pleistocene archaeological sequence in South America: Toca do Sítio do Meio, Piauí, Brazil. PaleoAmerica 2(4):286–302
Bordes F (1961) Typologie du Paleolithique Ancien et Moyen. Impriméries, Delmas, Bordeaux
Borrazzo K (2006) Tafonomía lítica en dunas: una propuesta para el análisis de los artefactos líticos. Intersec Antropol 7:247–261
Borrazzo K (2011a) Tafonomía lítica en la estepa patagónica: experimentación y registro arqueológico de superficie. In: Borrero LA, Borrazzo K (eds) Bosques, Montañas y cazadores: investigaciones arqueológicas en Patagonia Meridional. CONICET-IMHICIHU, Buenos Aires, pp 127–153
Borrazzo K (2011b) Tafonomía lítica y pseudoartefactos: el caso de la península El Páramo (Tierra del Fuego, Argentina). Intersec Antropol 12:155–167
Borrazzo K (2013) Tafonomía lítica y modelo de la dinámica eololacustre del norte de la bahía San Sebastián (Tierra del Fuego, Argentina). Rev Comechingonia 17(1):149–169

Borrazzo K (2016) Lithic taphonomy in desert environments: contributions from Fuego-Patagonia (Argentina). Quat Int 422:18–29
Borrazzo K, Borrero LA (2015) Taphonomic and archaeological perspectives from Northern Tierra del Fuego, Argentina. Quat Int 373:96–103
Borrazzo K, Weitzel C (eds) (2014) Taphonomic approaches to the archaeological record. Intersec Antropol 15(3)
Borrero LA (2001) Regional taphonomy. Background noise and the integrity of the archaeological record. In: Kuznar LA (ed) Ethnoarchaeology of Andean South America. Contributions to archaeological method and theory. International Monographs in Prehistory, Ann Arbor, pp 243–254
Borrero LA (2011) La función transdisciplinaria de la arqueozoología en el siglo XXI: restos animales y más allá. Antípoda 13:267–274
Borrero LA (2014) Multi-service taphonomy. Shells, garbage and floating palimpsests. Intersec Antropol 15(3):13–20
Borrero LA (2015) Con lo mínimo: los debates sobre el poblamiento de América del Sur. Intersec Antropol 16:5–38
Borrero LA (2016) Ambiguity and debates on the early peopling of South America. https://doi.org/10.1080/20555563.2015.1136498
Breuil H, Lantier R (1965) The men of the old stone age (Palaeolithic and Mesolithic). St. Martin's Press, New York
Carranza ME (2017) Análisis tafonómico de conjuntos líticos de superficie en la costa norte del golfo San Matías (Rio Negro, Argentina). Intersec Antropol 18(1):91–101
Carranza Elola JJ (2015) Tecnología y tafonomía lítica del sitio quebrada Quereo: metodología para abordar conjuntos líticos ambiguos del Pleistoceno tardío en la costa de Los Vilos. Unpublished undergraduate thesis, Universidad de Chile, Santiago
Carranza Elola JJ, Méndez C (2017) Lithic taphonomy at the Quebrada de Quereo site: addressing the problem of anthropogenic ambiguity in Late Pleistocene assemblages. In Alberti J, Borrazzo K, Buscaglia S, Castro Esnal A, Elías A, Franco N (eds) 11th international symposium on knappable materials "From toolstone to stone tools", Book of abstracts, IMHICIHU-CONICET, Buenos Aires, p 199
Castro Esnal A, Pérez de Micou CB, Casanueva ML (2017a) Early Holocene occupation of the forest-steppe ecotone of Southern South America: evidence from Casa de Piedra de Roselló Cave (Chubut, Patagonia Argentina). Paleoamerica. https://doi.org/10.1080/20555563.2017.1330102
Castro Esnal A, Stern C, Pérez de Micou C (2017b) Geochemical studies of archaeological obsidian artifacts from both stratigraphic and surface contexts in Aldea Beleiro Village, SW Chubut (Patagonia, Argentina). Magallania 45(1):123–135
Chlachula J, Le Blanc R (1996) Some artifact-diagnostic criteria of quartzite cobble-tool industries from Alberta. Canadian J Archaeol 20:61–74
Cotterell B, Kamminga J (1987) The formation of flakes. Am Antiq 52(4):675–708
Cruz I (2007) Avian taphonomy: observations at two Magellanic Penguin (Spheniscus magellanicus) breeding colonies and their implications for the fossil record. J Archaeol Sci 34:1252–1261
Demeter F, Edoumba E, Duringer P, Bacon A-M, Sytha P, Bano M, Laychour V, Cheangleng M, Sari V (2010) Reinterpretation of an archaeological pebble culture from the Middle Mekong River Valley, Cambodia. Geoarchaeology 25(1):75–95
Domínguez-Rodrigo M, Fernández-López S, Alcalá L (2011) How can taphonomy be defined in the XXI Century? J Taph 9:1–13
Dorren LK (2003) A review of rockfall mechanics and modeling approaches. Prog Phys Geogr 27(1):69–87
Duvall JG, Venner WT (1979) A statistical analysis of the lithics from the calico site (SBCM 1500A), California. J Field Archaeol 6(4):455–462
Eren MI, Durant A, Neudorf C, Haslam M, Shipton C, Bora J, Korisettar R, Petraglia M (2010) Experimental examination of animal trampling effects on artifact movement in dry and water saturated substrates: a test case from South India. J Archaeol Sci 37:3010–3021

Fariña RA, Tambusso PS, Varela L, Czerwonogora A, Di Giacomo M, Musso M, Bracco R, Gascue A (2014) Arroyo del Vizcaíno, Uruguay: a fossil-rich 30-ka-old megafaunal locality with cut-marked bones. P Roy Soc B (Biol Sci) 281. https://doi.org/10.1098/rspb.2013.2211
Fiedel SJ (2017) Did monkeys make the pre-clovis pebble tools of Northeastern Brazil? PaleoAmerica 3(1):6–12. https://doi.org/10.1080/20555563.2016.1273000
Garvey R, Mena F (2016) Confronting complexities of artifact-geofact debates: re-analysis of a coarse volcanic rock assemblage from Chilean Patagonia. Lithic Technol 41:114–129
Gifford DP (1981) Taphonomy and paleoecology: a critical review of archeology's sister discipline. Adv Archaeol Method Theor 4:365–438
Gifford-Gonzalez D (1989) Ethnographic analogues for interpreting modified bones: some cases from East Africa. In: Bonnichsen R, Sorg M (eds) Bone modification. University of Maine, Institute for Quaternary Studies, Orono, pp 179–246
Gillespie JD, Tupakka S, Cluney C (2004) Distinguishing Between naturally and culturally flaked cobbles: a test case from Alberta, Canada. Geoarchaeology 19(7):615–633
Goodman ME (1944) The physical properties of stone tool materials. Am Antiq 9:415–433
Gutiérrez MA, Miotti L, Barrientos G, Mengoni Goñalons G, Salemme M (2007) Taphonomy and zooarchaeology in Argentina. BAR International Series, Oxford
Haynes CV (1973) The calico site: artifacts or geofacts? Science 181:305–310
Hiscock P (1985) The need for a taphonomic perspective in stone artefact analysis. Queensland Archaeol Res 2:82–95
Hosfield R, Chambers J (2003) Flake modifications during fluvial transportation: three cautionary tales. Lithics 24:57–65
Johnson L, Behm JA, Bordes F, Cahen D, Crabtree DE, Dincauze DF, Hay CA, Hayden B, Hester TR, Katz PR, Knudson R, McManamon FP, Malik SC, Müller-Beck H, Newcomer MH, Paddayya K, Price-Beggerly P, Ranere AJ, Sankalia HD, Sheets PD (1978) A history of flint-knapping experimentation, 1838-1976 [and comments and reply]. Curr Anthropol 19(2):337–372
Kowalewski M (1999) Actuopaleontology: the strength of its limitations. Acta Paleontol Pol 44(4):452–454
Kowalewski M, Labarbera M (2004) Actualistic taphonomy: death, decay, and disintegration in contemporary settings. Palaios 19:423–427
Lahaye C, Hernandez M, Boëda E, Felice GD, Guidon N, Hoeltz S, Lourdeau A, Pagli M, Pessis AM, Rasse M, Viana S (2013) Human occupation in South America by 20,000 BC: the Toca da Tira Peia site, Piauí, Brazil. J Archaeol Sci 40:2840–2847
Lin S, Rezek Z, Dibble H (2017) Experimental design and experimental inference in stone artifact archaeology. J Archaeol Method Theory. https://doi.org/10.1007/s10816-017-9351-1
Lopinot N, Ray J (2007) Trampling experiments in the search for the earliest Americans. Am Antiq 72(4):771–782
Luedtke B (1986) An experiment in natural fracture. Lithic Technol 15(2):55–60
Lyman RL (1984) Broken bones, bone expediency tools and bone pseudotools: lessons from the blast zone around Mount St. Helens, Washington. Am Antiq 49(2):315–333
Lyman RL (1994) Vertebrate taphonomy. Cambridge University Press, Cambridge
Mallol C, Bertran P (eds) (2010) Geoarchaeology and taphonomy. Quat Int 214
Marean C (1995) Of taphonomy and zooarchaeology. Evol Anthropol 4(2):64–72
Mason RJ (1965) Makapansgat limeworks fractured stone objects and natural fracture in Africa. S Afr Archaeol B 20(77):3–16
Massigoge A, González M (ed) (2012) Taphonomic approaches in archaeology: current topics and methods from Southern South America. Quat Int 278
McBrearty S, Bishop L, Plummer T, Dewar R, Conard N (1998) Tools underfoot: human trampling as an agent of lithic artifact edge modification. Am Antiq 63(1):108–122
Méndez Muñoz V (2015) Historias Depositacionales de conjuntos líticos en la Transición Pleistoceno-Holoceno en el Sitio Valiente, Provincia del Choapa. Unpublished undergraduate thesis, Universidad de Chile, Santiago

Nash DT (1993) Distinguishing stone artifacts from naturefacts created by rockfall processes. In Goldberg P, Nash DT, Petraglia MD (eds) Formation processes in archaeological context, Monographs in world prehistory, vol 17. Prehistory Press, Madison, pp 125–138

Parenti F (2015) Old and new on the same site: putting Vale de Pedra Furada into a wider context. A comment on Lahaye et al. 2015. Quat Chronol 30:48–53

Pobiner BL, Brown DR (2005) Applying actualism: considerations for future research. J Taph 3(2):57–65

Prentiss AM, Walsh MJ, Barnett KD, Murphy MM, Kuenstle J (2015) The coarse volcanic rock industry at Rio Ibáñez 6 West, Aisén Region, Patagonian Chile. Lithic Technol 40(2):112–127

Prentiss AM, Barnett KD, Walsh MJ (2016) The coarse volcanic industry at the Río Ibáñez 6 West Site, Chilean Patagonia: assessing geogenic versus anthropogenic processes. Lithic Technol 41(2):130–138

Ritter MDN, De Francesco CG, Erthal F, Hassan GS, Tietze E, Martínez S (2016) Manifesto of the South American School of (Actualistic) Taphonomy. Palaios 31:20–24

Suárez R., Borrero LA, Borrazzo K, Ubilla M, Martínez S, Perea D (2014) Archaeological evidences are still missing: comment on Fariña et al. Arroyo del Vizcaíno Site, Uruguay. P Roy Soc B (Biol Sci) 281:20140449. https://doi.org/10.1098/rspb.2014.0449

Thiébaut C, Coumont M-P, Averbouh A (2010) The taphonomic approach: an archaeological necessity. In Thiébaut C, Coumont M-P, Averbouh A (eds) Mise en commun des approaches en taphonomie. Actes du workshop Nº 16, X Congress International de I'UISPP, pp 21–28

Valin L, Masson B, Caspar JP (2001) Taphonomy at Hermies, France: a mousterian knapping site in a Loessic Context. J Field Archaeol 28(3/4):419–436

Warren SH (1914) The experimental investigation of flint fracture and its application to problems of human implements. J Roy Anthropol Inst Great Britain Ireland 44:412–450

Weitzel C, Borrazzo K, Ceraso A, Balirán C (2014) Trampling fragmentation potential of lithic artifacts: an experimental approach. Intersec Antropol 3:97–110

Wiśniewski A, Badura J, Salamon y, Lewandowski J (2014) The alleged Early Palaeolithic artefacts are in reality geofacts: a revision of the site of Kończyce Wielkie 4 in the Moravian Gate, South Poland. J Archaeol Sci 52:189–203

Yeshurun R, Bar-Oz G, Kaufman D, Weinstein-Evron M (2014) Purpose, permanence, and perception of 14.000-year-old architecture contextual taphonomy of food refuse. Curr Anthropol 55(5):591–618

Chapter 13
Perceptions on Actualistic Paleontology in Four Distinct Areas of Energy Supply Along the Coast of the States of Rio de Janeiro, São Paulo, and Paraná, Brazil

Renato Pirani Ghilardi, Maria Clara Silva Pinto, Thays Paganella Marcondes and Sabrina Coelho Rodrigues

Abstract The study of actuopaleontology is fundamental to observe patterns in the current record and to correlate them to the fossil record. Different taphonomic signatures in mollusks bioclasts were observed (articulation, fragmentation, bioerosion, encrustation, abrasion, and dissolution) and related to four different points in the south coast of Brazil that have different geological aspects. The main objective is to compile the taphonomic data already collected for the Brazilian coast and correlated and search patterns of taphonomic signatures of these points: 1—Cananéia-SP has an open coastline with influence of river discharge, 2—Ubatuba-SP presents jagged relief with characteristics of little energetic movement, 3—Macaé-RJ presents open coast, but with influence of coastal upwelling and 4—Paranaguá-PR is a bay with strong fluvial influence. In general, disarticulation and fragmentation were the predominant characteristics in the four different localities and their genesis is related to the different factors. In Paranaguá (CEP) and Ubatuba is possible that the fragmentation is biofacilitated, especially in the location where the highest incidence of bioerosion was found. Environmental factors that contributed to the development of surface textures in the bioclasts studied, through a combination of bioerosion, encrustation, abrasion, and dissolution process include: (I) geographic distance—proximity to an island and/or upwelling area, whereas both provide a greater availability of nutrients; (II) hydrodynamics—factors that promote less disturbance of the environment/shells/sediment, such as geographical protection from an island or an indented coast, as well as a greater depth; (III) fluvial influence with reduction of salinity, such as in the area of an estuary. The preservation of macroinvertebrates shells from the siliciclastic bottoms of the southeastern continental subtropical climate shelf of Brazil are related in a complex way and associated with taphonomic signatures.

R. P. Ghilardi (✉) · M. C. S. Pinto · T. P. Marcondes
Faculdade de Ciências, Universidade Estadual Paulista, UNESP, Av. Edmundo Coube, Bauru, SP 17033-360, Brazil
e-mail: renato.ghilardi@unesp.br

S. C. Rodrigues
Instituto de Ciências Exatas e Naturais do Pontal, Universidade Federal de Uberlândia, Rua Vinte, 1600, Ituiutaba, MG 38304-402, Brazil

S. Martínez et al. (eds.), *Actualistic Taphonomy in South America*,
Topics in Geobiology 48, https://doi.org/10.1007/978-3-030-20625-3_13

Keywords Siliciclastic bottom · Mollusk taphonomy · Coast Brazil · Paranaguá · Ubatuba · Macaé · Cananéia · Taphonomic signatures

13.1 Actuopaleontology of Macroinvertebrates: The Search for Patterns in the Sedimentary Registry

The term taphonomy, as originally coined by Efremov (1940), refers to the study of the 'laws' that govern the transition of organic remains from the biosphere to the lithosphere, including the action of the biostratinomic and diagenetic processes. Several biotic and abiotic processes act in a complex way on the skeletal remains of the most different groups of organisms, which also interfere in the patterns of preservation in the various sedimentary environments, altering even the information contained in the fossil record. Thus, Behrensmeyer and Kidwell (1985) extended the definition of the concept of taphonomy, which refers to the study of preservation processes and how they affect the information in the fossil record. Recognizing such processes in taphonomic signatures in bioclasts is quite useful for (paleo)environmental and (paleo)ecological analyses, since, "in theory", taphonomic signatures are products of specific sedimentary/biological elements, such as (a) sedimentation rates, (b) energy levels of the environment, (c) depth, (d) primary productivity, and (e) rates of bioerosion and encrustation, to name a few examples (e.g., Fürsich and Oschmann 1993).

Actuopaleontology is a branch of Paleontology that aims to improve the understanding of the information contained in the sedimentary record, based on studies of sedimentary patterns and processes that operate on recent accumulations. Even if not all of the trends observed in recent accumulations show equal potential to be preserved in the sedimentary record in deep time, the actualistic approach contributes greatly to hypothesis testing with regard to the fossil record (Kowalewski 1999).

Actualistic studies (e.g., Kowalewski et al. 1994; Kidwell et al. 2001) and Paleozoic marine deposits (e.g., Speyer and Brett 1988, 1991; Simões et al. 2000) have demonstrated the usefulness of the concept of taphonomic facies to distinguishing environmental gradients. These, in most cases, can be recognized based on litho- and biofaciological criteria. Based on this perspective, the analysis of taphofacies may amplify the resolution and add important evidence, complementary to the analyses of lithofacies and biofacies. Since a particular taphonomic feature can provide unequivocal evidence for a range of environmental conditions (biotic/abiotic), its recognition also implies the identification of specific environmental characteristics (Brett and Baird 1986). Thus, a great effort has been made in the last three decades to develop a methodology for studies of comparative taphonomy and recognition of taphonomic facies (i.e., bodies of rocks identified by their quantifiable taphonomic characteristics), both for rocks deposited in the marine environment and for bioclastic accumulations in recent sediments (e.g., Speyer and Brett 1988, 1991; Kowalewski et al. 1994; Simões et al. 2000; Kidwell et al. 2001, Rodrigues et al. 2009). Although some taphonomic signatures are due to the action of specific environmental conditions, generalizations for the application of the signatures in environmental analyses are not always feasible. This is because each taphonomic signature produced in a particular bioclast in a given environment at a given time interval reflects the individual history of the bioclast in the taphonomically active zone (TAZ, Davies et al. 1989), which may be unique to that environment. Moreover, among the factors less understood in the taphonomy of marine environments are the effects of climate and latitude on the preservation of the bioclasts in the varied depositional environments (Best and Kidwell 2000a; Simões et al. 2005; Rodrigues and Simões 2010). Predominantly siliciclastic environments that are influenced by the currents are typically more abrasive due to the presence of iron and other metals, and they may present less saturated interstitial waters compared to carbonate environments (e.g., Aller 1982). However, among the environments of the different climatic zones, the way physical, chemical, and biological factors act on the preservation of bioclasts is, to a certain extent, poorly understood. For example, the interstitial water acidity that promotes the dissolution of calcium carbonate bioclasts in temperate climates (Aller 1982) is expected to be more intense in the tropics (Best and Kidwell 2000a). Also, in tropical environments, bioerosion is among the most effective taphonomic processes (Best and Kidwell 2000a; Simões and Kowalewski 2002), and is rare in polar environment bioclasts (Cerrano et al. 2001). On the other hand, encrustation rates are comparatively higher in polar environments, with greater diversity and an abundance of encrustation fauna, as made evident by the accumulations of brachiopod shells in the bays of Ubatuba and Picinguaba, Brazil (*Bouchardia rosea*), and Admiralty Bay, Antarctica (*Liothyrella uva*) (Simões et al. 2005).

Several examples of taphonomic generalizations are available in the scientific literature, but the conviction that each case is unique (Holz and Simões 2002, p. 189) grows among the taphonomic community. Nevertheless, knowing the particularities of each sedimentary environment and how the environmental variables influence the marks of their processes in the bioclastic accumulations is still the most reliable way to recognize the preservation patterns within Actuopaleontology. Also, as recently demonstrated by Ritter et al. (2018) several taphonomic signatures either covary with depth or are directly influenced by depth. It is based on such knowledge that tests of hypotheses should be performed for the Phanerozoic sedimentary record.

13.1.1 Taphonomy of Recent Accumulations of Brazilian Mollusks and Brachiopods: A Brief History

Several groups of recent marine invertebrates have been the target of taphonomic studies. For example, echinoids (e.g., Nebelsick 1999), crinoids (e.g., Llewellyn and Messing 1993), corals (e.g., Pandolfi and Greenstein 1997; Perry 2000), foraminifera (e.g., Martin and Liddell 1991), Bryozoa (e.g., Smith and Nelson 1994), crustaceans (e.g., Stempien 2005), brachiopods (e.g., Simões and Kowalewski 2003; Tomašových and Rothfus 2005) and, more extensively, mollusks (e.g., Fürsich and Flessa 1987;

Davies et al. 1989; Best and Kidwell 2000a, b; Kidwell et al. 2001; Callender et al. 2002; Parsons-Hubbard 2005). Among the different types of skeletons of marine invertebrates (solid, arborescent, univalve, and multielement, Holz and Simões 2002), bivalve shells stand out because they present a wide range of taphonomic signatures that can be quantified (e.g., abrasion, disintegration, fragmentation, bioerosion, encrustation, and dissolution) (Speyer and Brett 1988). Although a growing number of taphonomic studies have emerged in the last three decades (see Rodrigues et al. 2009, for a review of this subject), no methodological consensus has emerged to guide research in the field of taphofacies recognition. In taphonomic and faciological studies of both current skeletal accumulations and the geological record, methodological disparities are observed at a wide variety of stages, including the sampling phases of the material (e.g., type of sampler), collection (e.g., granulometric fraction considered, number of observers), and data analysis (e.g., restrictive analyses). Moreover, the similarity between the taphonomic signatures used in the various studies is low and little attention has been paid to issues such as the establishment and quantification of the degree of taphonomic alteration (Kidwell et al. 2001). Thus, an enormous amount of data has been generated for taphonomic facies, from the Paleozoic to the Modern periods, but these data are not comparable to each other. In general terms, protocols for taphonomic analysis have been the subject of discussions in the scientific literature (Kidwell et al. 2001; Kowalewski and Hoffmeister 2003; Rothfus 2004; Yesares-García and Aguirre 2004) due to the need for a better understanding of the effects that methodological decisions can have on the results.

In Brazil, great methodological contributions to the taphonomic analysis of marine macroinvertebrates have come out of macroinvertebrate studies of the southeastern continental shelf of Brazil (e.g., Simões et al. 2000, 2004a, b). In the past decade, the researchers of these studies have examined recent accumulations of brachiopods (*Bouchardia rosea*) and bivalve mollusks, sampled from distinct sedimentary environments located in the bays of Ubatuba and Picinguaba (state of São Paulo), including the beaches, and the inner and outer continental shelf. This pioneering research has produced important contributions to the development of actuopaleontology at the national level and has made an impact on international science. Data and discussions have been the basis for the establishment of taphonomic analysis protocols (e.g. Rodrigues et al. 2009), the recognition of taphofacies and contributions to the origin of signatures (e.g. Rodrigues and Simões 2010), including experimental taphonomy (e.g. Torello 2004; Simões et al. 2005, 2007) and developments in paleobiology, such as studies on encrustation (e.g. Rodland et al. 2014), predation (e.g. Simões et al. 2007), parasitism (e.g. Rodrigues et al. 2009), and the paleobiology of conservation (e.g. Simões et al. 2009). Although a large quantity of taphonomic and paleobiological data has been gathered from recent accumulations of bivalve mollusk and brachiopod shells, there is still much research to be done to before we understand the complex relationships between taphonomic processes and products, with a view toward application to the sedimentary record.

Thus, several lessons for actuopaleontology have arisen from such studies. In general terms, there is a consensus that there is a need to establish a methodological protocol for carrying out careful taphonomic analyses right from the beginning of any and all taphonomic studies. Furthermore, in spite of recent studies on taphofacies carried out with deposits generated in a variety of depositional environments (e.g. a siliciclastic bottom along the shallow continental shelf, Fürsich and Flessa 1987; a carbonate bank, Llewellyn and Messing 1993), presenting the potential to serve as a tool for (paleo)environmental and (paleo)ecological interpretations, they vary in a complex way along a given bathymetric gradient (e.g. Rodrigues and Simões 2010). For example, studies of taphofacies in recent mollusk accumulations generated in intertidal habitats and the shallow continental shelf (e.g., Meldahl and Flessa 1990) show that each lithofacies holds indications of the energetic conditions of its environment and thus the taphonomic signatures of the associated skeletal accumulations reflect the post-mortem history of the organisms. Thus concentrations characterized by bioerosion and encrustation are, in general, found in low-energy environments and/or deeper waters, particularly those marked by a low sedimentation rate (Brett and Baird 1986). Skeletal accumulations generated in high-energy environments in shallow water conditions are characterized by high rates of abrasion and fragmentation (Brett and Baird 1986). These ideas permeate the taphonomic literature of the 1990s when the conceptual framework of taphonomic facies was being constructed. However, taphonomic studies from the 2000s on skeletal accumulations of mollusks (Best and Kidwell 2000a, b; Callender et al. 2002; Staff et al. 2002; Parsons-Hubbard 2005) and living brachiopods (Simões and Kowalewski 2002), indicate that the above relationships are not as evident or predictable when examined in a variety of environments (e.g., different locations along the shallow and deep water continental shelf). Furthermore, not all taphonomic signatures are equally sensitive to environmental conditions and at times occur in a generalized manner along a specific bathymetric gradient.

Still, in the context of taphonomic studies, understanding the role of these processes in the genesis of shell accumulations, as well as the possible tendencies introduced by the biostratinomic and diagenetic agents is fundamental. Accumulations rich in shells are notable features in the sedimentary record of the Phanerozoic and constitute a source of paleoecological and evolutionary information. The genesis of such accumulations is governed by complex taphonomic factors (e.g., susceptibility to transport and differential destruction) and environmental factors (e.g., sea level, sedimentation rate) that do not always occur in an intuitive manner (see Paleozoic and Mesozoic examples in Speyer and Brett 1988, 1991; Brett 1995; Brett and Baird 1997; Brett et al. 1997; McFarland et al. 1999; Tomašových 2004). In this sense, current and experimental taphonomic studies, both in the laboratory and in the field (e.g. Lescinsky et al. 2002; Messina and LaBarbera 2004; Torello 2004), contribute to the understanding of the genesis of accumulations, be they current or fossil. This branch of taphonomy (i.e. experimental taphonomy) allows geologists and taphonomists to simulate the role of biostratigraphic and diagenetic processes (e.g., transport, abrasion, dissolution), which also assists in the understanding of agents that are active in the taphonomically active zone (TAZ, Davies et al. 1989) under different environmental conditions.

Furthermore, taphonomic signatures, in addition to being useful for determining environmental conditions, are also worthy of study from a biological and evolutionary point of view. This is because, in many cases, they result from established biotic interactions between living organisms (e.g., predation, parasitism, commensalism, competition) that have affected the structure of communities in the past and still do so today. Perforations and other marks left on biomineralized skeletons are among the more reliable evidence of biological interactions that are likely to be preserved in the metazoan fossil record (e.g. Hoffman et al. 1974; Vermeij et al. 1980; Kitchell et al. 1981; Vermeij 1987; Allmon et al. 1990; Anderson et al. 1991; Bengtson and Zhao 1992; Kelley and Hansen 1993, 1996; Harper et al. 1998; Leighton 2001; Kowalewski and Kelley 2002; Kelley et al. 2003).

Finally, actuopaleontology contributes to the paleobiology of conservation, as bioclastic accumulations contain relevant ecological information about the faunistic composition of a wide variety of marine and continental environments. This is because, due to taphonomic processes and the temporal mixing phenomenon, the compositional, ecological and spatial characteristics of taphocenoses do not always faithfully reflect living communities (biocenosis) in the skeletal remains that have been preserved. Palaeobiology of conservation is a new, socially relevant branch of the geosciences (Kowalewski 2017) that emerged from the observation that the sub-recent fossil record (from previous centuries and millennia) may serve as an unprecedented guide to studying the causes and rates of environmental changes at time scales not accessible to the methods employed by neontological studies. Not only quantitative comparisons between living, death, and fossil assemblages are important not only in strictly taphonomic studies, but have grown a leading tool for conservation paleobiology analysis, but also in studies of trophic webs of a target taxa (Ritter and Erthal 2013; Simões et al. 2009). This unique perspective can provide rigorous criteria for evaluating the extent of environmental change in a marine or continental ecosystem. This, in turn, allows for a realistic assessment of anthropogenic changes, providing effective bases for mitigation efforts.

13.2 Actuopaleontological Perspectives on Macroinvertebrates: Case Studies

During the last decade, the majority of actuopaleontological studies on marine bivalves were conducted on the northern coast of the state of São Paulo (Rodrigues 2006; Rodrigues et al. 2009; Rodrigues and Simões 2010). More recently, great contributions on coastal lagoons mollusk assemblages (e.g., Ritter et al. 2013) and recent river mollusk accumulations (e.g., Martello et al. 2018) have also lead to a broad taphonomic approach. In marine realm, these studies carried out a taphonomic analysis of bivalve mollusks and brachiopods collected in different environments and at various depths: (a) coastal, covering different states, dissipative and reflective; (b) bay, low energy, and clayey bottom; (c) bay, low energy and sandy bottom, 10–20 m deep and (d) inner continental shelf, sandy bottom, depths of more than 20 m. Recently, taphonomic studies have branched out to examine occurrences in other regions of Brazil, such as the studies on bivalve mollusk shells obtained from the continental shelf of the State of Rio Grande do Sul (e.g., Soberon et al. 2007; Erthal 2012). Erthal (2012) made the association between these taphonomic signatures and environmental factors on a large scale, using concentrations of shells from the southern and northeastern coasts of Brazil, two geographically remote and environmentally distinct areas. Still, in the southern region of the country, Ritter et al. (2013) carried out actuopaleontological studies on clusters of mollusks with a focus on estuarine-lagoon systems.

The coastline of the southeastern coast of Brazil is an excellent observatory for actuopaleontological studies, due to its variety of coastal morphologies. Unlike the coast of São Paulo, the coast of Paraná has a small coastline with oceanic features, including a wide coastal plain characterized by long, exposed sandy beaches (Bigarella et al. 1978), separated by two well-developed estuaries: the Guaratuba Bay and the Paranaguá Estuary Complex (CEP).

Although advances in taphonomic studies of recent macroinvertebrate accumulations and the great efforts made by various groups of researchers have garnered much attention, the gap in our understanding of the complex interactions between sedimentary processes is still quite large. For example, there are few actuopaleontological works examining bioclasts of gastropod shells. The analysis here presented adds to the body of work examining the taphonomy of mollusks (bivalves and gastropods) and seeks to contribute to a better knowledge of taphonomic signatures found in distinct coastal regions of Brazil (Fig. 13.1). The collection stations of the bioclasts for the study of presented cases are the same as those cited by Marcondes (2013) and Pinto (2017).

13.2.1 Environmental Characterization of the Sites in the Study: Paranaguá (Paraná, Brazil), Cananéia (São Paulo, Brazil), Ubatuba (São Paulo, Brazil) and Macaé (Rio de Janeiro)

13.2.1.1 Paranaguá

The Paranaguá Estuary Complex (CEP for its abbreviation in Portuguese) is located in the north-central portion of the coast of Paraná State. This system has two mouths that flow into the Atlantic Ocean: one considered the north mouth and another considered the south, based on their geographical position in relation to the Ilha do Mel island (Lana et al. 2001; Lamour 2007).

The inner area of the CEP is bordered by mangroves, marshes and tidal plains, while the adjacent oceanic area and the areas around the mouths are composed of extensive sandy beaches and rocky coastline (Angulo 1992). The CEP area is composed of several bays and beaches. It is a very jagged area and composed of

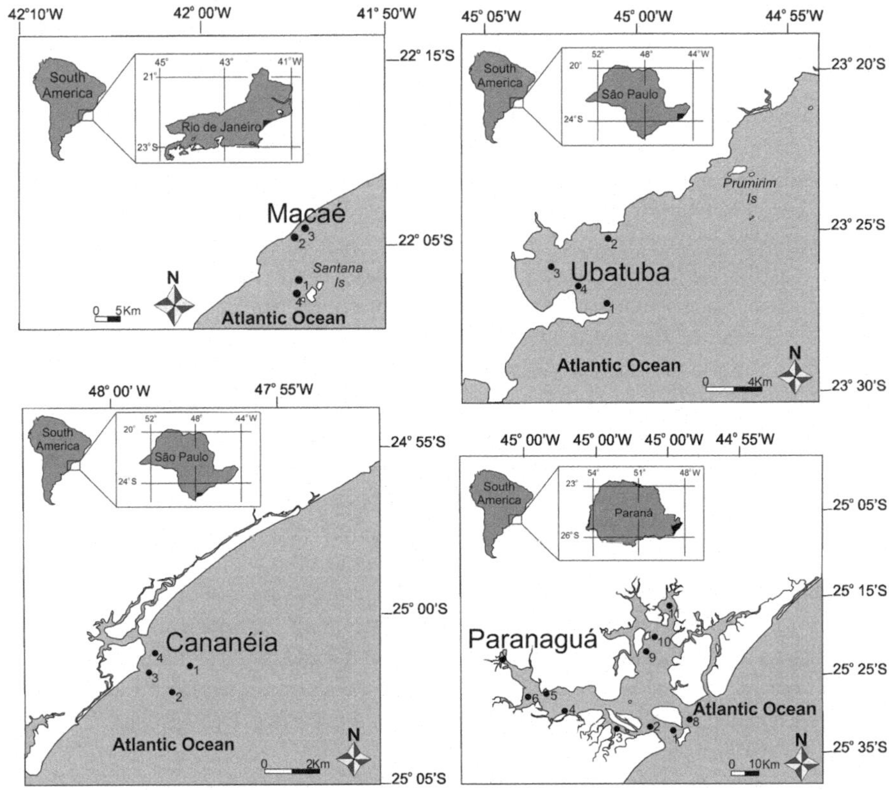

Fig. 13.1 Location maps of the different study sites on the southern coast of Brazil. The black circles represent the specific points of analysis of the bioclasts

numerous islands: Ilha Rasa da Cotinga, Ilha das Cobras, Ilha das Peças, Ilha do Mel, Ilha do Superagui, and Ilha do Teixeira, among others (Angulo 1992).

Its areas with calm waters and weak currents favor the deposit of sediment and the formation of shoals and tide plains. The shallows or shallow depths do not exceed 2 m (6 ft) in depth and may emerge from the water during low tides. Its formations include sandy clay banks, marsh vegetation, and marine grasses. In the Bay of Paranaguá there are extensive shoals, such as Baixio do Perigo, Baixio do Bagre, and Baixio do Saco do Limoeiro (Angulo 1992).

In general, fine sand constitutes the bottom sediments that predominate on the shallow continental shelf of the Paraná coast. The sediments of Paranaguá Bay are composed of fine silts and fine sand (Lamour et al. 2004), being thinner and not sorted by size for the interior of the bay (Bigarella et al. 1978). At the southern mouth, fine to medium sand is predominant, and in the north, fine to coarse sand is predominant. These sands are moderately well sorted in the south and moderately to poorly sorted in the north (Lamour 2007).

Table 13.1 Sites of gathering of the different geographical points with their respective bathymetry and the number of specimens analyzed

Location	Site	Depth (m)	Total number of specimens
Macaé	M1	5–10	21
	M2	5–10	31
	M3	10–15	86
	M4	10–15	105
Ubatuba	U1	10–15	151
	U2	10–15	155
	U3	5–10	57
	U4	10–15	68
Cananéia	C1	10–15	109
	C2	10–15	15
	C3	5–10	43
	C4	5–10	54
Paranaguá	P1	3	150
	P2	10	150
	P3	14	150
	P4	9	67
	P5	3	15
	P6	4	150
	P7	3	150
	P8	8	150
	P9	0.5	150
	P10	1.5	150
	P11	4	150

In total, 1438 bivalve mollusk bioclasts were used in the taphonomic analyzes, from the collection of the siliciclastic bottom of the locations in the inner area of the CEP and the locations at the outlets of the CEP. Sampling was performed from 11 collection stations that covered the different sub-environments, within a depth of 0.5–13.6 m (1.5–45 ft) (Table 13.1).

13.2.1.2 Macaé

This region is located on the southern coast of the state of Rio de Janeiro and is characterized by wide open spaces. Surrounding it is the Santana Archipelago, composed of the Ilha de Santana, Ilha do Francês, and Ilhote Sul islands, as well as a variety of crags and slabs. It is located near the upwelling area of Cabo Frio, which promotes a high inflow of nutrients. It is also influenced by the waters of the Paraíba do Sul river delta located just to the north (Sancinetti 2011).

In general, the coastal morphology is characterized by high wave energy and this is related to a profile of erosive behavior. As the beaches are subject to greater wave energy (i.e., they are reflective), their composition is that of thicker sediments and gravel and an absence of silt. They are intensely oxygenated due to their high energy (Oliveira and Muehe 2013).

Macaé receives a great deal of influence from the South Atlantic Central Water (SACW) in relation to Ubatuba and later Cananéia. This is due to its geographical proximity to the region of Cabo Frio-RJ. This influence from coastal upwelling results in low temperatures, high salinity and a large number of nutrients throughout the year, with greater intensity in the summer (Valentin 1984). There is a predominance of sand, with terrigenous sediments interspersed by deposits of sand or mud, throughout almost all the inner and intermediate continental shelf (in shallower waters). At greater depths toward the open ocean, up to the edge of the continental shelf, there are thicker sediments, such as coarse sand and gravel (carbonates) and less influence from terrigenous inputs (Figueiredo and Madureira 2004).

A total of 243 gastropod mollusk bioclasts from the siliciclastic bottoms of locations along the continental shelf were used in the taphonomic analysis. Sampling was carried out from four collection stations, covering proximal (5–10 m or 16–33 ft deep) and distal (15–20 m of 49–66 ft deep) sub-environments (Table 13.1).

13.2.1.3 Ubatuba

The Ubatuba Bay, located on the northern coast of the State of São Paulo, is characterized by a jagged relief, due to its proximity to the Serra do Mar system of mountain ranges, with small, isolated massifs and the promontories of the terminal spurs (Ab'Saber 1955). This results in an almost total absence of coastal plains, although there are small bays and beaches. Most of these beaches can be classified as protected (dissipative) or intermediate due to this physical protection from the most energetic wave trains, consequently the bottom is sporadically churned (Mahiques and Souza 1999).

The shelf sediment is predominately characterized by fine and very fine sand, with segments of mud and very thick sand and the sparse appearance of pelites, indicating a small influx of modern sediments (Muehe 1996).

Among the samples, 431 individual gastropod mollusks were recovered from the siliciclastic bottoms of locations on the continental shelf at four collection stations, covering proximal (5–10 m or 16–33 ft deep) and distal (15–20 m of 49–66 ft deep) sub-environments (Table 13.1).

13.2.1.4 Cananéia

The estuary-lagoon complex of Cananéia-Iguape is located at the southern end of the coast of São Paulo, with the municipality of Iguape to the north, the islands of Cananéia and Cardoso to the south, the Ilha Comprida Island to the east and the Serra do Mar to the west. The complex is linked to the ocean by means of the Barras de Icapara, Cananéia and Ararapira (Mendonça and Katsuragawa 2001). Of these connections to the estuary system, Barra de Cananéia is the Cananéia-Iguape complex's main access to the ocean and the access with the largest opening. Thus the highest volume of water exchanged is between the lagoon region and the sea. And, for this reason, it is traditionally used by the water traffic in the region.

Here it is characterized by an area wide open to the ocean without any coves and is formed by a series of meanders from a canal with a bathymetry superior to 4 m (13 ft), which expands from the mouth of Cananéia to the northeast of the Ilha de Cananéia island. The area of the outlet of Cananéia is formed by alkaline massifs of the São João Hill in Cananéia and of Morrete in Ilha Comprida, which provides this region with an accumulation of sediments. The swiftest currents (1.2 m/s or 4 ft per second) take place in this area of the outlet during the ebb tide, which is also the location of the greatest variation in bathymetric profiles and where the features of the bottom of the inner areas create mega waves and those of the outer areas, close to the ocean, form sandy waves.

Here the beaches are mostly dissipative, though there are also intermediate beaches. Its intertidal zone, surf zone, and the shallow continental shelf are very similar to each other in terms of sedimentation, with more than 80% of their sediment being fine and very fine sand (Herrera 2017).

Two hundred and twenty one bioclasts of gastropods were analyzed in Cananéia, collected from four collection stations located on siliciclastic bottoms, including in proximal (5–10 m deep) and distal (15–20 m deep) sub-environments (Table 13.1).

13.2.2 Looking for Patterns, Understanding the Limitations

The taphonomic analysis of bivalve mollusks and gastropods from collection stations at Paranaguá, Cananéia, Ubatuba, and Macaé was carried out individually for each bioclast using a stereomicroscope (10x magnification). When applicable, a taphonomic grade was established for the signatures as follows: (1) articulation (present/absent, for bivalve mollusks); (2) fragmentation (present/absent, for bivalve mollusks and gastropods); (3) bioerosion (present/absent, for bivalve mollusks and gastropods, with notes on the producer of the trace); (4) encrustation (present/absent, for bivalve mollusks and gastropods, with notes on the encrustation organism); (5) abrasion (present/absent, for bivalve mollusks) and (6) dissolution (present/absent, for bivalve mollusks and gastropods) (Fig. 13.2).

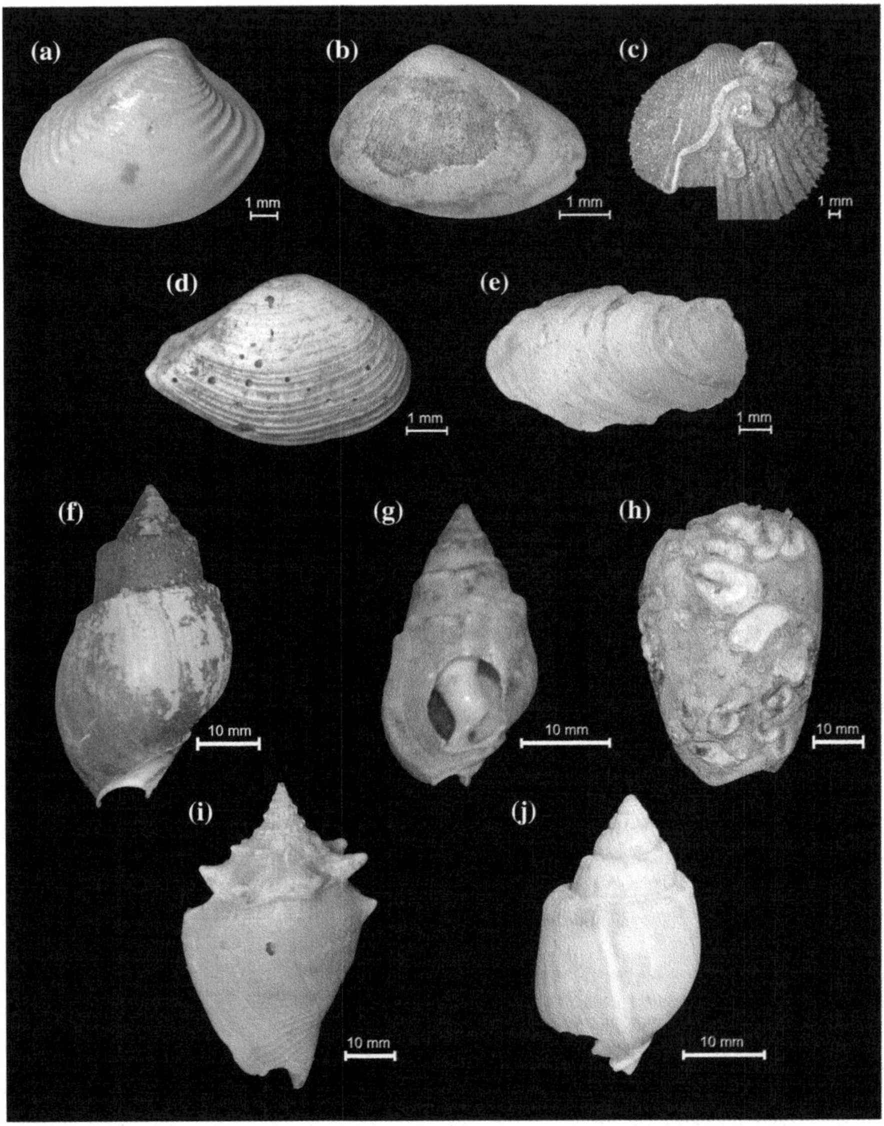

Fig. 13.2 Shells showing different taphonomic signatures. Bivalves: **a** *Anomalocardia brasiliana* with dissolution; **b** Mollusk indet. with abrasion; **c** *Trachycardium* sp. with encrustation **d** *Corbula* sp. with bioerosion and **e** Bivalve mollusk fragmentation. Gastropods **f** *Buccinanops gradatus* with dissolution; **g** *Buccinanops gradatus* with abrasion; **h** *Olivancellaria urceus* with encrustation; **i** *Strombus pugilis* with bioerosion and **j** *Buccinanops gradatus* with fragmentation

13.2.2.1 Articulation

In the Paranaguá Estuary Complex region, there was a predominance of disarticulation of mollusk valves (88–97%), not only at the collection stations near the outlet where the energetic conditions are stronger, but also at the collection stations of the more interior areas of the complex, which are protected and the energy conditions are low. Disarticulation was also found to be predominant among the bivalve mollusks studied by Rodrigues (2006) in the Ubatuba region. This is because, according to Rodrigues (2006), disarticulation of mollusk valves can take place in low energy environments due to important factors such as the amount of time the bioclasts have been in the final interment zone and their exposure or re-exposure to the taphonomically-active zone, in addition to other physical factors (such as the waves generated by ships, in this case) and biological factors (such as predation).

13.2.2.2 Fragmentation

Fragmentation was found in the bioclasts from all of the locations studied, whether in bivalves or gastropods. Thus, a high incidence of fragmentation (64–99%) was observed both at the collection stations in the outer area of the CEP and the inner areas, in addition to all of the places where samples were collected in Macaé (94%), Ubatuba (79%) and Cananéia (78%).

Fragmentation is a complex process that can be related to various factors beyond physical processes in high energy environments. According to Erthal (2012), fragmentation may be the result of a combination of several ecological processes (pre-death) or taphonomic (postmortem) processes, with predation being the main ecological factor responsible for fragmentation. Corroborating the fact that fragmentation can be caused by various different processes in different sub-environments, Rodrigues and Simões (2010) found a high incidence of fragmentation from different causes among remains of the brachiopod *Bouchardia rosea* in different environments in Ubatuba Bay. The authors verified incidents of fragmentation caused by transport and the reworking of shells in coastal environments of high energy, as well as incidents of fragmentation among bioclasts in calmer environments along the inner continental shelf, caused in this case by attacks from durophagous predators. There is a third case, in an environment along the outer continental shelf, where fragmentation is facilitated by the intense dissolution.

According to Zuchin et al. (2003), as the causes of bioclast fractures are diverse, information based on distinct fragmentation patterns greatly increase the value of a fragment to the analysis of taphofacies, among other analyses. One of the main sources of fragmentation in shells is the feeding habit of various durophagous organisms. The effects of these activities are diverse and can only weaken the shell, even to the point of its total destruction (Zuchin et al. 2003). According to the authors, the main cases of bivalve mollusk shells being weakened are caused by predation, either through perforation or marginal breakage. In the CEP region, the result of predation via perforation was observed at a low frequency at all collection points. This being the case, in spite of many bioclasts presenting various indications of fragmentation via predation, it is not the only factor related to fragmentation. Dissolution and erosion can be found on the highest parts of the ornamentation of the shells. However, the characteristic that dominates is an irregular or even rounded edge, which is an indication of physical damage. The cause of fragmentation, therefore, is probably due to more than one factor, but predation may be the biggest factor in this case.

Macaé is an example of a high-energy location with its reflective beaches (Oliveira and Muehe 2013). In this case, high fragmentation rates can be related to the conditions of a high-energy environment with high sedimentation rates, in which damage caused by abrasion and fragmentation is intense (Brett and Baird 1986; Smith and Nelson 2003). Although they showed signs of a harsh environment in terms of sedimentation and hydrodynamics, the predominant condition of the shells from both the bioclasts sampled from Cananéia and Ubatuba was fragmented.

13.2.2.3 Bioerosion

In the inner and outer areas of the CEP, bioerosion is found in low percentages (7.3–9.5%). The types of bioerosion registered were mainly those caused by sponges, followed by bioerosion from algae and/or fungi. Similar rates were observed in Ubatuba (6–11%), with the only exception of station Ubatuba 4, whose bioclasts were found to present a rate of bioerosion of up to 35%. The frequency of bioerosion was found to be even lower in the rest of the locations (3–5%) or even completely absent (such as was the case at three of the four collection station at Cananéia).

However, the lack of damage caused by bioerosion in Macaé and Cananéia suggest that bio-erosive agents are probably more sensitive to high environmental energy and high sedimentation rates (Brett and Baird 1986). This can be better understood if it is taken into consideration that bioerosion occurs after the death of the organisms in a bioclast and, if it is submitted to high energy environmental conditions, the substrate will be unstable, which is unfavorable to bioerosion. Moreover, this low frequency or absence of bioerosion may be related to the low exposure of the bioclasts in the sediment-water interface, that is, the boundary between the water column and the sediment (Young and Nelson 1985). In fact, the low frequency of bioerosion found in Cananéia can be related to the short exposure time of the bioclasts to the sediment-water interface (Young and Nelson 1985). Furthermore, low bioerosion rates may be related to the greater instability of the bioclasts near the sediment-water interface, as these locations correspond to more energetic environments. Moreover, in warmer (tropical) aquatic environments, bioerosion damage is more common than in colder environments, such as polar environments, which are quite rare. This corroborates with the lower incidence of bioerosion in both Macaé and Cananéia, where the average temperature during the collection period was lower than in Ubatuba.

In Cananéia, the incidence of signatures of biological origin was low among the three collection regions. With only 17% of the collected bioclasts encrusted and 1% bioeroded. Such damage, caused by bioeroding organisms like sponges, and encrustation organisms such as Bryozoa, which are organisms typical to marine environments. Thus such damage is rarely seen or never occurs in lagoon environments or environments fed with fresh water, which in turn corroborate these low incidences of bioerosion and encrustation (Parsons-Hubbard 2005). Ritter et al. (2013), for example, have reported the absence of signs of biological impacts on the shells examined in studies on bivalve mollusks from the Tramandaí lagoon complex. The authors conclude that the taphonomic signatures of estuary/lagoon environments are mainly produced by dissolution and not by biological factors (Ritter et al. 2013).

Ubatuba was the only region that had a significant incidence of bioerosion with 12% of the shells collected there being bioeroded. As with encrustation, the higher incidences are associated with a bathymetry of 10–15 m (33–49 ft), which are deeper and more distant from shallow and more agitated waters, as low levels of energy influences this damage (Brett and Baird 1986). Bioclasts found in accumulations in these environments can represent more stable substrates, which are more favorable to bioerosion. Another factor related to bioerosion is the warmer water of Ubatuba. Bioerosion is stimulated by nutrients and occurs more rapidly in more productive waters, which contribute to and increase this kind of damage to shells (Santos and Mayoral 2008). Bioerosion is directly dependent on the primary productivity of the marine environment. That is to say, the higher the marine productivity, the higher the bioerosion. Furthermore, bioerosion rates are also directly or indirectly controlled by the rate interment (sedimentation rate) and depth of the water column (Santos and Mayoral 2008). This suggests that the bioeroded bioclasts from this collection region may have spent more time exposed to the sediment-water interface and were only buried for a short time (Young and Nelson 1985).

13.2.2.4 Encrustation

The areas that were studied in the CEP showed low incidences of encrustation (2–23%), except in one location (35.3%). Most of the encrustation was caused by algae, followed by Bryozoa, bivalves, crustaceans, annelids and, occasionally, gastropods. Another peculiar pattern is found in Macaé, where the physiographic relief has few indentations, making it open to a large amount of energy (reflective beaches) (Oliveira and Muehe 2013). Under these conditions, high concentrations of biological damage are not expected to be observed, as it is more commonly found in low-energy and/or deeper water environments particularly marked by low sedimentation rates (Brett and Baird 1986). However, high encrustation rates appear in Macaé, at 67%. The locations with a shallower bathymetry, from 5 to 10 m (16–33 ft), had the lowest incidences of this kind of damage. Incidences of encrustation were higher in the bioclasts sampled from deeper bathymetries, 10–15 m (33–49 ft), probably due to less environmental stress (lower energy). An interesting fact is that spot geographically closer to the Santana Islands showed a higher incidence of encrustation, which is promoted by the "island effect" (Rodrigues 2006).

Thus, the high incidence of encrusted shells in Macaé, despite its more energetic characteristics, is probably due to its geographic location and the intense influence of the South Atlantic Central Water upwelling area in Cabo Frio, which decreases the temperature but increases the local primary productivity (Valentin 1984) and therefore reflects the action of encrustation organisms (Santos and Mayoral 2008).

However, in Ubatuba, a higher incidence of encrustation was found in the bioclasts studied, with a rate of 69%, followed by Macaé (67%) and Cananéia (17%). Damages of biological origin are more commonly found in the Ubatuba collection region, as it is in a low energy and/or deeper water environment, which has lower sedimentation rates (Brett and Baird 1986), and where low energy environments are attributed to coves (Rodrigues et al. 2002). This fact corroborates with the higher incidences found in deeper bathymetries, at 10–15 m (33–49 ft) and the lower incidences found in bathymetries of 5–10 m (16–33 ft), which, because they are closer to shallower waters, is probably the reason for the lower incidences. This elevated incidence of encrustation in Ubatuba, found in at least more than half of the shells at all of the collection points, is probably related to the fact that encrustation organisms are more common in warmer waters and where the primary productivity is high (Santos and Mayoral 2008). This corroborates with the high incidence of encrustation found in Ubatuba since this location had higher temperatures than the other two regions. And, notably, both Ubatuba and Macaé are strongly affected by the upwelling currents from the South Atlantic Central Water. Because the SACW, in addition to increasing the availability of nutrients and the oxygenation of the water, increases primary productivity and thus favors the propagation of epibionts (encrustation organisms) (Silva 1968). Therefore, the geographic distance of this upwelling area also seems to influence the incidence of damage of biological origin.

In this way, the low energy associated with the higher temperatures of Ubatuba is related to the greater wealth and abundance of species, as well as the greater frequency of the biological damage in this region.

The elevated incidence of encrustation (73%) in a region of geographic proximity to Bom Abrigo Island, reflects the protection pattern of this island that acts as a physical barrier to the hydrodynamic action that takes place between the Cananéia River and the tide, making this environment less energetic compared to the others, which would also explain the low rate of abrasion already discussed above. Furthermore, the island's proximity seems to favor the propagation of epibionts (encrustation organisms), or rather, the "island effect" (Rodrigues 2006). Thus, these details may be related to the low incidence of abrasion, high incidence of encrustation and high incidence of dissolution of this region of Cananéia. In this way, the presence of encrustation organisms seems to indicate specific environments, limits to temperature variation, salinity, fluvial contribution, and sedimentation rates (Erthal et al. 2016).

Encrustation and bioerosion are practically absent in freshwater environments since most of the organisms that cause this kind of damage live in marine habitats (Parsons-Hubbard 2005). Ritter et al. (2013), when investigating taphonomic signatures in bivalve mollusks from the Tramandaí lagoon complex, did not observe signs of biological activities in the shells. For them, taphonomic signatures caused by biological activities in marine environments are produced by dissolution in the estuary/lagoon environments. Thus, transitional environments would present a pattern of taphonomic damage more similar to fluvial environments (Ritter et al. 2013).

According to Farinati et al. (2008), the low degree or absence of bioerosion can also be indicative of low exposure time at the sediment-water interface.

In the case of the CEP, it is believed that the low frequency of biological damage cannot be attributed to low exposure of the bioclasts to the TAZ. This is due to the high frequency of signatures such as disarticulation and fragmentation, which are related to constant re-exposure of the bioclasts to the TAZ (Rodrigues 2006), and to the fact that biological damage play an important role in weakening shells, which also favors types of damages such as disarticulation and fragmentation (Zuchin et al. 2003). As shown, the Paranaguá Estuary Complex, CEP, is a very diverse environment, which may result in a wide combination of signatures in the bioclasts.

13.2.2.5 Abrasion

Macaé is an example of a high-energy location with its reflective beaches (Oliveira and Muehe 2013). In this case, high incidences of fragmentation can be related to the conditions of a high-energy environment with high sedimentation rates, in which damage caused by abrasion and fragmentation is intense (Brett and Baird 1986; Smith and Nelson 2003). The incidence of abrasion at Macaé was 14% of the sampled bioclasts. The collection points with the deepest bathymetry, between 10 and 15 m (33–49 ft), were highlighted by both the incidences of fragmentation and abrasion, 15 and 18%, respectively.

Ubatuba presented an insignificant incidence of abrasion (1%) probably because of a protected, indented coastal environment characterized by low energy and dissipative beaches. This pattern is consistent with the geomorphological features of this location. However, a high incidence of fragmentation (79%) was observed. The lowest incidence (44%) of this kind of damage occurred at the shallowest point, a depth of between 5 and 10 m (16 and 33 ft), close to the shallower waters, which tend towards higher energy and where this kind of damage most commonly occurs (Brett and Baird 1986; Smith and Nelson 2003).

Cananéia is the location with the highest incidence of abrasion, with 26% of the bioclasts exhibiting this signature. In addition, this abrasion signature is more notable when considering the incidences of each collection point. The higher incidences in proximal collection points are probably due to the influence of the Ribeira do Iguape river on the hydrodynamics of this estuarine region. The waves influence hydrodynamic transport, but in Cananéia this is predominantly affected by the currents, in spite of the fact that Bom Abrigo Island creates a geographic barrier against waves from the south and south-east of the area of the Cananéia outlet. Thus it is the tide that is the most effective agent for the transport of sediment, the mixing of waters and the input of nutrients (Miyao and Harari 1989).

The highest rates of abrasion at Cananéia occurred in the shallowest bathymetries, from 5 to 10 m (16–33 ft). Furthermore, the bioclasts from point 3 probably have greater abrasion damage because it is closer to the shore of Cardoso Island, where there is a greater tendency toward erosive marine activity, mainly the action of waves.

This is due to extratropical cyclones that increase the energy of the waves that reach the coast.

13.2.2.6 Dissolution

In the case of dissolution, Macaé presents the highest incidence (35%) among its bioclasts, especially in points of deeper bathymetry, between 10 and 15 m (33–49 ft). These points are further from shallower and more energetic waters, and there are points with greater protection due to their proximity to the Santana Archipelago and the resulting influence from the "island effect," which possibly contributes to the stability of the shells/sediment. This fact suggests that the shells from the points with a high incidence of dissolution suffered more damage because they were incorporated in the substrate, where the pore water or percolating water (*pore water*) is unsaturated in $CaCO_3$ (Alexandersson 1979). Another factor that may have contributed to the higher frequency of dissolution damages in Macaé is that low temperatures and salinity are involved in modulating the solubility of carbonate material (Alexandersson 1979), such as is the case in this location.

In Ubatuba, dissolution was found in up to 20% of the bioclasts collected. This fact may also be related to the calmer and deeper waters (10–15 m or 33–49 ft), further away from the shallower and more agitated waters, where consequently the sediments are finer and muddier (fine sand, silt, and clay) and because these bioclasts were possibly more incorporated into the substrate where the water is more unsaturated.

Dissolution was observed less frequently in Cananéia, at an incidence of 8%. However, this signature seems to be an important cause of damage in the Cananéia bioclasts, because of the influence of the estuary on this location, which leads to a lower incidence of damages of biological origin, such as encrustation and bioerosion. Dissolution tends to be present in the estuarine environment due to the presence of local vegetation and its roots, where there is great oxidation from organic matter, constant rainfall and the influence of freshwater (Marshall et al. 2008). The dissolution occurs in these environments because the acidity of the water is increased by the high rate of CO_2 produced by the large quantity of living heterotrophic organisms there. Thus, the lower pH (acid) increases the solubility of calcium carbonate (Marshall et al. 2008). The low temperature contributes to dissolution (Alexandersson 1979; Aller 1982). The highest incidence of dissolution in Cananéia occurred near Bom Abrigo Island, where the location receives greater influence from this island than the other points, which possibly makes it a less energetic and more stable environment.

In a study on the damages present in mollusks and brachiopods from the coves of Ubatuba and Picinguaba, Rodrigues (2006) found that articulation, fragmentation, dissolution, structural modification, bioerosion, and encrustation are signatures that showed a homogeneous distribution in the accumulations studied. According to the author, this may be due to the fact that the bioclasts are subjected to a complex taphonomic history, with exposure or re-exposure to the taphonomically active zone, leading to a prolonged residence in the final interment zone and leaving them constantly exposed to physical, chemical, and biological processes prior to their complete destruction or likely final incorporation into the substrate. In this way, Rodrigues (2006) verified that the surface texture signature on a shell allows for a better environmental interpretation since it is a product of the other taphonomic signatures. According to Rodrigues (2006), the distribution of the signatures abrasion, dissolution and bioerosion among the sampled environments correlate well with the incidences on the surface textures.

Rodrigues (2006) found that the shells that did not have texture changes were typical to clay bottom cove environments, where abrasion, dissolution, and bioerosion are rare. A polished texture was more commonly found in the bioclasts of the beach environments and rare in the other environments. Precisely in the beach environment, the most energetic among those studied, is where the highest rates of abrasion were recorded.

The granular texture was more frequently found in the sandy bottom of inner continental shelf environments (more than 50%), which is where the highest incidences of dissolution were recorded (low-energy sandy bottom bays and sandy bottom on the inner continental shelf) and incidences of bioerosion were also high. A high incidence of granular textures was also observed in the bioclasts from the sandy bottom bay environments. However, the dominant textures found in these environments were biogenic, which were found in all the environments studies, including the beach environments, although with lower incidences.

In the CEP, the most commonly registered surface textures were polished and granular, followed by perforated. In the same way observed by Rodrigues (2006), the incidences of the surface textures correlated well with the incidences of dissolution, bioerosion, and abrasion at the collection points: the stations that presented the highest incidences of abrasion and dissolution are also the stations that presented higher incidences of polished textures and, to a lesser extent, the granular textures. However, these are the collection stations of the innermost areas of the CEP. The stations near the outlet of the CEP had lower incidences of dissolution and higher incidences of a natural type texture. Polished and granular textures were also found at these stations, but to a lesser degree than the group of innermost collection stations.

Ninety-six percent of the bioclasts from collection stations in the inner area the CEP presented a polished surface texture. This type of texture is identified when the bioclast undergoes sporadic loss of its superficial layers and is considered a type of physical damage. Though it is a location with muddier sediments and lower levels of energy, the high incidence of this texture is due to the loss of superficial layers of bivalve mollusks, especially in the umbo region as a consequence of their sessile nature, as they spend their lives attached to a rock.

A greater diversity of surface textures is found in the most distal collection stations due to the greater variety of physical, chemical and biological types of damage observed. Because it is an environment with higher energy levels, in addition to the disturbance caused by the various ships that come and go at the Paranaguá port, the bioclasts here are constantly exposed to physical damage and have a greater exposure to the sediment-water interface, which is also favorable for biological attacks, as found at the stations close to the outlet of the CEP.

13.3 Final Comments

In general terms, the conditions that were predominant in the four locations studied were disarticulation and fragmentation, both condition attributed to more than one factor. In the case of Paranaguá (CEP), it is possible that the fragmentation is biofacilitated, being under greater influence of the biological types of damage, especially predation. In Ubatuba, it is also possible that some of the fragmentation was biofacilitated, especially in the location where the highest incidence of bioerosion was found.

Bioerosion was not observed to a notable degree. However, only the collection stations near the mouth of the CEP presented this type of damage. Encrustation appeared at all collection stations, also to a small degree.

As in previous studies (Rodrigues and Simões 2010), the signature of a change in the external surface of a shell correlated positively with the signatures of dissolution, abrasion and bioerosion: the stations with the highest incidence of dissolution were also the ones with the highest incidences of polished or granular changes to the surface texture, and the stations that presented bioerosion were the ones that also presented biogenic alterations to the surfaces, with a perforated texture being the most common. Alteration of the external surface, however, did not present a correlation with the environment, as found in other works. That is to say, the innermost stations and those with calmer waters were the stations that presented the highest incidence of dissolution and polished and granular textures.

In this sense, environmental factors that contributed to the development of surface textures in the bioclasts studied, through a combination of bioerosion, encrustation, abrasion, and dissolution process include: (I) geographic distance—geographic proximity to an island and/or upwelling area, whereas both provide a greater availability of nutrients; (II) hydrodynamics—factors that promote less disturbance (greater stability) of the environment/shells/sediment, such as geographical protection from an island or an indented coast, as well as a greater depth, as this would put the bioclasts farther from the shallow and more agitated waters; (III) fluvial influence—which promotes the reduction of salinity, such as in the area of an estuary, for example.

In this way, the factors listed above are related in a complex way and are possibly associated with taphonomic signatures and are therefore relatable and useful to the identification of taphonomic variations and, therefore, capable of being applied to paleoenvironmental and paleoecological comparative analyses of macroinvertebrates from the siliciclastic bottoms of the southeastern continental shelf of Brazil and its subtropical climate.

References

Ab'Saber AN (1955) Contribuição à geomorfologia do litoral paulista. Rev Bras Geogr 17(1):3–48

Alexandersson ET (1979) Marine maceration of skeletal carbonates in the Skagerrak, North Sea. Sedimentology 26(6):845–852

Aller RC (1982) Carbonate dissolution in nearshore terrigenous muds: the role of physical and biological reworking. J Geol 90(1):79–95

Allmon WD, Nieh JC, Norris RD (1990) Drilling and peeling of turritelline gastropods since the Late Cretaceous. Palaeontology 33:595–611

Anderson LC, Geary DH, Nehm RH, Allmon WD (1991) A comparative study of naticid gastropod predation on *Varicorbula caloosae* and *Chione cancellata*, Plio-Pleistocene of Florida, U.S.A. Palaeogeogr Palaeoclimatol Palaeoecol 85:283–290

Angulo RJ (1992) Geologia da planície costeira do Estado do Paraná, 334 p. Ph.D. thesis, Instituto de Geociências, Universidade de São Paulo

Behrensmeyer AK, Kidwell SM (1985) Taphonomy's contributions to paleobiology. Paleobiology 11(1):105–119

Bengtson S, Zhao Y (1992) Predatorial borings in Late Precambrian mineralized exoskeletons. Science 257:367–369

Best MMR, Kidwell SM (2000a) Bivalve taphonomy in tropical mixed siliciclastic-carbonate settings. I. Environmental variation in shell condition. Paleobiology 26:80–102

Best MMR, Kidwell SM (2000b) Bivalve taphonomy in tropical mixed siliciclastic-carbonate settings. II. Effect of bivalve life habits and shell types. Paleobiology 26:103–115

Bigarella JJ, Becker RD, Matos DJ, Werner A (1978) A Serra do Mar e a porção oriental do Estado do Paraná: um problema de segurança ambiental e nacional. Secretaria de Estado de Planejamento/ADEA, Curitiba, p 249

Brett CE (1995) Sequence, stratigraphy, biostratigraphy, and taphonomy in shallow marine environments. Palaios 10:597–616

Brett CE, Baird GC (1986) Comparative taphonomy: a key to paleoenvironmental interpretation based on fossil preservation. Palaios 1(3):207–227

Brett CE, Baird GC (1997) Epiboles, outages, and ecological evolutionary bioevents: taphonomy, ecology, and biogeographic factors. In: Brett CE, Baird GC (eds) Paleontological events: stratigraphic, ecological, and evolutionary implications. Columbia University Press, New York, pp 249–284

Brett CE, Baird GC, Speyer SE (1997) Fossil Lagerstätten: stratigraphic record of paleontological and taphonomic events. In: Brett CE, Baird GC (eds) Paleontological events: stratigraphic, ecological, and evolutionary implications. Columbia University Press, New York, pp 3–40

Callender WR, Staff GM, Parsons-Hubbard KM, Powell EN, Rowe G, Walker SE, Brett CE, Raymond A, Carlson DD, White S, Heise EA (2002) Taphonomic trends along a forereef slope: Lee Stocking Island, Bahamas. I. Location and water depth. Palaios 17:50–65

Cerrano C, Bavestrello G, Calcini B, Cattaneo-Vietti R, Chiantote M, Guidetti M, Sarà A (2001) Bioerosive processes in Antartic Seas. Polar Biol 24:790–792

Davies DJ, Powell EN, Stanton RJ Jr (1989) Taphonomic signature as a function of environmental process: shells and shell beds in a hurricane-influenced inlet on the Texas coast. Palaeogeogr Palaeoclimatol Palaeoecol 72:317–356

Efremov JA (1940) Taphonomy: new branch of paleontology. Pan Am Geol 74:81–98

Erthal F (2012) Assinaturas Tafonômicas em Bivalves marinhos recentes na costa do Brasil e seu significado paleoambiental. Ph.D. thesis, Instituto de Geociências, Universidade Federal do Rio Grande do Sul, Porto Alegre

Erthal F, Ritter MN, Kotzian CB (2016) Assinaturas tafonômicas em moluscos recentes e seu significado paleoambiental. Terra e Didática 13(1):4–29

Farinati EA, Spagnuolo J, Aliotta S (2008) Tafonomía de bivalvos holocenos en la costa del estuario de Bahía Blanca, Argentina. Geobios 41:61–67

Figueiredo AG Jr, Madureira LSP (2004) Granulometria e composição de sedimentos. In: Topografia, composição, refletividade do substrato marinho e identificação de províncias sedimentares na região Sudeste-Sul do Brasil. Instituto Oceanográfico da USP, São Paulo (Série Documentos Revizee – Score Sul), pp 46–55

Fürsich FT, Flessa KW (1987) Taphonomy of tidal flat molluscs in the northern Gulf of California: paleoenvironmental analysis despite the perils of preservation. Palaios 2:543–559

Fürsich FT, Oschmann W (1993) Shell beds as tools in basin analysis: the Jurassic Kachchh, Western India. J Geol Soc London 150:169–185

Harper EM, Forsythe GTW, Palmer T (1998) Taphonomy and the Mesozoic marine revolution: preservation state masks the importance of boring predators. Palaios 13:352–360

Herrera DR (2017) Distribuição, estrutura populacional, reprodução e crescimento de *Callinectes danae* Smith, 1869 e *Callinectes ornatus* Ordway, 1863 (Decapoda, Brachyura, Portunidae) na região de Cananéia, litoral sul paulista. Ph.D. thesis, Instituto de Biociências de Botucatu, Universidade Estadual Paulista

Hoffman A, Pisera A, Ryszkiewski M (1974) Preation by muricid and naticid gastropods on the Lower Tortonian mollusks from the Korytnica clays. Acta Geol Pol 24:249–261

Holz M, Simões MG (2002) Elementos fundamentais de tafonomia. Editora da Universidade/UFRGS, Porto Alegre

Kelley PH, Hansen TA (1993) Evolution of the naticid gastropod predator-prey system: an evaluation of the hypothesis of escalation. Palaios 8:358–375

Kelley PH, Hansen TA (1996) Recovery of the naticid gastropod predator-prey system from the Cretaceous-Tertiary and Eocene-Oligocene extinctions. Geol Soc Spec Publ 102:373–386

Kelley PH, Kowalewski, M, Hansen TA (eds) (2003) Predator-prey interactions in the fossil record. Topics in geobiology, vol 20. Kluwer Academic/Plenum Publishers, New York, 464 pp

Kidwell SM, Rothfus TA, Best MM (2001) Sensitivity of taphonomic signatures to sample sizes, sieve size, damage scoring system, and target taxa. Palaios 16:25–52

Kitchell JA, Boggs CH, Kitchell JF, Rice JA (1981) Prey selection by naticid gastropods: experimental tests and application to the fossil record. Paleobiology 7:533–552

Kowalewski M (1999) Actuopaleontology: the strength of its limitations. Acta Paleont Pol 44(4):452–454

Kowalewski M (2017) The youngest fossil record and conservation biology: holocene shells as eco-environmental recorders. In: Dietl GP, Flessa KA (eds) Conservation paleobiology: science and practice, pp 7–21

Kowalewski M, Hoffmeister AP (2003) Sieves and fossils: effects of mesh size on paleontological patterns. Palaios 18:460–469

Kowalewski M, Kelley PH (eds) (2002) The fossil record of predation. Paleontological society special papers 8. Paleontological Society, Yale Printing Service, New Haven, CT, 398 pp

Kowalewski M, Flessa KW, Aggen JA (1994) Taphofacies analysis of recent shelly cheniers (beach ridges), Northeastern Baja California, Mexico. Facies 31:209–242

Lamour MR (2007) Morfodinâmica sedimentar da desembocadura do complexo estuarino de Paranaguá – PR. Ph.D. thesis, Universidade Federal do Paraná, Curitiba

Lamour MR, Soares CR, Carrilho JC (2004) Mapas dos parâmetros texturais dos sedimentos de fundo do Complexo Estuarino de Paranaguá – Pr. Bol Paran Geoc 55:77–82

Lana PC, Marone E, Lopes RM, Machado EC (2001) The subtropical estuarine complex of Paranaguá Bay, Brazil. In: Seeliger U, Lacerda LD, Kjerfve B (eds) Coastal marine ecosystems of Latin America. Ecological studies, vol 144. Springer Verlag, Basel, pp 131–145

Leighton LR (2001) New example of Devonian predatory boreholes and the influence of brachiopod spines on predator success. Palaeogeogr Palaeoclimatol Palaeoecol 165:53–69

Lescinsky HL, Edinger E, Risk MJ (2002) Mollusc shell encrustation and bioerosion rates in a modern epeiric sea: taphonomy experiments in the Java Sea, Indonesia. Palaios 17:171–191

Llewellyn G, Messing CG (1993) Compositional and taphonomic variations in modern crinoid-rich sediments from the deep-water margin of a carbonatic bank. Palaios 8:554–573

Mahiques MM, Souza LAP (1999) Shallow seismic reflectors and upper Quaternary sea level changes in the Ubatuba region, São Paulo State, Southeastern Brazil. Rev Bras Oceanogr 47(1):1–10

Marcondes TP (2013) Actuopaleontologia em moluscos bivalves no Complexo Estuarino Paranaguá: implicações paleoambientais e paleoecológicas em diferentes áreas de aporte energético. M.Sc. dissertation, Faculdade de Filosofia, Ciências e Letras de Ribeirão Preto, Universidade de São Paulo

Marshall DJ, Santos JH, Leung KMY, Chak WH (2008) Correlations between gastropod shell dissolution and water chemical properties in a tropical estuary. Mar Env Res 66(4):422–429

Martello AR, Kotzian CB, Erthal F (2018) The role of topography, river size and riverbed grain size on the preservation of riverine mollusk shells. J Paleolimn 59:309–327
Martin RE, Liddell WD (1991) Taphonomy of foraminifera in modern carbonate environments: implications for the formation of foraminiferal assemblages. In Donovan SK (ed) The processes of fossilization. Belhaven Press, pp 179–193
McFarland S, Westrop SR, Cheel RJ (1999) Allogenic versus autogenic processes in the genesis of Middle Ordovician brachiopod-rich shell beds, Verulam Formation, Ontario. Palaios 14:282–287
Meldahl KH, Flessa KW (1990) Taphonomic pathways and comparative biofacies and taphofacies in a recent intertidal/ shallow shelf environment. Lethaia 23:43–60
Mendonça JT, Katsuragawa M (2001) Caracterização da pesca artesanal no Complexo Estuarino-Lagunar de Cananéia – Iguape, Estado de São Paulo, Brasil (1995-1996). Acta Scient 23(2):535–547
Messina C, LaBarbera M (2004) Hydrodynamic behavior of brachiopod shells: experimental estimates and field observations. Palaios 19(5):441–450
Miyao SY, Harari J (1989) Estudo preliminar da maré e das correntes de maré na região estuarina de Cananéia (25°S e 048°W). (SP), Brasil. Bol Inst Oceanogr 37(2):107–123
Muehe D (1996) Geomorfologia costeira. Geomorfologia: uma atualização de bases e conceitos 4:191–238
Nebelsick JH (1999) Taphonomy of *Clypeaster* fragments: preservation and taphofacies. Lethaia 32:241–252
Oliveira JF, Muehe DCEH (2013) Identificação de áreas de sedimentos compatíveis na plataforma continental interna para recuperação de praias entre as cidades de Niterói e Macaé – Rio de Janeiro, Brasil. Rev Gest Cost Int 13(1):89–99
Pandolfi JM, Greenstein BJ (1997) Taphonomic alteration of reef corals: effects of reef environment and coral growth form. I. The great barrier reef. Palaios 12:27–42
Parsons-Hubbard K (2005) Molluscan taphofacies in recent carbonate reef/lagoon systems and their application to sub-fossil samples from reef cores. Palaios 20(2):175–191
Perry CT (2000) Factors controlling sediment preservation on a North Jamaica fringing reef: a process-eased approach to microfacies analysis. J Sed Res 70:633–648
Pinto MCS (2017) Assinaturas Tafonômicas de Gastrópodes atuais do litoral do Rio de Janeiro e São Paulo: Biodiversidade e implicações Ambientais. M.Sc. dissertation, Universidade Estadual Paulista Júlio de Mesquita Filho
Ritter MN, Erthal F (2013) Fidelity bias in mollusk assemblages from coastal lagoons of Southern Brazil. Rev Bras Paleont 16(2):225–236
Ritter MN, Erthal F, Coimbra JC (2013) Taphonomic signatures in molluscan fossil assemblages from the Holocene lagoon system in the northern part of the coastal plain, Rio Grande do Sul State, Brazil. Quat Int 305:5–14
Ritter MN, Erthal F, Coimbra JC (2018) Depth as an overarching environmental variable modulating preservation potential and temporal resolution of shelly taphofacies. Lethaia https://doi.org/10.1111/let.12289
Rodland D, Krause RA, Kowalewski M, Simões MG (2014) Stowing away on ships that pass in the night: sclerobiont assemblages on individually dated bivalve and brachiopod shells from a subtropical shelf. Palaios 29:170–183
Rodrigues SC (2006) Tafonomia de moluscos Bivalves e Braquiópodes das Enseadas de Ubatuba e Picinguaba, Norte do Estado de São Paulo: implicações do uso de assinaturas tafonômicas no reconhecimento de gradientes ambientais. Ph.D. thesis, Instituto de Geociências, Universidade de São Paulo, São Paulo
Rodrigues SC, Simões MG (2010) Taphonomy of *Bouchardia rosea* (Rhynchonelliformea, Brachiopoda) shells from Ubatuba Bay, Brazil: implications for the use of taphonomic signatures in (paleo)environmental analysis. Ameghiniana 47(3):373–386
Rodrigues M, Mahiques MM, Tessler MG (2002) Sedimentação atual nas enseadas de Ubatumirim e Picinguaba, região norte de Ubatuba, Estado de São Paulo, Brasil. Rev Bras Oceanogr 50:27–45

Rodrigues SC, Simões MG, Pires-Domingues RA (2009) Viés analítico em estudos tafonômicos com macroinvertebrados: implicações (paleo)ambientais e (paleo)ecológicas. Rev Inst Geoc USP 9(1):101–114

Rothfus TA (2004) Clams and Brachiopods: chips that pass out of sight. Palaios 19(5):507–513

Sancinetti GS (2011) Distribuição espaço-temporal e estrutura populacional do camarão *Artemesia longinaris* Bate, 1988 (Crustacea: Decapoda: Penaeidae) no litoral de Macaé, RJ. M.Sc. dissertation, Instituto de Biociências de Botucatu, Universidade Estadual Paulista

Santos A, Mayoral E (2008) Bioerosion versus colonisation on Bivalvia: a case study from the Upper Miocene of Cacela (southeast Portugal). Geobios 41(1):43–59

Silva PMC (1968) O fenômeno ressurgência na costa meridional brasileira. Pub Inst Pesq Marin 24:1–31

Simões MG, Kowalewski M (2002) Taphonomy of recent articulate brachiopods from subtropical, siliciclastic-carbonate environments of southern Brazil. In: Geological Society of America, abstracts with program, 34, p 357

Simões MG, Kowalewski M (2003) Modern accumulations of brachiopod shells in unconsolidated surficial beach deposits, northern coast of São Paulo State, Brazil: taphonomic implications for the genesis of skeletal concentration. In: 3rd Latin American Congress of Sedimentology, Belém, Brazil, Abstracts, pp 220–223

Simões MG, Kowalewski M, Mello LHC, Rodland DL, Carroll M (2000) Present-day terebratulid brachiopods from the southern Brazilian shelf: paleontological and biogeographic implications. Geol Soc Am Bull 32(7):14

Simões MG, Kowalewski M, Mello LHC, Rodland DL, Carroll M (2004a) Present-day terebratulid brachiopods from the southern Brazilian shelf: paleontological and biogeographic implications. Palaeontology 47:515–533

Simões MG, Kowalewski M, Carroll M, Barbour Wood SL, Krause RA, Rodrigues SC, Wehmiller JF (2004b) Taphonomy and time averaging as a tool in (paleo)oceanography: a case study on a shallow subtropical shelf from Brazil. 42 Congresso Brasileiro de Geologia, Anais:S27–696

Simões MG, Rodrigues SC, Petti MAV, Nonato V, Kowalewski M, Rodland D (2005) Patterns of brachiopod shell encrustation: a comparison of tropical (Ubatuba Bay, Brazil) and polar (Admiralty Bay, Antarctica) environments. In: 2nd international meeting Taphos'05, Abstract volume: 147–148

Simões MG, Rodrigues SC, Kowalewski M (2007) Comparative analysis of drilling frequencies in recent brachiopod-mollusk associations from the southern Brazilian shelf. Palaios 22(2):143–154

Simões MG, Rodrigues SC, Kowalewski M (2009) *Bouchardia rosea*, a vanishing brachiopod species of the Brazilian platform: taphonomy, historical ecology and conservation paleobiology. Hist Biol 21(3–4):123–137

Smith AM, Nelson CS (1994) Selectivity in sea-floor processes: taphonomy of bryozoans. In: Hayward PJ, Ryland JS, Taylor PD (eds) Biology and paleobiology of bryozoans: proceedings 9th international bryozoology conference, 1992, pp 177–180

Smith AM, Nelson C (2003) Effects of early sea-floor processes on the Taphonomy of temperate shelf skeletal carbonate deposits. Earth-Sci Rev 63(1):1–31

Soberon G, Lacerda MA, Erthal F, Kotzian CB, Rodrigues SC, Simões MG (2007) Assinaturas Tafonômicas em Tanatocenoses de Bivalves (Mollusca) Recentes de duas Localidades da Costa Sul (RS) do Brasil. Boletim Informativo da Sociedade Brasileira de Paleontologia 57:24–25

Speyer SE, Brett CE (1988) Taphofacies models for epeiric sea environments: middle Paleozoic examples. Palaeogeogr Palaeoclimatol Palaeoecol 63:225–262

Speyer SE, Brett CE (1991) Taphofacies controls background and episodic processes in fossil assemblage preservation. In: Allison PA, Briggs DEG (eds) Taphonomy: releasing the data locked in the fossil record. Topics in geobiology, vol 9, pp 501–545

Staff GM, Callender WR, Powell EN, Parsons-Hubbard KM, Brett CE, Walker SE, Carlson DD, White S, Raymond A, Heise EA (2002) Taphonomic trends along a forereef slope: Lee Stocking Island, Bahamas. II. Time. Palaios 17:66–83

Stempien JA (2005) Brachyuran taphonomy in a modern tidal-flat environment: preservation potential and anatomical bias. Palaios 20:400–410
Tomašových A (2004) Effect of extrinsic factors on biofabric and brachiopod alteration in a shallow intraplatform carbonate setting (Upper Triassic, West Carpathians). Palaios 19:349–371
Tomašových A, Rothfus TA (2005) Differential taphonomy of modern brachiopods (San Juan Islands, Washington State): effect of intrinsic factors on damage and community-level abundance. Lethaia 38:271–292
Torello FF (2004) Tafonomia Experimental do Fóssil-Vivo *Bouchardia rosea* (Brachiopoda- Terebratellidae) e suas implicações paleontológicas. Ph.D. thesis, Instituto de Geociências, Universidade de São Paulo
Valentin JL (1984) Analyses des parameters hydrobiologiques dans la remontée de Cabo Frio (Brésil). Mar Biol 82(3):259–276
Vermeij GJ (1987) Evolution and escalation. Princeton University Press, Princeton, New Jersey, p 527
Vermeij GJ, Zipser E, Dudley EC (1980) Predation in time and space: peeling and drilling in terebrid gastropod. Paleobiol 6:352–364
Yesares-García J, Aguirre J (2004) Quantitative taphonomic análisis and taphofacies in Lower Pliocene temperate carbonate-siliciclastic mixed platform deposits (Almería-Nijar basin, SE Spain). Palaeogeogr Palaeoclimatol Palaeoecol 207(1–2):83–103
Young HR, Nelson CR (1985) Biodegradation of temperate-water skeletal carbonates by boring sponges on the Scott shelf, British Columbia, Canada. Mar Geol 65(1–2):33–45
Zuchin M, Stachowitsch M, Stanton RJ Jr (2003) Patterns and processes of shell fragmentation in modern and ancient marine environments. Earth-Sci Rev 63:33–82

Index

S. Martínez et al. (eds.), *Actualistic Taphonomy in South America*, Topics in Geobiology48, https://doi.org/10.1007/978-3-030-20625-3

Printed by Printforce, the Netherlands